GEOLOGIC EVOLUTION OF THE EASTERN UNITED STATES

Field Trip Guidebook
NE–SE GSA 1991

Edited by Art Schultz and Ellen Compton-Gooding

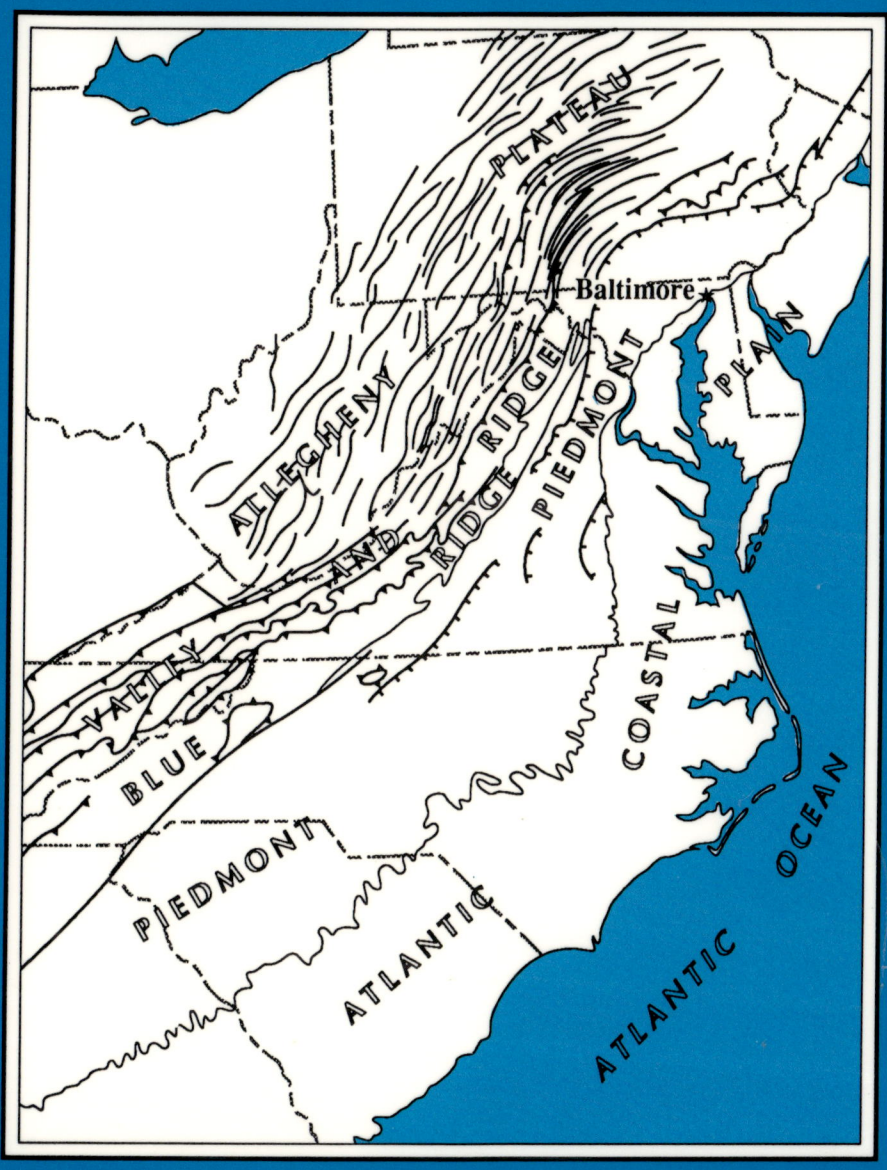

Virginia Museum of Natural History
Guidebook Number 2

Geologic Evolution of the Eastern United States

GEOLOGIC EVOLUTION OF THE EASTERN UNITED STATES

**Field Trip Guidebook
NE–SE GSA 1991**

Edited by Art Schultz and Ellen Compton-Gooding

Virginia Museum of Natural History
Guidebook Number 2
Martinsville, 1991

Copyright 1991 by the Virginia Museum of Natural History
All rights reserved
Printed and bound in the United States of America
Virginia Museum of Natural History, Martinsville, Virginia 24112

ISBN 0-9625801-2-0

Cover map modified from Figure 1, King, P.B., 1964, Further thoughts on tectonic framework of southeastern United States, in Lowry, W.D., Tectonics of the Southern Appalachians, VPI Department of Geological Sciences, Memoir 1.

 This book is printed on recycled paper.

Geology
QE
78.3
.G45
1991

CONTENTS

1. Terranes and Tectonics of the Maryland and Southeast Pennsylvania Piedmont
   Alexander E. Gates, Peter D. Muller, and David W. Valentino — 1

2. Stratigraphy and Structure of the Great Valley and Valley and Ridge, West Virginia
   Peter Lessing, Stuart L. Dean, and Byron R. Kulander — 29

3. Tectonic History of the Blue Ridge Basement and Its Cover, Central Virginia
   Mervin J. Bartholomew, Sharon E. Lewis, Scott S. Hughes, Robert L. Badger, and A. Krishna Sinha — 57

4. Taconic Collision in the Delaware-Pennsylvania Piedmont and Implications for Subsequent Geologic History
   Mary Emma Wagner, LeeAnn Srogi, C. Gil Wiswall, and J. Alcock — 91

5. Stratigraphy of Upper Proterozoic and Lower Cambrian Siliciclastic Rocks, Southwestern Virginia and Northeastern Tennessee
   Dan Walker and Edward L. Simpson — 121

6. Tertiary Lithology and Paleontology, Chesapeake Bay Region
   Lauck W. Ward and David S. Powars — 161

7. Bottom Sediments of the Chesapeake Bay: Physical and Geochemical Characteristics
   Jeffrey P. Halka and James M. Hill — 203

8. Sedimentology and Sequence Stratigraphic Framework of the Middle Devonian Mahantango Formation in Central Pennsylvania
   Anthony R. Prave and William L. Duke — 207

9. Geology of the Robertson River Igneous Suite, Blue Ridge Province, Virginia
   Richard P. Tollo, Tamara K. Lowe, Sara Arav, and Karen J. Gray — 229

10. Late Proterozoic Sedimentation and Tectonics in Northern Virginia
    Stephen W. Kline, Peter T. Lyttle, and J. Stephen Schindler — 263

11. Sideling Hill Road Cut and Visitors Center: An Educational Opportunity Combining Outcrop and Classroom
    Kenneth A. Schwarz — 295

PREFACE

The eleven chapters in this guidebook bring together a wealth of data on a wide variety of topics. Today, such a volume is rarely found. Most collections of papers are focused compendiums; this volume spans much of the entire framework of Appalachian and eastern United States geology. As such, it is of interest to both the specialist and the regional geologist. As a tool for new students of Appalachian geology, this collection is an excellent starting point. The locations of important outcrops, the detailed site descriptions, and the comprehensive reference lists are invaluable.

I thank the authors for their time and concerted efforts in the preparation of the guidebook chapters. Special thanks go to Lauck W. Ward and the Virginia Museum of Natural History for their generous publication support. I also thank Ellen Compton-Gooding for her thorough edit of the final stages of the guidebook and Rick Boland for help in text preparation.

Art Schultz
Reston, Virginia
January 21, 1991

1
TERRANES AND TECTONICS OF THE MARYLAND AND SOUTHEAST PENNSYLVANIA PIEDMONT

Alexander E. Gates
New York Geological Survey
3136 Cultural Education Center, Albany, NY 12230

Peter D. Muller
Department of Earth Sciences
SUNY College at Oneonta, Oneonta, NY 13820

David W. Valentino
Pennsylvania Geologic Survey
P.O. Box 2357, Harrisburg, PA 17120

ABSTRACT

The Baltimore terrane, the Potomac terrane and the Westminster terrane in the Maryland and southeast Pennsylvania Piedmont contain markedly different metasedimentary rocks of the same age. In the Baltimore terrane, anticlines and domes contain Grenville gneiss cores with a Cambrian-Ordovician age metasedimentary cover. The Setters, Cockeysville, and Loch Raven Formations are shelf deposits that have equivalent units of the same age in the Valley and Ridge Province. In the Potomac terrane, the Cambrian-Ordovician Liberty Complex is a polygenetic melange composed of the Morgan Run and Sykesville Formations. The Morgan Run Formation is a metavolcanic-sedimentary unit that may be a melange from within the forearc prism and possibly part of the arc itself. The Morgan Run Formation contains clasts of mafic and ultramafic material in pelite and plagioclase-dominated, amphibole-bearing metagraywackes. The overlying Sykesville Formation exhibits one less deformation than the Morgan Run Formation and is interpreted as an olistostrome derived from the Morgan Run Formation. The boundaries of the Westminster terrane are redefined in the area of the Susquehanna River. The Late Proterozoic(?)-Cambrian Peters Creek Formation is proposed to be part of the Westminster terrane. It contains graded beds that are interpreted as turbidite deposits and conglomerate dominated by blue quartz and K-feldspar. The Peters Creek Formation is proposed to be rift related with a Grenville basement sediment source. The 490 Ma State-Line complex of the Baltimore mafic complex is in faulted intrusive contact with the Peters Creek Formation. The State-Line complex is a layered mafic intrusion and likely also rift related.

INTRODUCTION

In this fieldtrip, we will define several of the terranes of the Maryland and southeastern Pennsylvania Piedmont, suggest modification of their areal extent and stratigraphic subdivisions and propose a revised tectonic model for the evolution of the area. The concept of tectonostratigraphic terranes is new to the central Appalachian Piedmont. Early workers described a continuous Late Proterozoic/Cambrian to Ordovician(?) stratigraphy resting unconformably on Grenville basement (Knopf and Jonas, 1923; Hopson, 1964; Southwick and Fisher, 1967; Fisher, 1970). These studies explained local sequences and stratigraphic relations but could not satisfactorily account for the relations between sequences. Within the context of plate tectonic theory, the Piedmont was reevaluated and consequently subdivided into a volcanic arc sequence, an ocean floor sequence, and a North American sequence (Crowley, 1976; Fisher et al., 1979; Muller and Chapin, 1984). Much of the North American sequence was later determined to consist of tectonic melange that was exotically derived and transported to its present position during the Taconic orogeny (Drake and Morgan, 1981; Drake, 1985; Higgins and Conant, 1986; Drake et al., 1989; Muller et al., 1989). Even the Grenville basement rocks may have been fragmented microcontinents of North American crust that were reattached during subsequent orogenies (Fisher et al., 1979).

The concept of tectonostratigraphic terranes was first applied to western North America (Coney et al., 1980) and quickly extended to the Appalachians (Williams and Hatcher, 1982; Zen, 1983). Horton et al. (1989) applied the terrane subdivision methods of Howell and Jones (1984) to the Appalachians and summarized all previous studies in a comprehensive compilation. The Maryland Piedmont is proposed to consist of five terranes which are (from east to west): the composite Chopawamsic terrane, the composite(?) Bel Air–Rising Sun terrane, the Baltimore terrane, the composite Potomac terrane, and the Westminster terrane (Figure 1). The Chopawamsic terrane (Williams and Hatcher, 1982) includes the James Run metavolcanic sequence of eastern Maryland and is considered a volcanic arc terrane. The Bel Air–Rising Sun terrane (Horton et al., 1989) is composed entirely of mafic and ultramafic plutonic rocks. The Baltimore terrane (Horton et al., 1989) is a continental basement terrane because it is floored by Grenville basement rocks. The Potomac terrane (Drake, 1985) is a disrupted composite terrane composed primarily of metasedimentary rocks with subordinate amounts of metavolcanic rock (Drake, 1989; Muller et al., 1989). The Westminster terrane (Muller et al., 1989) is a stratigraphic metasedimentary terrane composed predominantly of metamorphosed sandstone, siltstone, and shale with less extensive marble and greenstone. In this study we will redefine the boundaries of the Westminster and Potomac terranes and propose the abandonment of at least part of the Bel Air–Rising Sun terrane. It is important to bear in mind that although these terranes vary radically in lithology, they overlap in age.

BALTIMORE TERRANE

The Baltimore terrane is characterized by gneiss-cored anticlines and domes flanked by metasedimentary rocks (Figure 1, Table 1). Grenville-age Baltimore Gneiss is unconformably overlain by Cambrian(?) to Ordovician(?) quartzite, marble, calcsilicate schist, and metapelite. With the exception of some of the granitoid intrusives, all rock units in the terrane exhibit polyphase folding and middle to upper amphibolite facies mineral assemblages.

Baltimore Gneiss

The Baltimore Gneiss consists of layered, quartzo-feldspathic gneiss that is locally migmatitic and contains minor amphibolite and rare calcsilicate gneiss and quartzite (Hopson, 1964; Olsen, 1977; Crowley, 1976; Muller and Chapin, 1984). The Baltimore gneiss has been subdivided into at least four units (Crowley, 1976) with uncertain stratigraphic and/or intrusive relations. Texturally, the gneiss ranges from coarse-grained augen gneiss to medium-grained granofels. Samples of augen gneiss yield discordant zircon ages and Rb/Sr whole-rock ages of approximately 1.0 to 1.2 Ga, indicating a Grenville thermal event (Wetherill et al., 1968; Tilton et al., 1970; Grauert, 1973). The age and metamorphic grade of the Baltimore terrane indicate a tectonometamorphic affinity with gneisses of the Blue Ridge Province to the west (Rankin et al., 1989). There are, however, significant lithologic and structural differences between the Baltimore terrane and the Blue Ridge.

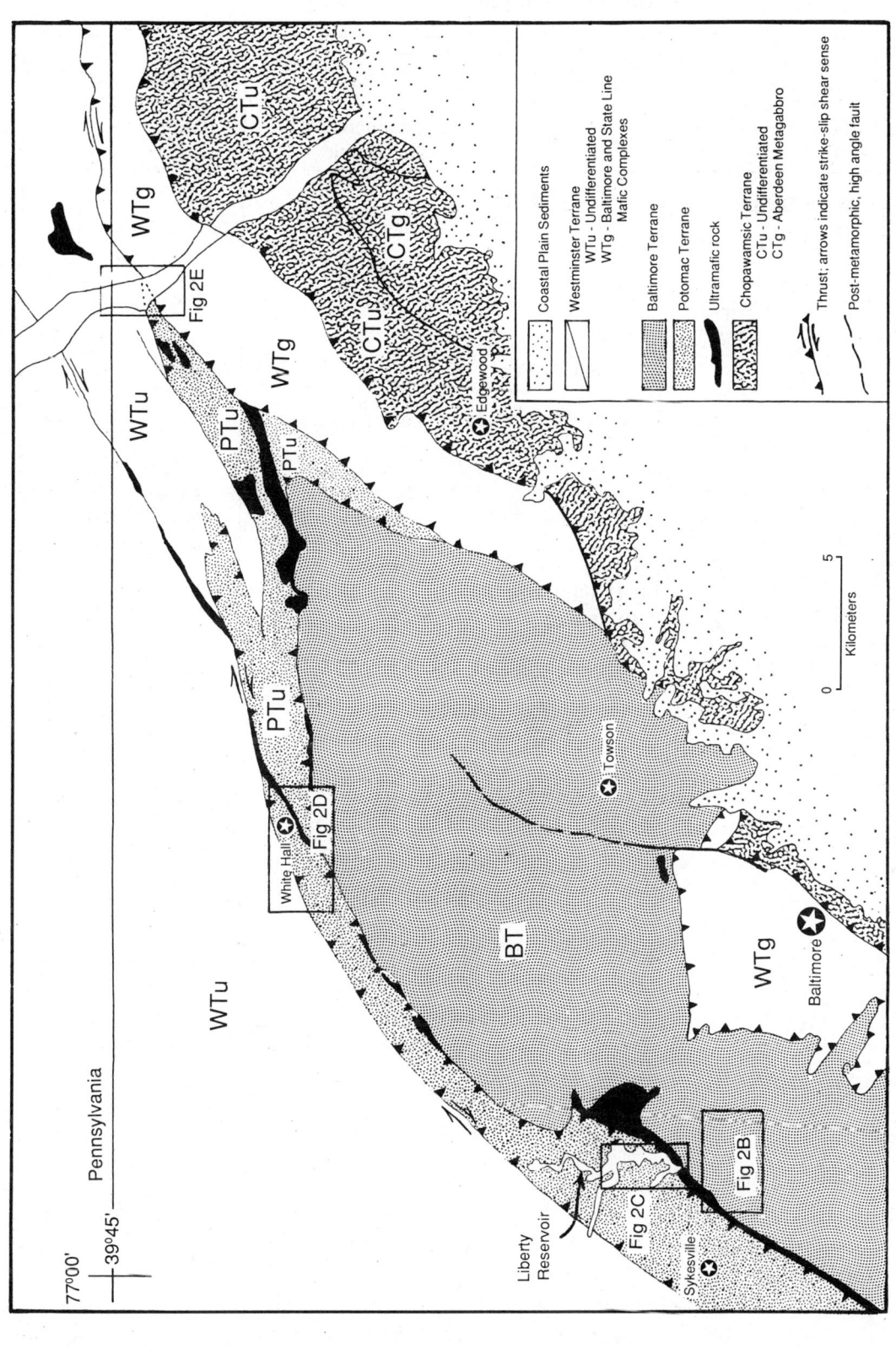

Figure 1. Simplified tectonic map of the Maryland and southeastern Pennsylvania Piedmont showing terranes and locations of Figures 2B–E. Baltimore and Potomac terranes are undifferentiated. WTg (Westminster Terrane gabbro)=Bel Air–Rising Sun Terrane of Horton et al. (1989) herein redefined.

Table 1. Tectonostratigraphic Terranes of the Central and Northern Maryland Piedmont

	Westminster North American rifted margin	Potomac forearc prism	Baltimore North American basement and shelf/bank	Chopawamsic * volcanic arc complex
Cambro-Ordovician	Peach Bottom Slate; Cardiff Meta-conglomerate; ~unconformity~; Peters Creek - Pleasant Grove Formation	Sykesville Fm; ~unconformity~; Morgan Run - Piney Run Fm; Soldiers Delight Mafic Complex (Liberty Complex)	Loch Raven Fm; Cockeysville Marble; Setters Fm; ~unconformity~	James Run - Port Deposit Complex
Proterozoic Z	Octoraro - Prettyboy Schist; Baltimore - State Line Mafic Complex		Baltimore Gneiss	* not included in field trip

∨∨∨∨∨ = unconformity

Baltimore Terrane Cover Sequence

The Baltimore Gneiss is unconformably overlain by a thick sequence of dominantly siliciclastic metasedimentary rocks that were, until recently, grouped together into a single stratigraphic package designated the Glenarm Supergroup (Crowley, 1976). Horton et al. (1989), Drake et al. (1989), and Muller et al. (1989) reinterpreted this stratigraphic package as comprising at least three distinct tectonostratigraphic terranes. We agree that the terms Glenarm Supergroup and Wissahickon Group be discontinued in Maryland, and we describe the Baltimore terrane cover sequence using individual formation names.

The basal Setters Formation is dominantly massive quartzite but ranges from muscovite-microcline quartzite with rare quartz-pebble metaconglomerate, to garnet $\pm$ kyanite $\pm$ staurolite mica schist. The Cockeysville Marble conformably(?) overlies the Setters Formation and is lithologically variable. It includes dolomitic phlogopite marble, calcite marble, and calcsilicate schist. The Cockeysville Marble is generally medium- to coarse-grained, and layering is well-developed except in some very coarse-grained, massive marble.

The overlying Loch Raven Formation consists of two dominant lithologies: 1) medium to coarse-grained garnet $\pm$ staurolite $\pm$ kyanite mica schist interpreted as metapelite; and 2) medium-grained quartz-mica-plagioclase $\pm$ garnet gneiss, interpreted as metagraywacke (Oella Formation of Crowley, 1976). Minor amphibolite, micaceous quartzite, and marble also occur locally. The Loch Raven Formation contains ilmenite-bearing pegmatitic segregations as well as more common quartz veins, pods, and stringers.

The Setters and/or Cockeysville Formation is locally absent and the Loch Raven Formation rests directly on the Baltimore Gneiss. The absence of these units appears to be the result of tectonic removal by synmetamorphic ductile shearing; however, some of the variation in thickness may reflect depositional distribution.

Granitoid Intrusions

A number of granitoid bodies are present in the Baltimore terrane. The composition of these plutonic rocks ranges from granodiorite to quartz monzonite (Hopson, 1964). They include strongly deformed, concordant bodies (Ellicott City granodiorite) and massive discordant stocks and dikes (Woodstock and Guilford). Pegmatite dikes and sills of alkali granite are associated with the margins of the gneiss domes. Most of the isotopic dating of the plutons has yielded ages of 400 to 520 Ma (Sinha et al., 1980). Zircons from the Gunpowder Granite, however, yielded a discordant lead age of 330 Ma (Grauert, 1973).

Tectonostratigraphic Interpretation

The Baltimore terrane has been interpreted as a fragment of North American continental crust composed of Grenvillian basement and a Late Proterozoic–Cambrian sedimentary cover (Rankin, 1975; Crowley, 1976; Fisher et al., 1979; Muller and Chapin, 1984; Horton et al., 1989; Muller et al., 1989; Rankin et al., 1989). The Setters Formation is interpreted as a high energy, shallow water deposit of a rift/drift margin (Fisher et al., 1979). The overlying Cockeysville Marble is interpreted as a shallow-water carbonate bank (Fisher et al., 1979). The position and stratigraphy of these units indicates they are Cambrian in age and correlative to the Valley and Ridge sequence to the west (Fisher, 1970; Crowley, 1976; Fisher et al., 1979; Muller and Chapin, 1984).

The Loch Raven Formation is interpreted to have been deposited in deep-water marine conditions (Fisher, 1970). The graded metagraywacke to subgraywacke and metapelite is interpreted to be a turbidity deposit composed of Cambro-Ordovician basinal mud and sand that was deposited on a foundered shelf sequence (Fisher, 1970).

POTOMAC TERRANE

Outcrops of Liberty Complex in the Potomac terrane, Maryland (Muller et al., 1989), will be visited on this fieldtrip. The Liberty Complex is a thick (2,000–5,000 m) assemblage of upper greenschist to lower amphibolite facies metaclastic rocks with intercalated mafic and ultramafic lenses (Figures 1, 2). It is composed of three units, the Morgan Run Formation, the Sykesville Formation, and the Soldiers Delight Ultramafic Complex (Table 1). At Liberty Lake, the Soldiers Delight Complex is overlain by the Morgan Run Formation, which in turn is overlain by the Sykesville Formation.

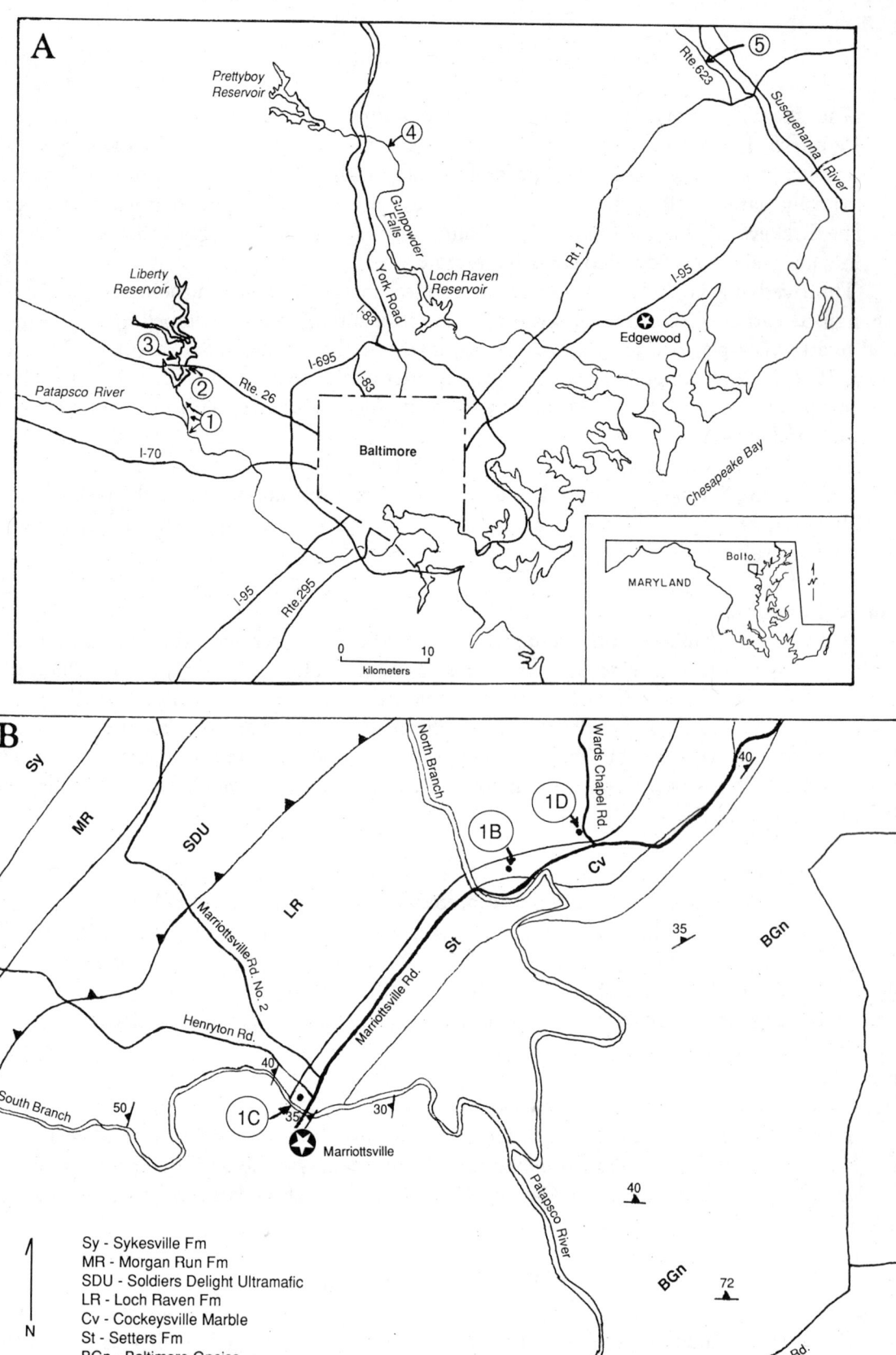

Figure 2. A (top) and B (bottom).

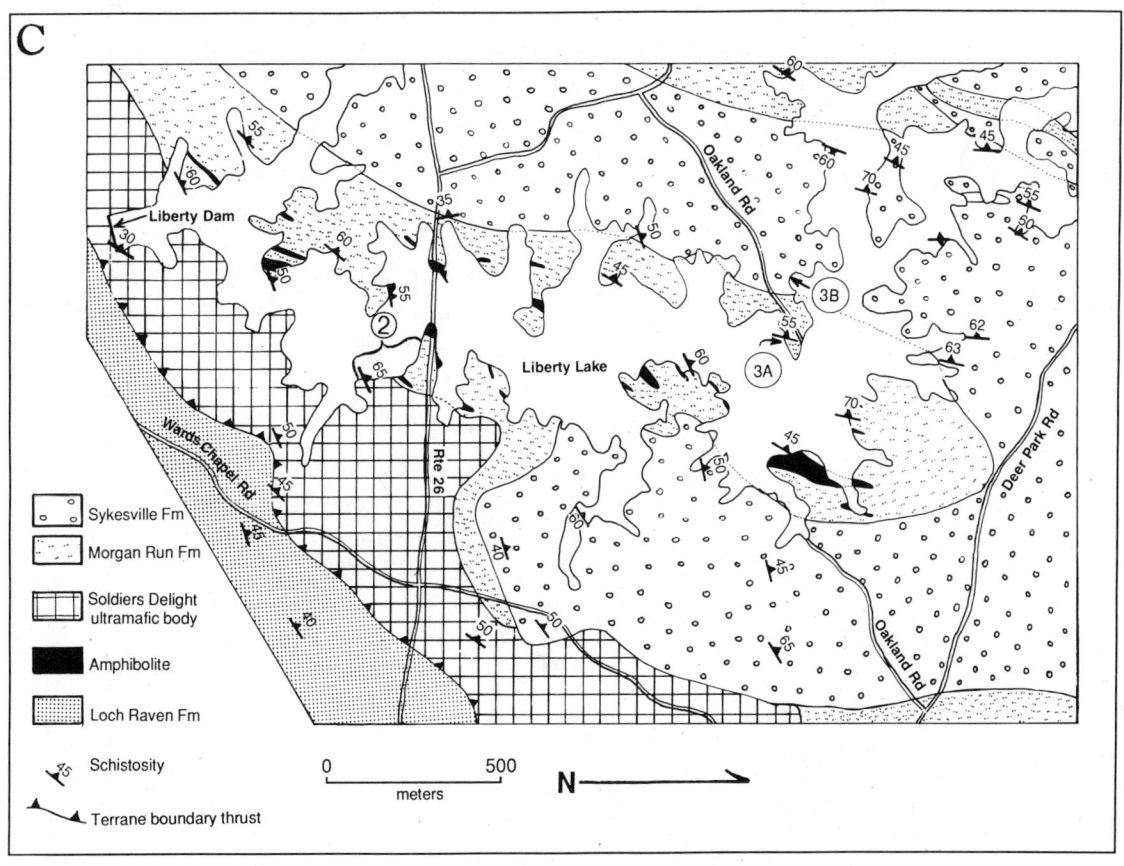

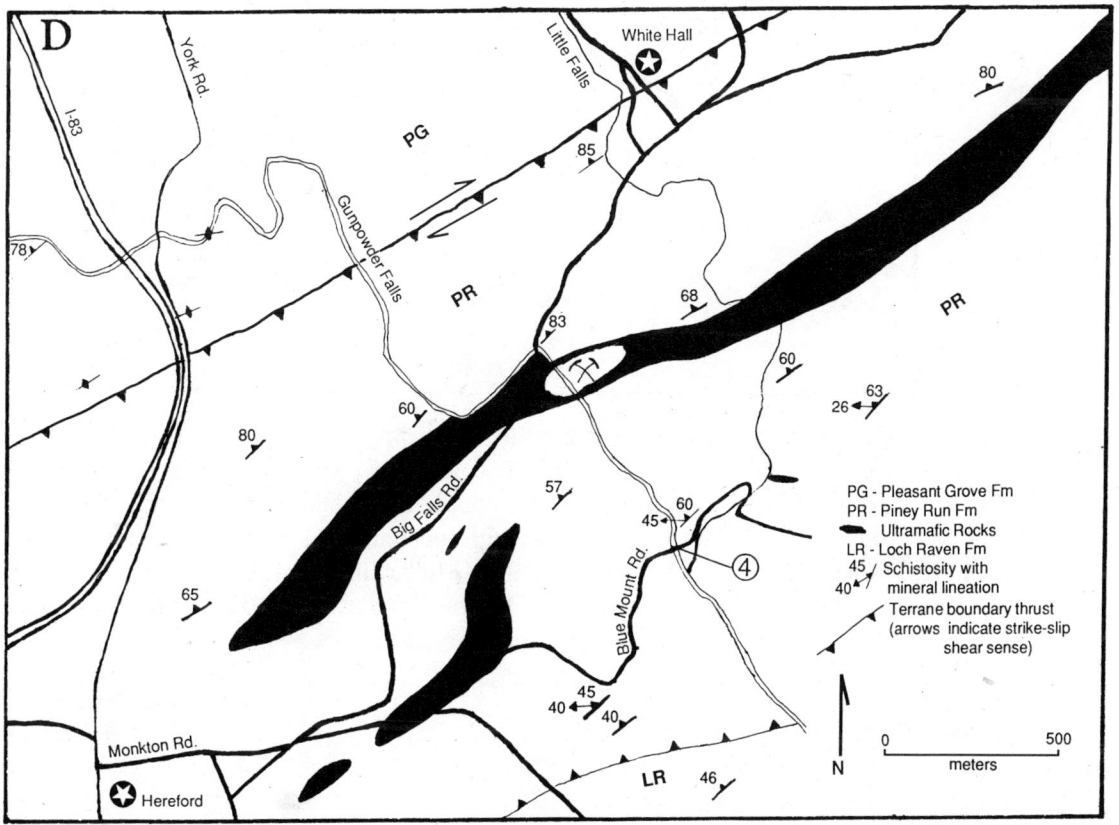

Figure 2. C (top) and D (bottom).

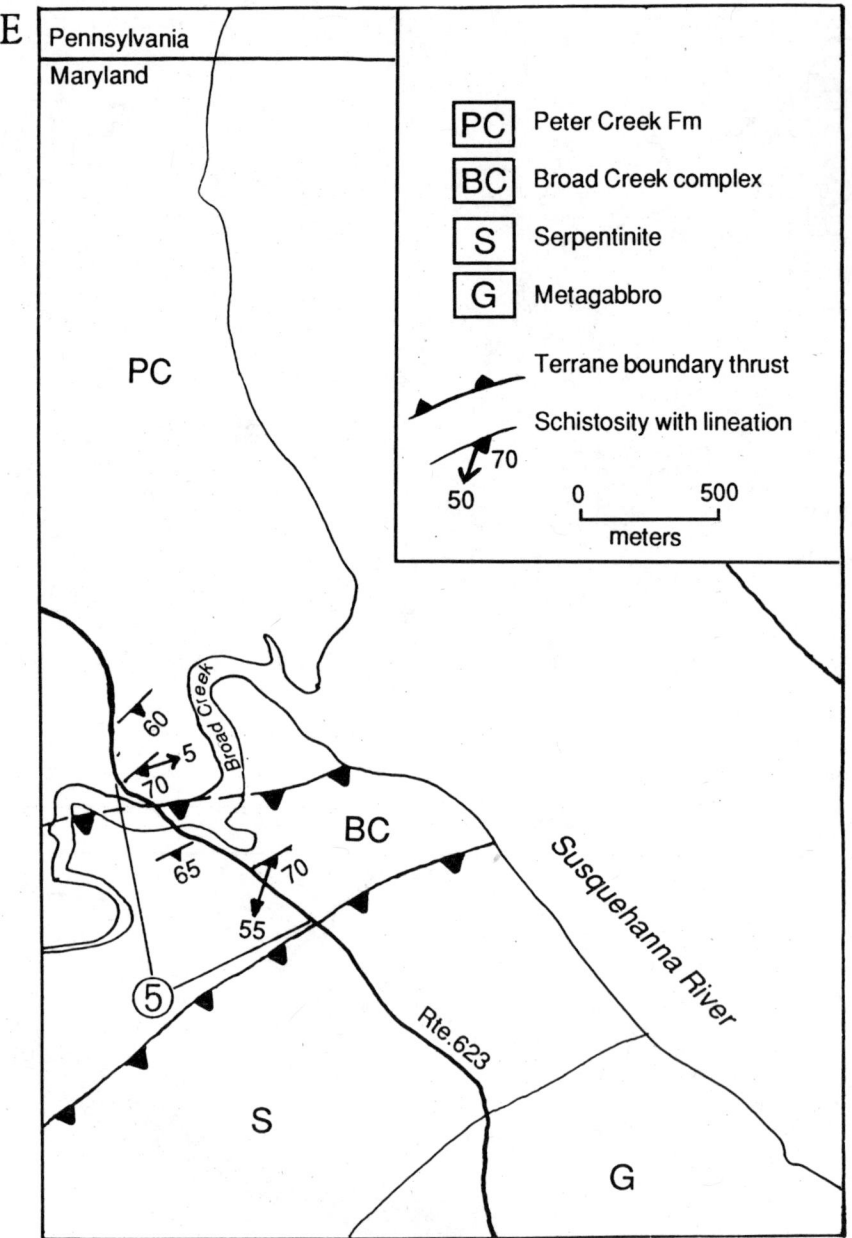

Figure 2. Location map for field stops 1–5 and geologic maps for field stop areas (locations on Figure 1). A, Location map for all field stops on day 1. B, Geologic map and outcrop locations for Stops 1A–D in the Baltimore Terrane (Muller, in press). Fine solid lines=contacts. Teeth on upper plate of thrust fault. C, Geologic map and outcrop locations of Stops 2 and 3 in the Liberty Complex, Potomac Terrane (Muller et al., 1989; Muller, in press). D, Geologic map and location of Stop 4 in the Liberty Complex, Potomac Terrane (Muller, 1985). E, Geologic map and location of Stop 5, Potomac–Westminster terrane boundary.

Soldiers Delight Complex

The Soldiers Delight ultramafic complex is composed of metamorphosed ultramafic rocks with lesser amounts of metagabbro and metapelite. It is primarily a serpentinite with variable amounts of chlorite and actinolite. Muller et al. (1989) considers the Soldiers Delight complex to be related to the smaller ultramafic bodies in the Morgan Run Formation, which appear to be blocks and clasts in a sedimentary unit.

Morgan Run Formation

The Morgan Run Formation (and correlative Piney Run Formation to the northeast) is composed of five rock types: 1) pelitic schist; 2) metagraywacke; 3) amphibole-epidote quartzite and quartzite granofels; 4) amphibolite and an inclusion-bearing metagabbro gneiss that appears to be metasedimentary; and 5) ultramafic rocks, including serpentinite and various chlorite-actinolite and chlorite-talc schist (Muller et al., 1989). In addition to these rocks, the Morgan Run contains minor amounts of foliated muscovite granite and several small ore deposits (Cu-Fe- Co-Ni-Zn mineralization) associated with detrital ultramafite lenses (Candela and Wylie, 1989; Candela et al., 1989).

The metagraywacke and quartzite of the Morgan Run Formation are devoid of alkali-feldspar and contain abundant epidote and amphibole. These rocks are characterized by high $(Ca+Na)/K$ and $(Fe+Mg)/Al$ ratios, reflecting the original immature nature of the sediment and a probable volcanic component of the source terrane(s). Also, the occurrence of detrital ultramafics with sea-floor exhalative-type mineralization and mafic metadiamictites indicates an oceanic source (Muller et al., 1989).

The Morgan Run Formation contains chaotically disrupted compositional layering and block-in-matrix structures exhibited in interlayered pelitic schist/metagraywacke and pelitic schist/ quartzite sequences (Muller et al., 1989). The chaotic layering has been interpreted as soft-sediment deformation. The block-in-matrix structure is thought to reflect pre- to early metamorphic, heterogeneous, layer-parallel shearing. Neither of these pre- to early-metamorphic deformational features are found in metasediments of the Westminster or Baltimore terranes.

The internal disruption of the pelitic schist-quartzite sequences exemplified by the block-in-matrix structures indicates that the Morgan Run Formation is transitional from a broken to a dismembered formation in the classification of Raymond (1984). The randomly interlayered lithologies further indicate that the entire Morgan Run Formation is a tectonic melange.

Sykesville Formation

The Sykesville Formation is a thick, poorly layered sequence of quartzofeldspathic schist and clast-bearing gneiss that was originally thought to be a xenolith-rich granitic intrusion (Stose and Stose, 1946). Hopson (1964) and subsequent workers (Southwick and Fisher, 1967; Fisher, 1970; Higgins, 1972) have shown that chemically and texturally the Sykesville Formation is a metasedimentary unit. It has been termed a metadiamictite by recent workers (Crowley, 1976; Fisher et al., 1979; Muller et al., 1989) and the coarse-grained clast-bearing units have been interpreted as mass-movement, debris flows.

Clasts in the Sykesville Formation range in size from millimeters to several meters and are angular to rounded. They are randomly oriented or aligned parallel to the matrix foliation. They are matrix-supported and many display internal foliations truncated by the clast margins. The medium-grained, biotite-epidote-plagioclase-quartz matrix is commonly schistose and rarely exhibits compositional layering. Clast types include vein quartz, metapelite schist, metagraywacke, quartzite and quartzite granofels, amphibolite, and ultramafic rocks.

Muller et al. (1989) consider the Morgan Run Formation to have been the primary source of the Sykesville Formation based on several lines of evidence as follows:
1) All clast types in the Sykesville Formation are identifiable as lithologies in the Morgan Run Formation.
2) All of the clast samples examined exhibit mineral assemblages that are characteristic of the Morgan Run Formation.
3) Numerous Sykesville clasts contain internal metamorphic foliations and veins both truncated by the clast margin and discordant to the matrix foliation.
4) The Sykesville and Morgan Run Formations are in close spatial association.
5) The large size of the clasts in the Sykesville Formation indicates proximity to the source area.

Tectonostratigraphic Interpretation

Muller et al. (1989) concluded that the Liberty Complex is an accretionary prism/subduction melange that was emplaced onto North America (Baltimore Terrane) during Taconic suturing of an island arc (Chopawamsic Terrane). The Morgan Run Formation is a mixture of volcanic arc-derived sediments deposited as trench or slope-basin turbidite deposits and oceanic crust which includes the Soldiers Delight Complex. These mixed rocks were tectonically juxtaposed in an accretionary prism structurally above an east-dipping subduction zone. As the arc-continent collision progressed, the Morgan Run Formation was rapidly uplifted from mid-crustal levels to the surface where it was eroded and redeposited as the Sykesville Formation. Final suturing of the arc terrane resulted in the deformation of the Morgan Run and Sykesville Formations and metamorphism to upper greenschist–lower amphibolite facies.

WESTMINSTER TERRANE

One objective of this fieldtrip is to show that the Peters Creek Formation of southeast Pennsylvania is of North American affinity and part of the Westminster terrane. We therefore describe the Peters Creek Formation as observed in the Susquehanna River drainage basin along the Maryland-Pennsylvania state line (Figure 3) and refer the reader to Muller et al. (1989) for descriptions of the Westminster terrane in other areas.

The Peters Creek Formation is composed of metagraywacke and mica-chlorite schist (interpreted as metapelite) with lesser amounts of metaconglomerate, quartzite, and greenstone. The rocks range from middle to upper greenschist facies and deformation ranges from moderate to intense.

A semi-continuous 3,200 m thick section of the Peters Creek Formation is exposed along the Susquehanna River from its northern contact with the Peach Bottom Formation at Peters Creek to its southern contact with the State-Line ultramafic complex (Bel Air–Rising Sun terrane). The lower section contains several 20-40 m thick massive quartzite layers separated by massive mica-chlorite schist. Most of the rest of the section contains 20-75 m thick packages of layered metagraywacke separated by sporadic 10–30 m thick quartzites. The layered metagraywackes (Figure 4) consist of 0.2 to 1 m graded beds of coarse sandstone-granule conglomerate to metapelite with rare and questionable ripple cross stratification. Grainsize and quartz content decrease and phyllosilicates and cleavage intensity increase towards the top of the beds. These younging criteria indicate an upright section with top to the southeast. Three greenstone layers of 10 m, 30 m, and 50 m thickness occur near the southeast contact of the unit.

Opposite: *Figure 3A.* Geologic map of the Susquehanna River–Octararo Creek area; Stops 6–10 in the Peters Creek Formation and State-Line complex of the Westminster Formation.

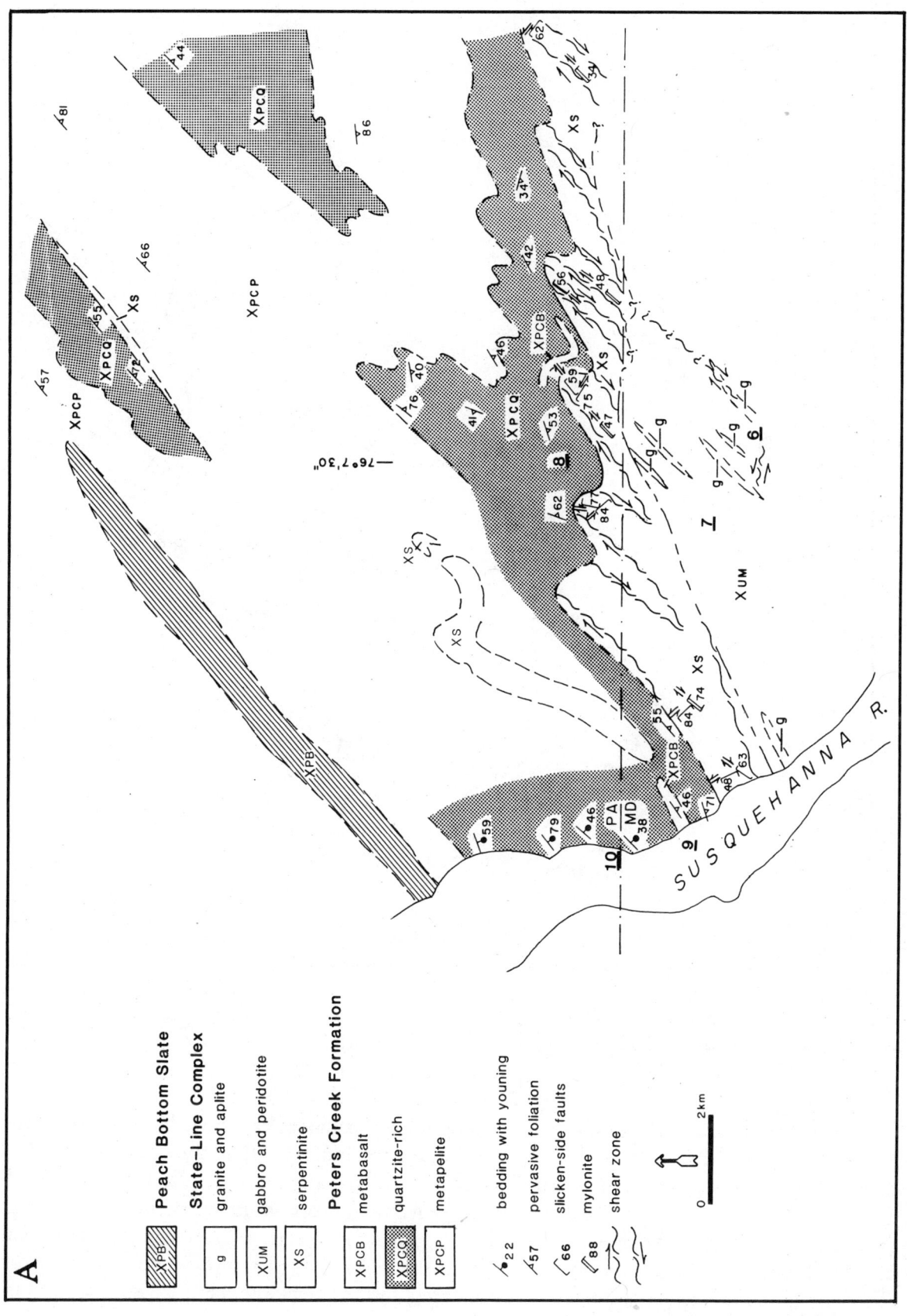

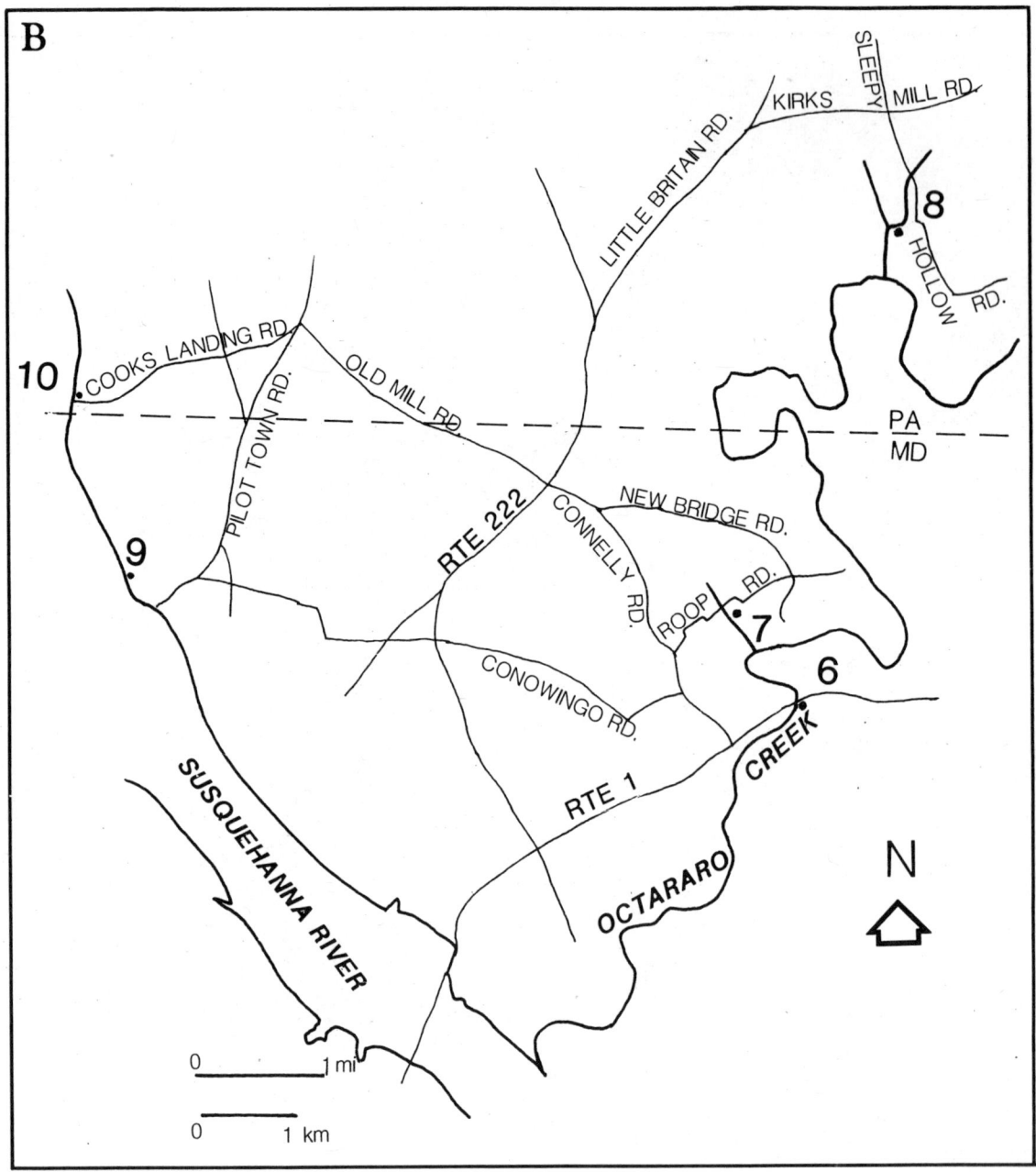

Figure 3B. Road map of the area in Figure 3A showing location of Stops 6–10 in the Peters Creek Formation and State Line complex of the Westminister Terrane.

Petrographic analysis of the coarse sandstone and conglomerate from both the Susquehanna River section and the entire Kirkwood-Rising Sun area indicates that they are feldspathic. The detrital grains (Figure 5) consist primarily of rutilated quartz, and perthitic alkali-feldspar with lesser amounts of oligoclase, apatite cemented quartzite fragments, and rounded zircon. The relict grains are enclosed in a recrystallized quartz, muscovite, chlorite, and biotite matrix. The greenstone consists of chlorite, epidote and albite with lesser amounts of amphibole, magnetite, and clinozoisite. Possible relict amygdules are composed of oval shaped intergrowths of epidote, clinozoisite and amphibole.

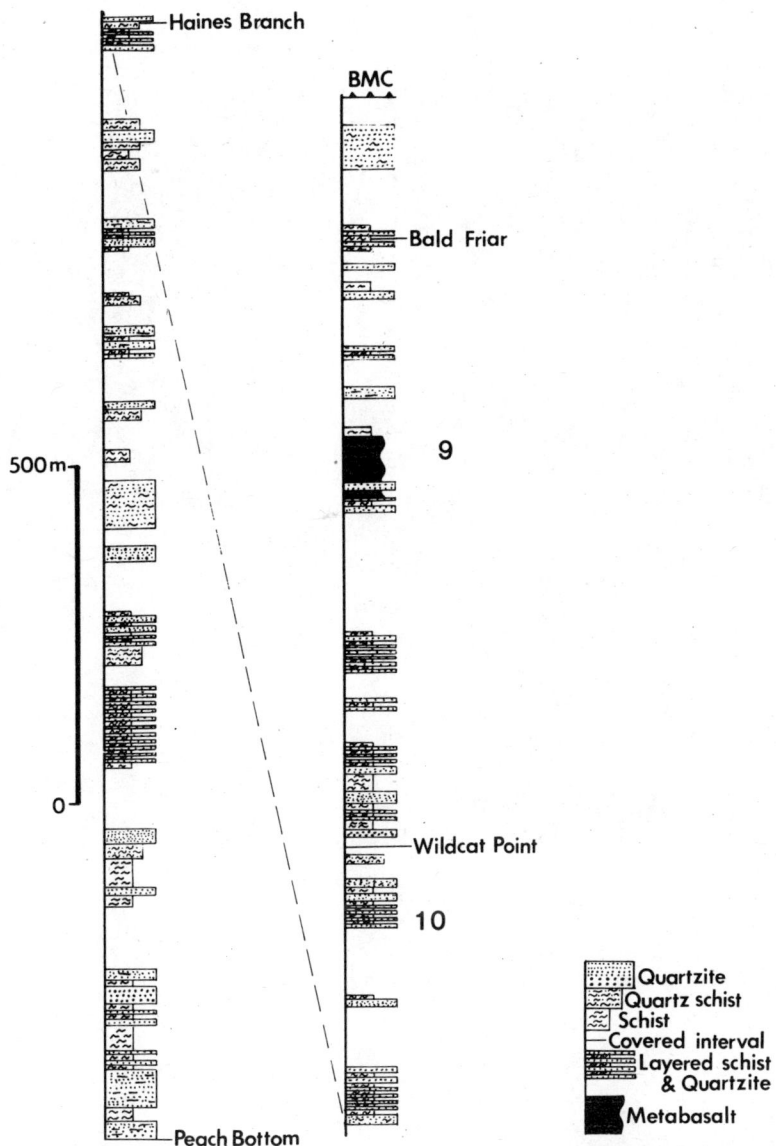

Figure 4. Stratigraphic section for Peters Creek Formation along Susquehanna River showing locations of Stops 9 and 10.

Tectonostratigraphic Interpretation

The Peters Creek Formation is interpreted as a proximal turbidite deposit based on the character and thickness of the graded beds, grainsize, abundance of pelitic material, and relation to the rest of the Westminster terrane. The paucity of coarse material indicates a considerable distance from the source and/or channels, but the thickness and composition of the beds support a more proximal position. Unfortunately, there are insufficient current indicators to determine flow direction.

We conclude that the Peters Creek Formation must have had a Grenville basement source based on the rutilated quartz, abundance of perthitic alkali-feldspar, low An content of the detrital plagioclase, rounded zircons, and the lack of volcanic or carbonate debris. The only period when there could have been an exclusive Grenville source was during the Late Proterozoic to Early Cambrian. Subsequent to this time, the Grenville massifs were blanketed by passive margin sediments.

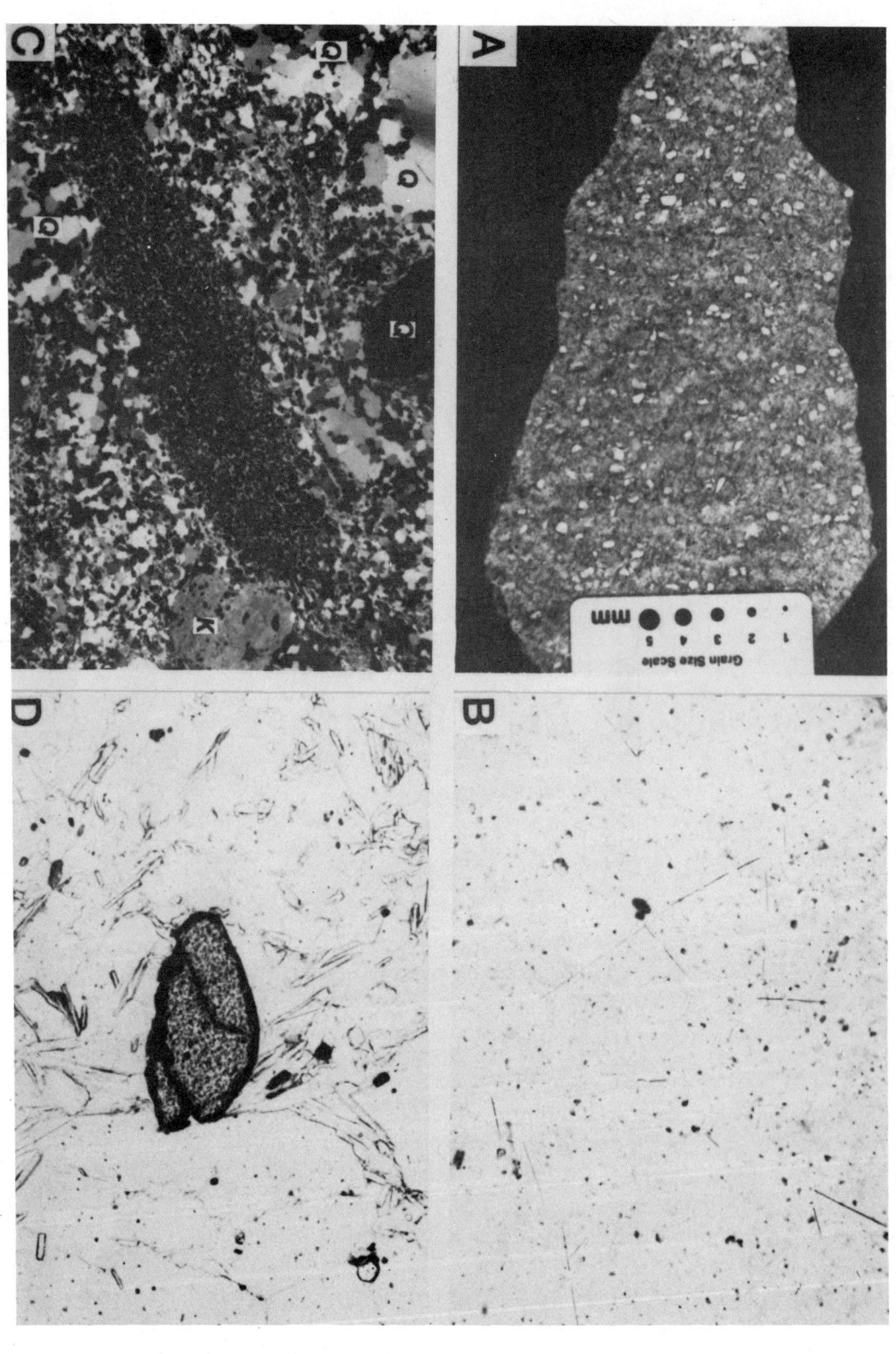

Figure 5. Metaconglomerates from the Peters Creek Formation. A, Photograph of granule metaconglomerate. B, Photomicrograph of rutile needles in quartz granule from metaconglomerate. Field of view = 0.2 mm. C, Photomicrograph of flattened apatite cemented quartzite granule and quartz (Q) and K-feldspar (K) detrital grains in a quartz-mica matrix from metaconglomerate. Field of view = 3 mm. D, Photomicrograph of rounded zircon grain in a quartz-mica matrix from metaconglomerate. Field of view = 0.5 mm.

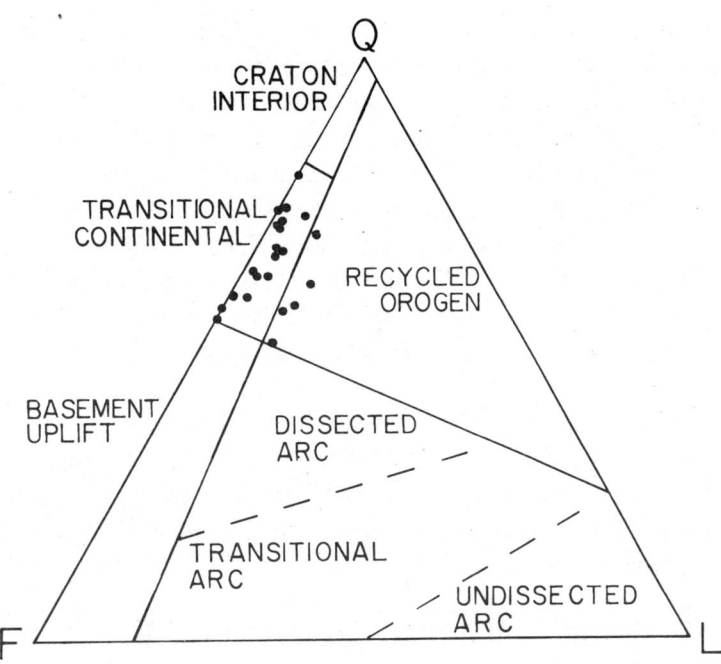

Figure 6. QFL discrimination plot with fields for sandstone provenance (Dickinson and Suczek, 1979). Points represent modal compositions of sandstones and conglomerates from the Peters Creek Formation.

Based on comparison of modal analyses of the conglomerate and sandstone with standard QFL discrimination diagrams (Figure 6) (Dickinson and Suczek, 1979; Dickinson et al., 1983), the provenance for these sediments is a continental block, transitional to a basement uplift. Such a provenance is consistent with the clast composition and provides further evidence for a Grenville basement source. The location of the point cluster on the QFL plot (Figure 6) is characteristic of rift-generated sediments (Dickinison et al., 1983). It is also nearly identical to the QFL plot for the rift portion of the Unicoi Formation of the Cambrian Chilhowee Group in the Valley and Ridge province (Simpson and Eriksson, 1989). The Peters Creek Formation is also similar to the turbidite portion of the rift-generated, Late Proterozoic Lynchburg Group of Virginia (Wehr, 1985). Neither the Lynchburg Group nor the Unicoi Formation extends into northern Virginia and Maryland (Wehr, 1985). The Peters Creek Formation may be the northern and more important, eastern continuation of the major Late Proterozoic/Cambrian rift basin in eastern North America.

The Peters Creek Formation could not have had a volcanic arc source as is proposed for the graywacke in the Liberty Complex of the Potomac terrane (Muller et al., 1989) because:
1) No volcanic detritus was found in the Peters Creek deposits.
2) The detrital plagioclase grains are oligoclase rather than andesine, as would be expected from a volcanic arc.
3) Most of the feldspar is perthitic alkali-feldspar.
4) The QFL plot does not support such a provenance.
5) No debris from Valley and Ridge shelf deposits occur in even the coarsest sandstone. These observations, however, do not preclude the possibility that the source for the Peters Creek deposits was the Baltimore terrane to the east.

Structural Geology

Southeast of the Peach Bottom Slate, the Peters Creek Formation exhibits less deformation than most rocks in the southeastern Pennsylvania Piedmont. Compositional layering in quartzite and metagraywacke is concordant with the pervasive S1 foliation. These foliations commonly strike northeast and dip 35-70SE along the Susquehanna River but strike more east-west to the north of the State-Line body. S2 and S3 crenulation cleavages are well developed to the north of and along the western side of the State-Line body. These foliations strike 025-045 and 005-030 respectively with moderate to steep southeast dips. The foliations are axial planar to mesoscopic and macroscopic tight folds of the same generation that are slightly to moderately asymmetric with northwest vergence. Along the Susquehanna River, S2 and S3 are rare to absent but variably developed S4 conjugate crenulation cleavages and associated open folds are common. The cleavages dip shallowly to moderately to the north and strike east-west to west-northwest. Conjugate box folds develop where both cleavages are well developed.

The contact between the Peters Creek Formation and State-Line complex to the southeast is interpreted to have been thrust faulted and later reactivated as a dextral strike-slip fault (Gates, in press). The deformation that produced the S2 and S3 foliations and associated folding is related to the strike-slip faulting. Other interpreted thrust faults in the Peters Creek Formation were reactivated as strike-slip faults.

BEL AIR–RISING SUN TERRANE

State-Line Complex

The Peters Creek Formation is bounded to the southeast by the State-Line mafic complex of the Bel Air–Rising Sun terrane (Figures 1, 3). We interpret the State-Line complex and the entire Bel Air–Rising Sun terrane to have intruded the Baltimore and Westminster terranes, and therefore propose the abandonment of the terrane designation for these bodies. The State-Line complex is a layered mafic intrusion (Hanan and Sinha, 1989; Gates, in press). The basal, northern portion is a serpeninite with relict peridotite including websterite and dunite as well as large chromite deposits (Pearre and Heyl, 1960). The serpentinite is proposed to be separated from the rest of the State-Line complex by a thrust fault (Higgins and Conant, 1986) and to be part of the Potomac terrane (Horton et al., 1989). No thrust faults were found between the serpentinite and main body but exposure is incomplete. The middle portion of the State-Line complex is gabbroic including massive and rhythmically layered portions. The upper portion is a quartz gabbro to diorite with small plagiogranite bodies, xenoliths, and roof pendants. The roof pendants and xenoliths include rutilated blue quartz quartzite and metapelite that are characteristic of the Peters Creek Formation. The quartz gabbro contains rutilated quartz with reaction rims around the grains. The igneous stratigraphy of the body indicates a southeast younging direction, similar to that in the Peters Creek Formation.

Interpretation

The State-Line complex and entire Baltimore Mafic complex have been interpreted as an ophiolite sequence (Crowley, 1976; Morgan, 1977; Muller and Chapin, 1984; Wagner and Srogi, 1987; Horton et al., 1989). Recent isotope geochemistry studies, however, have indicated that most of the Baltimore Mafic complex intruded through continental crust and was contaminated by it (Shaw and Wasserburg, 1984; Hanan and Sinha, 1989). The petrology of the complex indicates it is a layered mafic intrusion (Hanan and Sinha, 1989; Gates, in press). Such bodies are exclusively intracontinental intrusions. The rutilated quartz in the quartz gabbro and xenoliths are interpreted to be partially digested clasts from the Peters Creek Formation. We interpret the State-Line complex as an intrusion into the Peters Creek basin rather than a fragment of another terrane or an ophiolite. Geophysical studies of the body indicate that it is tabular and probably a sill, similar to the Palisades sill of the Mesozoic Newark rift basin (Sun, 1990). Intrusive ultramafic/mafic bodies are common in the Peters Creek Formation as well as the Lynchburg group (Wehr and Glover, 1984) and are characteristic of rift basins in general. The ultramafic bodies in the Liberty Complex are distinct from the rest of the Baltimore Mafic Complex both petrologically and by stratigraphic associations. They contain metagabbro and metapelite in addition to chlorite- and actinolite-rich ultramafic rocks whereas the State-Line serpentinite does not. The Liberty Complex ultramafic rocks are likely ophiolitic material.

TECTONIC MODEL

After the Grenville orogeny and the resultant consolidation of the eastern North American craton, there was a major Late Proterozoic/Early Cambrian rifting event (Figure 7a) (Rankin, 1975; Wehr and Glover, 1985; Simpson and Erickson, 1989). Extensive rift deposits formed along the southern Appalachian Blue Ridge. These pinch out northward and indicate dying out of rifting in this direction (Wehr and Glover, 1985). This apparent northern terminus of the rift basin coincides with a major Appalachian recess which was thought to result from the development of a transform margin during Late Proterozoic rifting (Rankin, 1975; Fisher et al., 1979). The identification of rift-generated sediments (Peters Creek Formation) to the north and especially east of the Lynchburg/lower Chilhowee basin supports the transform model and rifting in the central Appalachians.

The Peters Creek Formation was probably deposited on attenuated and possibly transitional continental crust. The bathymetric relief caused high energy deposits. Deposition probably lasted from Late Proterozoic(?) through Cambrian. Intrusion through the thinned crust occurred of mafic and ultramafic magma into the sedimentary pile either late syn- or post-deposition (Figure 7b). The Baltimore mafic complex has been dated at 490 Ma using Nd/Sm whole rock methods (Shaw and Wasserburg, 1984), indicating that rifting continued through the Cambrian in the Westminster terrane.

Stratigraphic studies in the Maryland and northeastern Virginia Piedmont suggest that a peninsular horst formed as a result of the Late Proterozoic rifting (Fisher et al., 1979; Muller and Chapin, 1984; Evans, 1984). The Baltimore terrane was proposed to have been detached from the Blue Ridge terrane to the west, and a deep water basin developed between them (Fisher et al., 1979). Sandstone that lies unconformably on the Grenville gneiss terranes is medium- to coarse-grained both to the west (Antietam Formation) and to the east (Setters Formation) but fine- to medium-grained between these areas. This repetition of areas of coarse material separated by fines indicates two sources, the continent and a horst (Fisher et al., 1979). The Conestoga and Frederick limestones are indicative of a deep water facies (Fisher et al., 1979) separating deep from shallow platform facies to the west. The apparent shallow water sandstone and limestone on the Baltimore gneiss domes to the east are therefore enigmatic. Fisher et al. (1979) suggest an embayment in the Early Cambrian carbonate bank between the Cockeysville Marble to the east and the Frederick limestone to the west. The embayment in the carbonate bank was thought to contain deposits of fine-grained pelitic material reflecting the paleobathymetry of the horst to continent transition.

An alternative hypothesis to the peninsular horst is the shuffling of terranes through offset on dextral transcurrent faults (Glover and Gates, 1987). The Baltimore terrane may have originated near Philadelphia, 100 km to the northeast. In this case, the shallow platformal Setters/Cockeysville sequence of the Baltimore terrane was part of the northern continuation of the Blue Ridge hinge zone in Pennsylvania and New Jersey. The entire Baltimore terrane may therefore be a fault block that moved southward relative to North America along strike-slip faults producing the apparent early Paleozoic horst/microcontinent. The many documented late dextral strike-slip faults throughout the

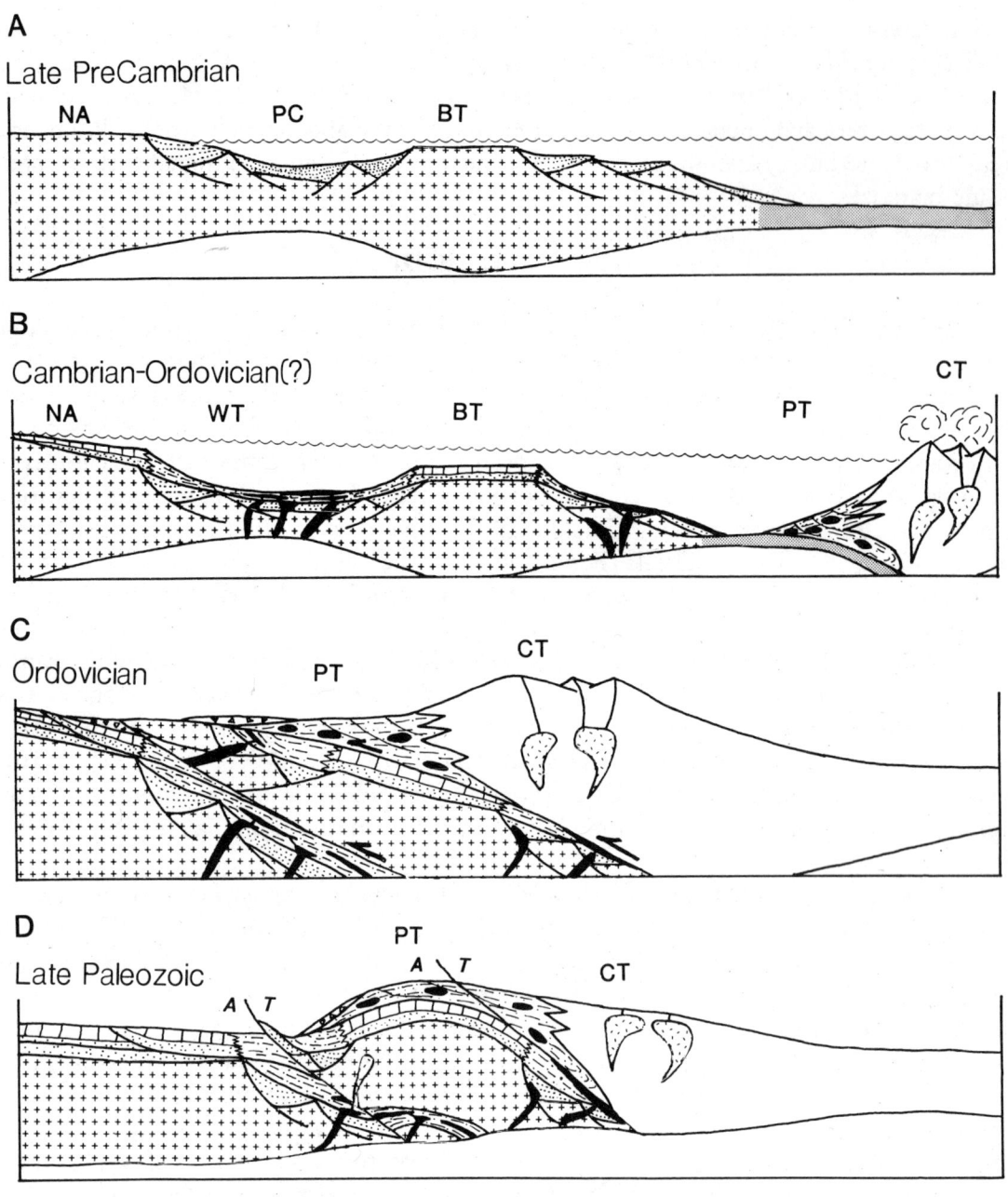

Figure 7. Tectonic model for the central Maryland through southeastern Pennsylvania Piedmont. A, Late Proterozoic–Cambrian rifting and deposition of high energy turbidite deposits with Grenville basement source in restricted basins. B, Cambrian-Ordovician(?) intrusion of mafic complexes through attenuated crust and extrusion of basalt. Deposition of passive margin deposits on basement highs, Baltimore Terrane and North American margin. Approach of composite Potomac-Chopawamsic terrane (island arc). C, Middle Ordovician Taconic Orogeny. Suturing of all terranes along thrust faults. Collapse of Peters Creek basin. D, Alleghanian(?) strike-slip faulting, granite intrusion and concurrent doming. NA = North America, PC = Peters Creek basin, BT = Baltimore terrane, WT = Westminster terrane, PT = Potomac terrane, CT = Chopawamsic terrane, A = away, T = towards.

area (Valentino, 1989; Krol et al., 1990; Gates, in press; 1990) support such a model.

Major differences in rock types and Paleozoic tectonic history exist between the Baltimore Gneiss (Muller and Chapin, 1984) and the Grenville gneisses of the Blue Ridge province (Rankin et al., 1989). The Baltimore Gneiss lacks the uppermost Proterzoic volcanic units or feeder dikes that are abundant in the Blue Ridge. The Baltimore terrane is geologically more similar to the Honey Brook Upland, Pennsylvania, than to the Blue Ridge. This connection supports the strike-slip model. On the other hand, the Baltimore terrane is also similar to the Goochland terrane (Farrar, 1984) of eastern Virginia. A connection between these two areas requires no strike-slip shuffling but does not preclude it.

During deposition of the Cambrian shelf sequence in the Baltimore terrane and the rift sequence in the Westminster terrane, the Chopawamsic terrane volcanic arc developed over an east-dipping subduction zone, in which the Iapetan ocean basin was consumed (review in Muller et al., 1989). The Potomac terrane is the forearc prism of this arc and therefore contains highly deformed volcaniclastics derived from the arc and fragments of ocean crust. The collision of the composite Chopawamsic/Potomac terrane arc with eastern North America occurred during the Taconic orogeny.

Problems with this model arise over the ages of the rocks in the Potomac and Chopawamsic terranes. Drake (1985; 1989) and Drake et al. (1989) proposed that the rocks of the Potomac terrane are pre-Arenig in age and were amalgamated from terrane fragments into a composite terrane. These rocks were deformed and intruded by the synkinematic Occoquan Granite, possibly during the Penobscottian orogenic event. The age of volcanism supports this model. The Chopawamsic volcanics are no younger than approximately 525 Ma (Higgins et al., 1977), some 50 to 75 Ma older than the Taconic orogeny. The composite Chopawamsic-Potomac terrane may therefore have been assembled prior to suturing to North America during the Taconic orogeny.

Regardless of the age of assembly, at the point of collision, the Westminster terrane was thrust westward onto the North American craton (Figure 7c). The Chopawamsic and Potomac volcanic arc was then thrust over the entire eastern margin. Therefore, at least two and possibly three coeval but vastly different packages of sedimentary rock (Westminster rift sequence, Potomac arc sequence, and Baltimore shelf sequence) were sutured together. The Baltimore terrane may have been added during later strike-slip faulting.

The Baltimore and Westminster terranes were laterally shuffled and uplifted, breaching the composite arc terrane probably during the Alleghanian orogeny (Figure 7d). The arc was divided into the Potomac terrane to the west of the Baltimore-Westminster terrane uplift and the Chopawamsic terrane to the east. Vertical movements and transpressional doming appear to have been common during Alleghanian dextral strike-slip faulting (Gates, 1987). The Chopawamsic/Carolina Slate Belt arc terrane was similarly breached through Alleghanian transpressional uplift of the Goochland terrane in Virginia (Farrar, 1984; Gates and Glover, 1989).

ACKNOWLEDGMENTS

Support for research was provided by the Maryland Geological Survey for Muller, by the Pennsylvania Geologic Survey for Valentino, and through a Rutgers University Research Fellowship and the Pennsylvania Geologic Survey for Gates. The comments of A. Drake and critical reviews of J. Gilotti and A. Schultz improved the text and are greatly appreciated.

FIELD TRIP LOG

DAY 1
STOP 1: BALTIMORE TERRANE

Stops 1A through 1D (Figures 2A, 2B) are intended to provide a brief look at a representative suite of metamorphic rocks from the Baltimore terrane. Because the focus of the field trip is on the Potomac and Westminster Terranes, we will spend only a brief time at these exposures.

STOP 1A Large hillslope outcrops in woods immediately south of railroad tracks, approximately 200 feet west of Old Court Road railroad crossing.

Layered, medium- to coarse-grained, biotite quartz-microcline-plagioclase Baltimore Gneiss containing granitic veins and pegmatite. Well-developed gneissic foliation contains isoclinal intrafolial folds. These folds are refolded by moderately southwest-plunging open folds. A narrow, steeply northwest-dipping ductile shear zone crosses foliation and displays a top-to-the-east shear sense. This exposure lies in the Woodstock dome.

STOP 1B Exposures in an abandoned quarry immediately north of Marriottsville Road. Medium-bedded, muscovite and K-feldspar quartzite of the Setters Formation dipping moderately to steeply to the north. Layering is folded by shallowly west-plunging gentle folds.

STOP 1C Small outcrops of weathered dolomitic Cockeysville Marble occur in the woods north of Marriottsville Road at edge of Patapsco River floodplain. Foliation in the marble dips moderately northwest. A small abandoned drift occurs in a pegmatite intruded along the Cockeysville–Loch Raven Formation contact. This outcrop of marble is in the same belt as the large, operating marble quarry approximately one mile to the northeast along Marriottsville Road.

STOP 1D Several partly overgrown roadcuts along the west side of Wards Chapel Road. Pelitic and semi-pelitic schist of the Loch Raven Formation. Coarse-grained, biotite $\pm$ garnet schist with abundant pegmatite pods. At the southern end of the exposures, schistosity is folded into open, shallowly west-plunging reclined folds. Interlayered micaceous quartzite occurs towards the northern part of the outcrop. WARNING: Heavy truck traffic.

STOPS 2 AND 3: LIBERTY COMPLEX

Stops 2 and 3A/3B (Figures 2A, 2C) are located along the southeast shore of Liberty Lake. Exposures are dependent on the water level being at least several feet below spillway elevation.

STOP 2 Series of outcrops along the shore of Liberty Lake south of the eastern abutment of the Route 26 bridge. Begin at the main promontory approximately 2000 feet south of Route 26, walking in from Route 26 on the dirt fire road.

The first exposure in the Soldiers Delight ultramafic body is a strongly sheared, inclusion-bearing, mafic gneiss. The inclusions are of two main types, chlorite-amphibole ultramafite and quartzofeldspathic granofels. The matrix is composed of chlorite-hornblende-plagioclase-quartz $\pm$ garnet. Muller et al. (1989) interpret the inclusions as sedimented clasts in a mafic metadiamictite. It is possible, however, that the inclusions are xenoliths and that the rock is a sheared metagabbroic intrusion. The gneiss exhibits a well-developed mineral stretching lineation that plunges obliquely down the northwest-dipping foliation. A down-dip stretching lineation is characteristic of the Liberty Complex in this area. If the water level is low enough, a prominent boudinaged pegmatite that was intruded parallel to foliation is exposed.

North along the lakeshore toward Route 26, mafic metadiamictite is structurally overlain by sheared talcose chlorite-amphibole ultramafites intruded by several-meter-thick sheets of leucocratic muscovite granite gneiss. Farther north are interlayered metagraywackes and metapelites containing lenses of amphibolite (Morgan Run Formation). The metagraywacke locally contains pebble-sized, relict detrital grains of quartz and quartz-plagioclase.

Rocks lying between the first exposure examined on the traverse and Liberty Dam to the south include various types of ultramafites, with minor intercalations of pelitic and mafic-matrix metasedimentary rocks.

STOP 3A A series of outcrops along the shoreface of Liberty Lake approximately 250 feet west of the intake for the Carroll County water treatment plant. The weathered exposures of the Morgan Run Formation consist mainly of garnet-mica-quartz schist containing phacoidal blocks of fine-grained, hornblende-bearing quartzite and stringers and pods of vein quartz. Muller et al. (1989) interpreted this block-in-matrix structure as evidence of stratal disruption during pre- to early- metamorphic shearing. There is also a 1 m thick amphibolite lens in this sequence.

Traversing east along the shore to the Oakland Road promontory, the northwest-dipping pelitic schist is underlain by locally pebbly metagraywacke, intruded by a muscovite granite vein. The vein cuts the main foliation, but is itself foliated and folded. Continuing north along the shoreline to the intersection of the old extension of Oakland Road and the lake, the shoreface is littered with float blocks of pelitic schist with prominent retrograde chlorite clots and fine-grained hornblende-bearing quartzite.

STOP 3B Outcrop and large slumped blocks along the shoreline north of Oakland Road entrance to the Baltimore City Water Department maintenance yard. The exposures are typical quartzofeldspathic metadiamictite of the Sykesville Formation. Centimeter to meter-sized clasts, including vein quartz, chlorite-biotite schist, biotite metagraywacke, and epidote-garnet $\pm$ hornblende quartzite are dispersed in a medium-grained, schistose to massive matrix of garnet-biotite-muscovite-plagioclase-quartz. Clasts range from thin chips to angular slabs and rounded cobbles. Some clasts have internal metamorphic foliations truncated by the clast margin. Foliation-compositional layering dips steeply northwest.

STOP 4: PINEY RUN FORMATION (LIBERTY COMPLEX)

A series of roadside and hillslope outcrops along Big Falls Road near the confluence of the Gunpowder Falls and Little Falls (Figures 2A, 2D). The Piney Run Formation (Morgan Run Formation equivalent) is a pelitic to semipelitic schist with intercalations of hornblende-bearing quartzite and amphibolite. Schistosity-compositional layering dips moderately to steeply northwest, and a west-plunging mineral lineation approximately parallels the axes of minor folds.

These exposures lie approximately 1000 m southeast of a large, elongate serpentinite body within the Piney Run outcrop belt. The Piney Run Formation is along strike of the Morgan Run Formation (stops 2 and 3A) to the south. The elongate ultramafic bodies in this area are similar to the Soldiers Delight complex to the south. The Sykesville Formation, the highest stratigraphic unit of the Liberty Complex, is not found in this area. Presumably its absence is the result of erosion, rather than nondeposition. The metamorphic grade of the Piney Run Formation at this location is slightly higher (staurolite zone) than that of the Morgan Run Formation exposed around Liberty lake (garnet zone). Therefore, in this area a slightly lower level of the Liberty Complex is exposed.

STOP 5: WESTMINSTER-POTOMAC TERRANE BOUNDARY

A series of roadcuts extends approximately 1000 m along route 623 from about 100 m north of Broad Creek to about 900 m south of Broad Creek (Figures 2A, 2E). These exposures cross the proposed boundary between the Westminster and Potomac terranes. Begin observations at the northernmost exposures and walk south along route 623. WARNING: There is a sharp curve just north of the narrow Broad Creek bridge. Please be alert to traffic while in this area of the traverse.

The first exposure contains gray-green, fine-grained, thinly layered quartzite of the Peters Creek Formation. The quartzite contains abundant layer-parallel vein quartz stringers and includes laminations and thin interlayers of chlorite-muscovite schist. Schistosity-compositional layering dips moderately to the southeast, and a well-developed northeast-trending, gently plunging intersection lineation is present.

South across the Broad Creek bridge, the first long exposure on the west side of Route 623 consists of quartzofeldspathic metadiamictite containing pebble- to cobble-sized clasts of vein-quartz, schist, and metagraywacke. Schistosity-compositional layering dips steeply to the southeast. Southeast-dipping shear bands cross the foliation and show consistent reverse motion. The Westminster-Potomac terrane boundary is interpreted to lie between these two outcrops.

Cross to the east side of Route 623 and continue walking south up the long straightaway. A series of small outcrops of quartzofeldpathic schist with interlayers of chlorite-rich schist occurs along the east side of the road. The schists are progressively finer-grained structurally towards the contact with serpentinite. There is a narrow zone of foliated talcose ultramafic rock at the serpentinite contact.

DAY 2

STOP 6: ROOF PORTION OF STATE-LINE COMPLEX

Park near Route 1 bridge over Octararo Creek, the outcrop is on east bank of the creek, south side of the bridge (Figure 3).

This outcrop provides a good example of the roof portion of the State-Line Complex. Unfoliated State-Line quartz diorite to quartz gabbro contains granite and aplite dikes and metapelite xenoliths. Granite dikes only intrude the upper portion of the body and are common in this area. The metapelite is a xenolith of country rock that was intruded by the complex and is likely Peters Creek Formation. The quartz diorite contains plagioclase, quartz (rarely rutilated), amphibole, minor chlorite, and relict pyroxene. The granite is medium-grained, equigranular, and plagioclase-rich, and cross-cuts the quartz diorite. The metapelite xenoliths are well-crenulated and composed of biotite-muscovite schist with quartz and plagioclase. The SE-dipping contact with the plagiogranite is sheared, exhibiting S-C fabric with consistent reverse sense. Locally, conjugate NW-dipping reverse faults are also present.

STOP 7: BLUE QUARTZ GABBRO, STATE-LINE COMPLEX

Drive into the valley, cross the stream, park on the right side of the road at the curve to the right. Hike south along the east side of the creek approximately 100 m (Figure 3).
NOTE: Permission is required. (Stop may be eliminated from trip.)

Rounded boulders in the stream and float on the eastern stream bank are blue quartz gabbro of the State-Line Complex. This is a very unusual lithology. Bright blue grains or clasts of rutilated quartz are contained within State-Line gabbro. Reaction rims of chlorite and amphibole occur around the quartz grains but the rock is dominated by clinopyroxene and plagioclase where not in contact with quartz. Rare occurrences of alkali-feldspar, epidote, and sphene have also been noted.

The blue quartz is interpreted to be xenocrysts from the country rock into which the State-Line body was emplaced. Rare clasts in the gabbro are quartzite, similar to those in the Peters Creek Formation. The quartzite appears to have been disaggregated and partially consumed by the melt. The reaction rims result from that interaction.

STOP 8: PETERS CREEK CONGLOMERATE

Drive up the hill, park on the left side at the field edge at the right-hand bend in the S-shaped curve (Figure 3). Cross the street and walk due east down the hill to the stream. Walk downstream, south and east, along the bank to the outcrops.
NOTE: Permission is required.

The first outcrops are composed of biotite-chlorite schist to phyllite, metagraywacke, and metapelite. The rocks are well-foliated with S2 and S3 crenulation cleavages. They commonly contain quartz veins.

Outcrops of metamorphosed granule to pebble conglomerate occur approximately 50 m downstream. The conglomerate is feldspathic with alkali-feldspar > plagioclase, quartz, rutilated quartz, and quartzite clasts in a metasandstone matrix (Figure 5). The plagioclase is well-twinned and of oligoclase composition. The alkali-feldspar is perthitic orthoclase and rarely occurs with plagioclase and/or quartz. The quartzite clasts are apatite cemented and ubiquitously flattened. Rounded zircon grains appear detrital.

The metaconglomerate is a L>S tectonite. The lineation is defined by quartz and feldspar rodding. Parallel to lineation and perpendicular to foliation, the granules and pebbles show shearing by rotated porphyroclasts and extended grains. Perpendicular to lineation, original grain boundaries and grain relations are commonly visible.

STOP 9: PETERS CREEK GREENSTONE

Park the cars along the railroad tracks and Susquehanna River. Walk north along the tracks approximately 0.5 km to the outcrop (Figure 3).

Peters Creek metabasalt is interlayered with Peters Creek schist and phyllite. The rock is well-foliated, green, and fine-grained. Individual layers are sporadic and range in thickness from 1-2 m to 25+ m. The mineralogy includes chlorite, albite, epidote, magnetite, and amphibole. Possible relict amygdules are ellipsoidal aggregates of epidote and chlorite.

STOP 10: PETERS CREEK GRADED BEDS

Park the cars at the railroad tracks and river; the outcrop is on the right side along the tracks (north) (Figure 3).

Outcrop contains graded beds of 0.2-1.5 m thickness. Grading is from coarse-grained sandstone, quartzose biotite schist to pelite, chlorite-biotite phyllite. Locally, probable ripple cross laminations occur. Deformation is intense. F4 folds and S4 crenulation cleavages are found in most beds.

We interpret this rock to be a proximal turbidite deposit. The clast lithology in the coarse-grained sandstones is identical to that in the Peters Creek conglomerate (Stop 8). These deposits are derived from Grenville basement massifs.

REFERENCES

Candela, P. A., and Wylie, A. G., 1989, Ultramafite-associated Cu-Fe-Co-Ni-Zn deposits of the Sykesville District, Maryland Piedmont: 28th International Geological Congress Field Trip Guidebook T241, American Geophysical Union, 10 p.

——, ——, and Burke, T. M., 1989, Genesis of the ultramafic rock-associated Fe-Cu-Co-Zn-Ni deposits of the Sykesville District, Maryland Piedmont: Economic Geology, v. 84, p. 663–675.

Coney, P. J., Jones, D. L., and Monger, J. W. H., 1980, Cordilleran suspect terranes: Nature, v. 288, p. 329–333.

Crowley, W. P., 1976, The geology of the crystalline rocks near Baltimore and its bearing on the evolution of the eastern Maryland Piedmont: Maryland Geological Survey Report of Investigations 27, 40 p.

Dickinson, W. R., Beard, L. S., Brakenridge, G. R., Erjavec, J. L., Ferguson, R. C., Inman, K. F., Knepp, R. A., Lindberg, F. A., and Ryberg, P. T., 1983, Provenance of North American sandstones in relation to tectonic setting: Geological Society of America Bulletin, v. 94, p. 222–235.

——, and Suczek, C. A., 1979, Plate tectonics and sandstone compositions: American Association of Petroleum Geologists Bulletin, v. 63, p. 2164–2182.

Drake, A. A., Jr., 1985, Sedimentary melanges of the central Appalachian Piedmont: Geological Society of America Abstracts with Programs, v. 17, p. 566.

——, 1989, Metamorphic rocks of the Potomac terrane in the Potomac Valley of Virginia and Maryland: International Geological Congress Field Trip Guidebook, T202, 22 p.

——, and Morgan, B. A., 1981, The Piney Branch Complex; A metamorphosed fragment of the central Appalachian ophiolite in northern Virginia: American Journal of Science, v. 281, p. 484–508.

——, Sinha, A. K., Laird, J., and Guy, R. E., 1989, The Taconic Orogen, in Hatcher, R. D., Jr., Thomas, W. A., and Viele, G. W., eds., The Appalachian-Ouachita Orogen in the United States: Geological Society of America, The Geology of North America, v. F-2, p. 101–177.

Evans, N. H., 1984, Latest Precambrian to Ordovician metamorphism and orogenesis in the Blue Ridge and western Piedmont, Virginia Appalachians: Unpublished Ph.D. dissertation, Virginia Polytechnic Institute and State University, Blacksburg, 324 p.

Farrar, S. S., 1984, The Goochland granulite terrane: Remobilized Grenville basement in the eastern Virginia Piedmont: Geological Society of America Special Paper, 194, p. 215–227.

Fisher, G. W., 1970, The metamorphosed sedimentary rocks along the Potomac River near Washington, D.C., in Fisher, G. W., ed., Studies of Appalachian Geology: Central and Southern: Wiley-Interscience, New York, p. 299–315.

——, Higgins, M. W., and Zeitz, I., 1979, Geological interpretations of aeromagnetic maps of the crystalline rocks in the Appalachians, northern Virginia to New Jersey: Maryland Geological Survey Report of Investigations no. 32, 40 p.

Gates, A. E., 1987, Transpressional dome formation in the southwest Virginia Piedmont: American Journal of Science, v. 287, p. 927–949.

——, in press, Shear zone control on mineral deposits in the State-Line serpentinite, Pennsylvania Piedmont, in Petruk, W., Hausen, D., and Vassiliou, A. H., eds., Process Mineralogy: Applications to Exploration: Elsevier, Amsterdam.

——, 1990, Complex structural development of the Towson dome, Maryland Piedmont: Geological Society of America Abstracts with Programs, v. 22, p. 19–20.

——, and Glover, L., III, 1989, Alleghanian tectono-thermal evolution of the dextral transcurrent Hylas zone, Virginia Piedmont, U.S.A.: Journal of Structural Geology, v. 11, p. 407–419.

Glover, L., III, and Gates, A. E., 1987, Alleghanian Orogeny in the Central and Southern Appalachians: Geological Society of America, Abstracts with Programs, v. 19, no. 2, p. 86.

Grauert, B. W., 1973, U-Pb isotopic studies of zircons from the Gunpowder Granite, Baltimore County, Maryland: Carnegie Institute of Washington yearbook, no. 72, p. 288–290.

Hanan, B. B., and Sinha, A. K., 1989, Petrology and tectonic affinity of the Baltimore mafic complex, Maryland: Geological Society of America Special Paper, 231, p. 1–18.

Higgins, M. W., 1972, Age, origin, regional relations and nomenclature of the Glenarm Series, central Appalachian Piedmont: A reinterpretation: Geological Society of America Bulletin, v. 83, p. 989–1026.

——, and Conant, L. C., 1986, Geologic map of Cecil County: Maryland Geological Survey, scale 1:62,500.

——, Sinha, A. K., Zartman, R. E., and Kirk, W. S., 1977, U-Pb zircon dates from the central Appalachian Piedmont: A possible case of inherited radiogenic lead: Geological Society of America Bulletin, v. 88, p. 125–132.

Hopson, C. A., 1964, The crystalline rocks of Howard and Montgomery Counties: Maryland Geological Survey, The Geology of Howard and Montgomery Counties, p. 27–215.

Horton, J. W., Jr., Drake, A. A., Jr., and Rankin, D. W., 1989, Tectonostratigraphic terranes and their Paleozoic boundaries in the central and southern Appalachians, in Dallmeyer, R. D., ed., Terranes in the Circum-Atlantic Paleozoic orogens: Geological Society of America Special Paper, 230, p. 213–245.

Howell, D. G., and Jones, D. L., 1984, Tectonostratigraphic terrane analysis and some terrane vernacular, in Howell, D. G., Jones, D. L., Cox, A., and Nur, A., eds., Proceedings of the Circum-Pacific Terrane Conference: Stanford University Publications in Geological Sciences, v. 18, p. 6–9.

Knopf, E. B., and Jonas, A. I., 1923, Stratigraphy of the crystalline schists of Pennsylvania and Maryland: American Journal of Science, v. 5, p. 40–62.

Krol, M. A., Onasch, C. M., and Muller, P. D., 1990, Kinematic analysis of the Pleasant Grove shear zone, eastern Maryland Piedmont: Geological Society of America Abstracts with Programs, v. 22, p. 28.

Morgan, B. A., 1977, The Baltimore Complex, Maryland, Pennsylvania and Virginia, in Coleman, R. G., and Irwin, W. P., eds., North American Ophiolites: Oregon Department of Geology and Mineral Industries Bulletin, 95, p. 41–49.

Muller, P. D., 1985, Geologic map of the Hereford, MD 7.5-minute quadrangle: Maryland Geological Survey, scale 1:24,000.

———, in press, Geologic map of the Finksburg, MD 7.5-minute quadrangle: Maryland Geological Survey, scale 1:24,000.

———, Candela, P. A., and Wylie, A. G., 1989, Liberty Complex; polygenetic melange in the central Maryland Piedmont, in Horton, J. W., Jr., and Rast, N., eds., Melanges and olistromes of the U.S. Appalachians: Geological Society of America Special Paper 228, p. 113–134.

———, and Chapin, D. A., 1984, Tectonic evolution of the Baltimore Gneiss anticlines Maryland: Geological Society of America Special Paper 194, p. 127–148.

Olsen, S., 1977, Origin of the Baltimore Gneiss migmatites at Piney Creek, Maryland: Geological Society of America Bulletin, v. 88, p. 1089–1101.

Pearre, N. C., and Heyl, A. V., 1960, Chromite and other mineral deposits in the serpentine rocks of the Pennsylvania upland, Maryland, Pennsylvania and Delaware: U.S. Geological Survey Bulletin, 1082-K, p. 707–833.

Rankin, D. W., 1975, The continental margin of eastern North America in the southern Appalachians; The opening and closing of the proto-Atlantic ocean: American Journal of Science, v. 275-A, p. 298–336.

———, Drake, A. A., Jr., Glover, L., III, Goldsmith, R., Hull, L. M., Murray, D. P., Ratcliffe, N. M., Read, J. F., Secor, D. T., Jr., and Stanley, R. S., 1989, Pre-orogenic terranes, in Hatcher, R. D., Jr., Thomas, W. A., and Viele, G. W., eds., The Appalachian-Ouachita Orogen in the United States: Geological Society of America, The Geology of North America, v. F-2, p. 7–100.

Raymond, L. A., 1984, Classification of melanges, in Raymond, L. A., ed., Melanges; their nature, origin, and significance: Geological Society of America Special Paper 198, p. 7–20.

Shaw, H. F., and Wasserburg, G. J., 1984, Isotopic constraints on the origin of Appalachian mafic complexes: American Journal of Science, v. 284, p. 319–349.

Simpson, E. L., and Eriksson, K. A., 1989, Sedimentology of the Unicoi Formation in southern and central Virginia: Evidence for Late Proterozoic to Early Cambrian rift-passive margin transition: Geological Society of America Bulletin, v. 101, p. 42–54.

Sinha, A. K., Hanan, B. B., Sans, J. R., and Hall, S. T., 1980, Igneous rocks of the Maryland Piedmont: Indicators of crustal evolution, in Wones, D. R., ed., The Caledonides in the USA: Virginia Polytechnic Institute, Department of Geological Sciences Memoir no. 2, p. 131–136.

Southwick, D. L., and Fisher, G. W., 1967, Revision of stratigraphic nomenclature of the Glenarm Series in Maryland: Maryland Geological Survey Report of Investigations 6, 19 p.

Stose, A. J., and Stose, G. W., 1946, The physical features of Carroll and Frederick Counties: Maryland Department of Geology Mines and Water Resources, p. 11–131.

Sun, D. F., 1990, A gravity study of the State-Line body in the Maryland-Pennsylvania Piedmont: unpublished M.S. thesis, Rutgers University, Newark, NJ, 65 p.

Tilton, G. R., Doe, B. R., and Hopson, C. A., 1970, Zircon age measurements in the Maryland Piedmont, with special reference to the Baltimore Gneiss problems, in Fisher, G. W., ed., Studies of Appalachian Geology: Central and Southern: New York, Wiley-Interscience, p. 429–434.

Valentino, D. W., 1989, Evidence for dextral strike slip shearing in the Pennsylvania Piedmont along the Susquehanna River: Geological Society of America Abstracts with Programs, v. 21, p. 73.

Wagner, M. E., and Srogi, L., 1987, Early Paleozoic metamorphism at two crustal levels and a tectonic model for the Pennsylvania-Delaware Piedmont: Geological Society of America Bulletin, v. 99, p. 113–126.

Wehr, F., 1985, Stratigraphy of the Lynchburg Group and Swift Run Formation, Late Proterozoic (730–570 Ma), central Virginia: Southeastern Geology, v. 25, p. 225–239.

——, and Glover, L., III, 1985, Stratigraphy and tectonics of the Virginia–North Carolina Blue Ridge: Evolution of a late Proterozoic–early Paleozoic hinge zone: Geological Society of America Bulletin, v. 96, p. 285–295.

Wetherill, G. W., Davis, G. L., and Lee-Hu, C., 1968, Rb-Sr measurements on whole rocks and separated minerals from the Baltimore gneiss, Maryland: Geological Society of America Bulletin, v. 79, p. 757–762.

Williams, H., and Hatcher, R. D., Jr., 1982, Suspect terranes and accretionary history of the Appalachian orogen: Geology, v. 10, p. 530–536.

Zen, E-an, 1983, Exotic terranes in the New England Appalachians: Limits, candidates and ages; A speculative essay: Geological Society of America Memoir, v. 158, p. 55–81.

2
STRATIGRAPHY AND STRUCTURE OF THE GREAT VALLEY AND VALLEY AND RIDGE, WEST VIRGINIA

Peter Lessing
West Virginia Geological Survey
P.O. Box 879, Morgantown, WV 26507

Stuart L. Dean
Department of Geology
University of Toledo, Toledo, OH 43606

Byron R. Kulander
Department of Geological Sciences
Wright State University, Dayton, OH 45435

INTRODUCTION

This field trip considers the geology of intensely folded and faulted Lower, Middle, and Upper Paleozoic sedimentary rocks in the central Appalachians of northeastern West Virginia. The trip begins at Harpers Ferry, West Virginia, on the western overturned limb of the Blue Ridge anticlinorium and proceeds westward across the Massanutten synclinorium (Great Valley) to its western border at the North Mountain fault. West of this location the structure and stratigraphy of eastern Valley and Ridge rocks will be examined in the context of modern concepts of thin-skinned deformation. Figure 1 is an index map for all 11 stops and Table 1 is the stratigraphic column, with abbreviations and stops for the entire field trip.

The field trip emphasizes new interpretations of the geology and regional tectonics that have evolved within the last decade, including those of Faill (1985), Mitra (1987), Kulander and Dean (1986), Dean et al. (1987), and Hatcher (1989). Detailed mapping by Dean et al.(1987) has greatly modified previous geological interpretations by Grimsley (1916) and Woodward (1941, 1943, 1949, 1951). This trip will hopefully foster a greater appreciation for the role of small-scale structures in the overall deformation and shortening of the Paleozoic section.

The tectonics of the region are dominated by the detached Massanutten–Blue Ridge overthrust sheet (Kulander and Dean, 1986). This allochthonous sheet was transposed westward on Upper Ordovician Martinsburg shales from a longitudinal ramp-zone located under the Blue Ridge anticlinorium, as shown by Mitra (1987). In essence, the Cambrian-Ordovician section has been doubled in thickness through emplacement of this detached sheet. Displacement of the Massanutten–Blue Ridge sheet is partly taken up at the North Mountain fault at the western border of the Great Valley. Additional displacement of this sheet appears to be translated westward into the Valley and Ridge on shales of the Martinsburg Formation, creating the upper detached Martinsburg sheet. Westward movement of the lower, or Waynesboro, sheet occurred on shale of the Lower Cambrian Waynesboro Formation, with longitudinal ramping and local imbrications that have rotated and accentuated pre-existing folds and faults in the upper Massanutten–Blue Ridge and Martinsburg sheets, as well as initiating new cover structures in the Martinsburg sheet.

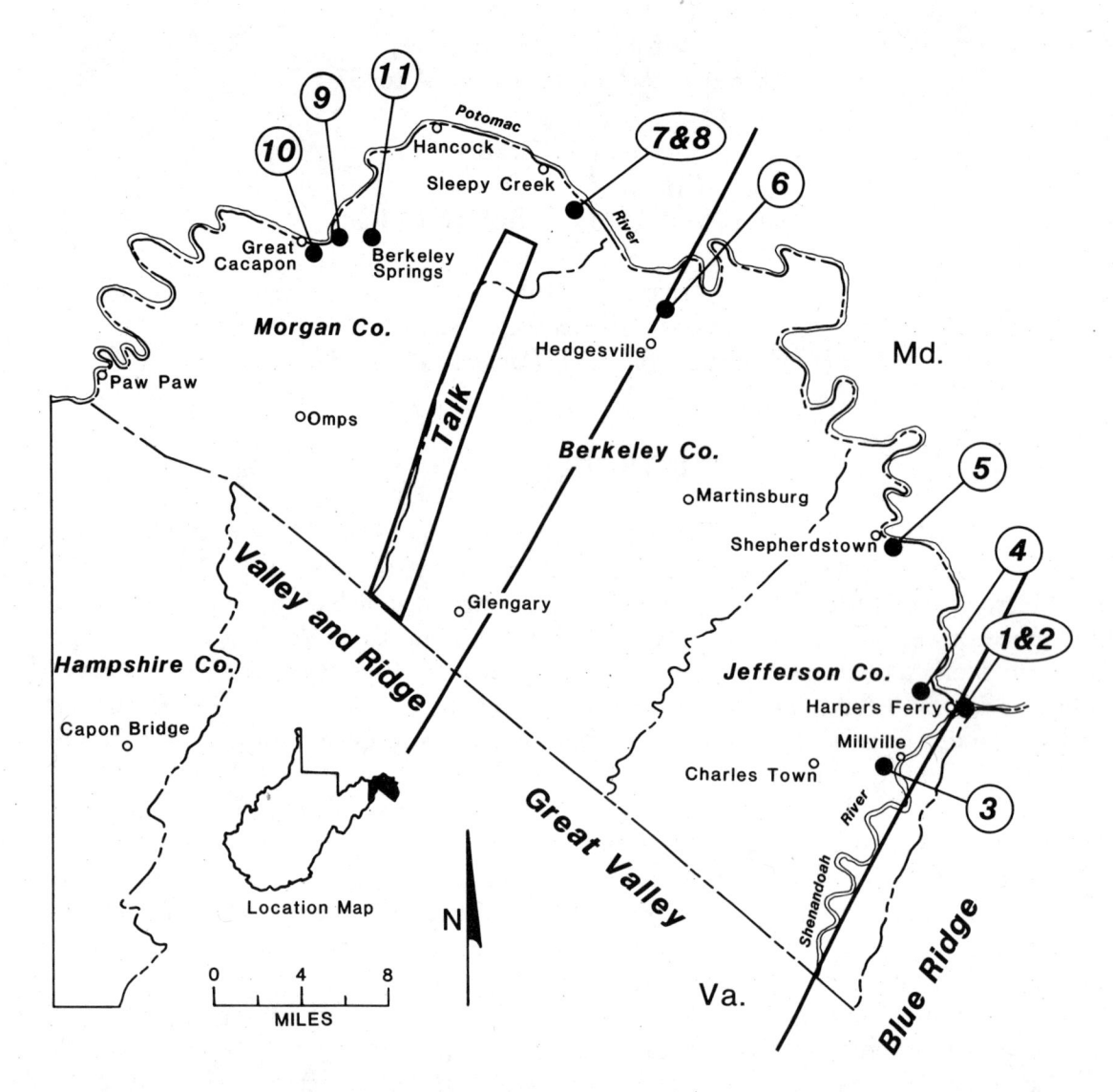

Figure 1. Index map of eastern West Virginia showing physiographic provinces and all stop locations.

TABLE 1. Stratigraphic column, abbreviations, and stops for the Great Valley and Valley and Ridge of eastern West Virginia.

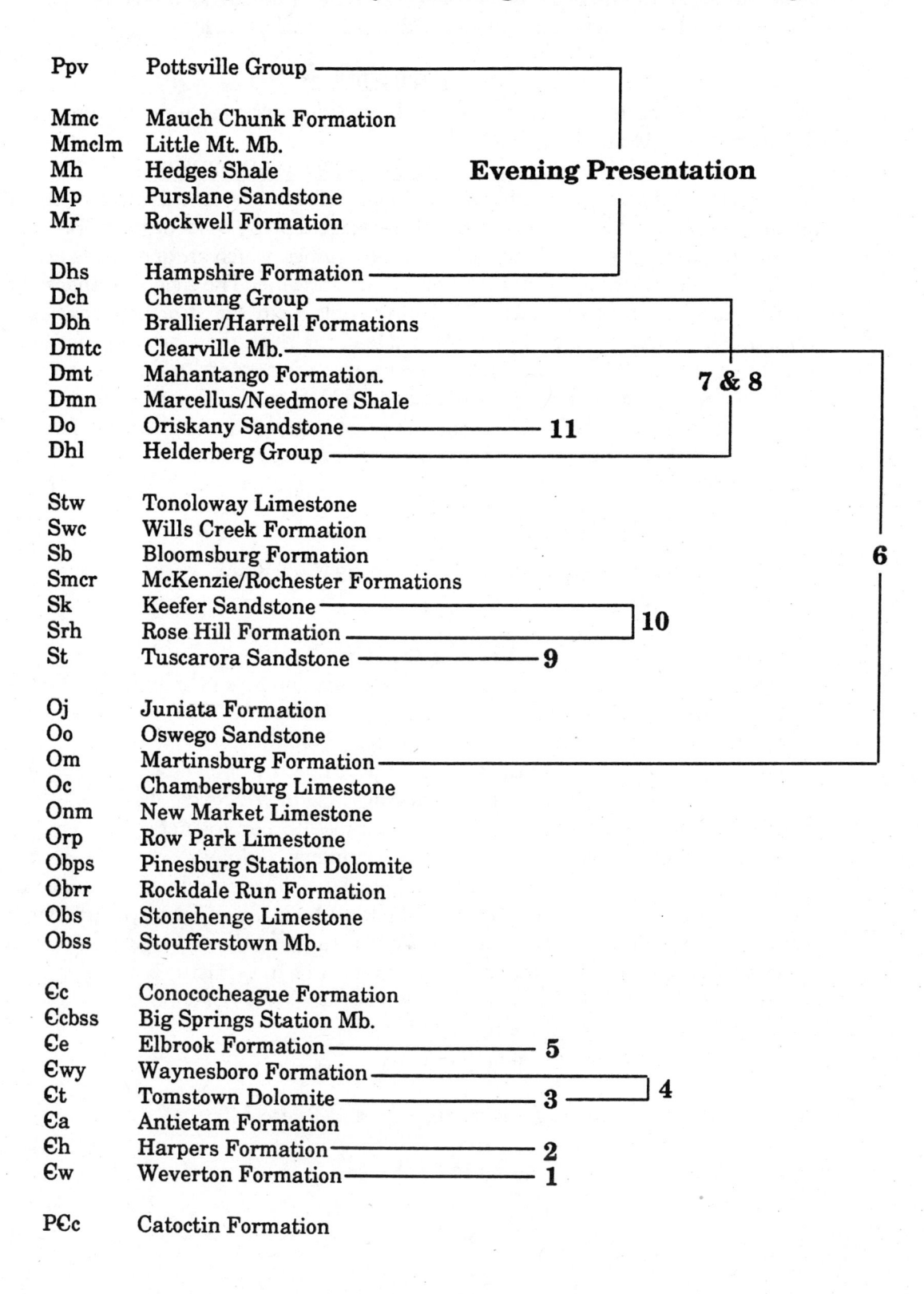

Ppv	Pottsville Group	
Mmc	Mauch Chunk Formation	
Mmclm	Little Mt. Mb.	
Mh	Hedges Shale	**Evening Presentation**
Mp	Purslane Sandstone	
Mr	Rockwell Formation	
Dhs	Hampshire Formation	
Dch	Chemung Group	
Dbh	Brallier/Harrell Formations	
Dmtc	Clearville Mb.	
Dmt	Mahantango Formation.	7 & 8
Dmn	Marcellus/Needmore Shale	
Do	Oriskany Sandstone — 11	
Dhl	Helderberg Group	
Stw	Tonoloway Limestone	
Swc	Wills Creek Formation	
Sb	Bloomsburg Formation	6
Smcr	McKenzie/Rochester Formations	
Sk	Keefer Sandstone	
Srh	Rose Hill Formation — 10	
St	Tuscarora Sandstone — 9	
Oj	Juniata Formation	
Oo	Oswego Sandstone	
Om	Martinsburg Formation	
Oc	Chambersburg Limestone	
Onm	New Market Limestone	
Orp	Row Park Limestone	
Obps	Pinesburg Station Dolomite	
Obrr	Rockdale Run Formation	
Obs	Stonehenge Limestone	
Obss	Stoufferstown Mb.	
Єc	Conococheague Formation	
Єcbss	Big Springs Station Mb.	
Єe	Elbrook Formation — 5	
Єwy	Waynesboro Formation	
Єt	Tomstown Dolomite — 3	4
Єa	Antietam Formation	
Єh	Harpers Formation — 2	
Єw	Weverton Formation — 1	
PЄc	Catoctin Formation	

31

ROAD LOG

Tuesday, March 12, 1991

Mileage

0.00   Road cut on south side of Route 340 with exposure of the metamorphic Precambrian Pedlar Formation, a hypersthene granodiorite with vertical chloritic dikes.

0.50   **STOP 1 (FIGURE 2)**

The Weverton Formation (west limb of Blue Ridge anticlinorium) at the Virginia–West Virginia border (Dean et al., 1990a).

CAUTION: HEAVY TRAFFIC AND MAD DRIVERS

The massive cliffs on the south side of Route 340 are overturned rocks of the Weverton Formation dipping 30–55 degrees SE with cleavage dipping 5–15 degrees SE. The rocks are cut by numerous quartz-filled en echelon tension gashes, which are apparently caused by fold-related fracturing on the overturned limb of the anticline. The sigmoidal shapes of many of the quartz-filled fractures indicate an earlier origin than the structural position they now occupy. Across the Potomac on the north side of the river, there are two concentric folds in the Weverton Fm. We are standing on the western overturned limb of the eastern anticline. Because the Weverton Fm. is approximately 500 ft thick, several folds are necessary to account for the 3,000 ft of this formation exposed here that forms the main part of the water gap. Figure 3 shows the structure of the north side of the Potomac River modified from Cloos (1951; 1958).

Proceed west on Route 340.

1.80   Cross Shenandoah River on Route 340; turn right off Route 340 onto Shenandoah St. at north end of bridge.

2.00   **STOP 2 (FIGURE 2)**

Group will leave bus at parking lot and walk along outcrops of the Harpers Fm. and return to bus. The Lower Cambrian Harpers Formation (Nichelsen, 1956) along this traverse in the National Park is overturned in most of the cuts. Deformation of subhorizontal cleavage has resulted in numerous chevron folds and the development of a weak S_3 crenulation foliation which in places crosses the original S_2 cleavage (Figure 4). S-pole diagrams of S_1 (bedding) and S_2 (chevron folds in foliation) have the same attitude for Beta (fold hinge of N20°E), implying that all folding here occurred as a continuum and not as separate events. Near the eastern end of this traverse, a small antiformal syncline is outlined by iron-oxide-stained bedding and is on an eastern overturned limb of the Harpers Ferry syncline. Figure 5, after Onasch and O'Conner (in press), shows the structure and the cleavage/bedding relationships at this location. Figure 6 shows our interpretation of the structure of this entire section in the Harpers Formation, as well as the Harpers Ferry syncline and Bolivar anticline immediately to the west (Dean et al., 1990a).

Proceed west along Route 340.

2.00   Turn right at intersection of Shenandoah St. and Route 340.

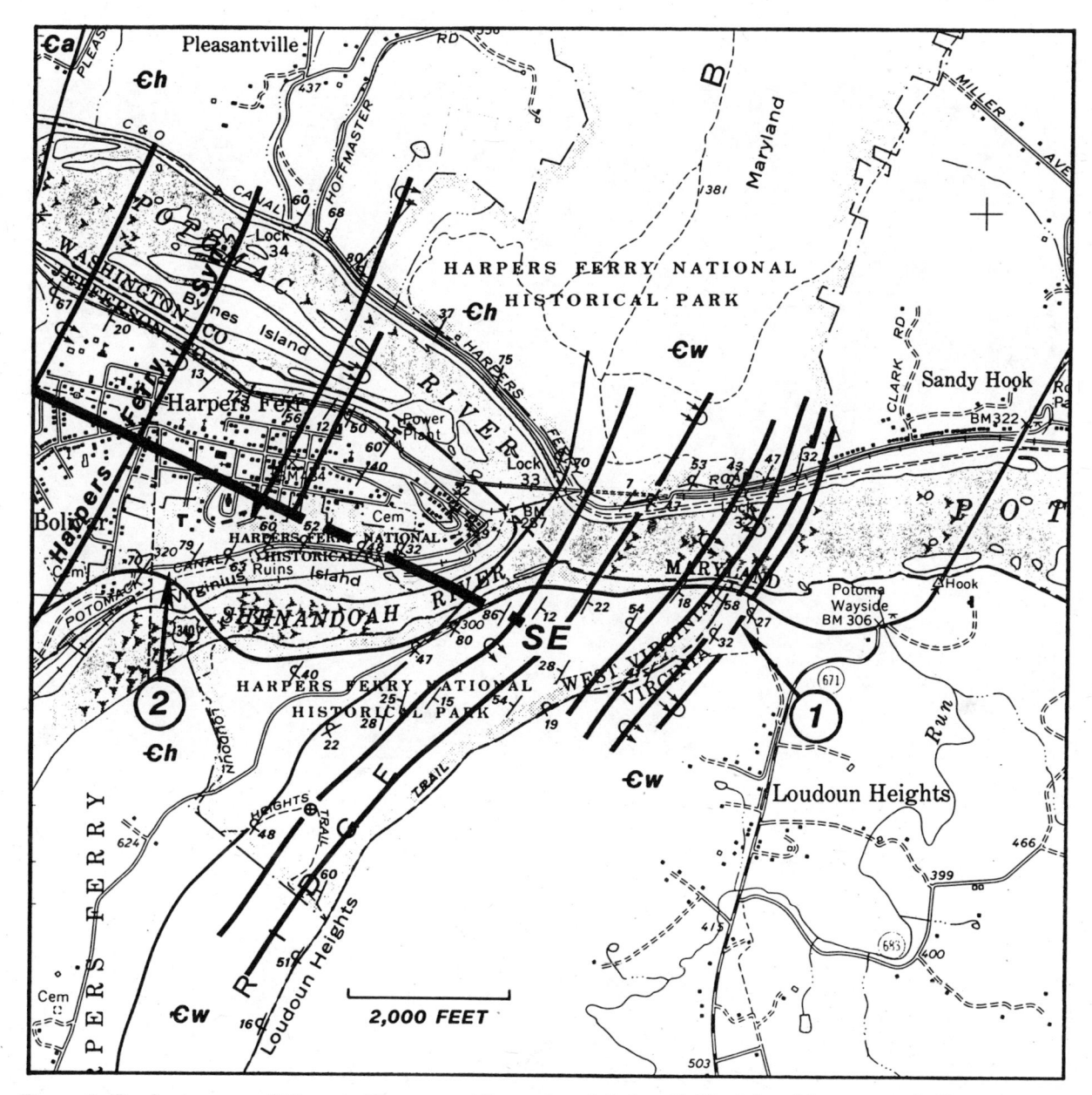

Figure 2. Geologic map of Harpers Ferry area; Stops 1 and 2 (see Table 1 for abbreviations). Cross section line (heavy black line), which continues on Figure 9, is illustrated as Figure 6.

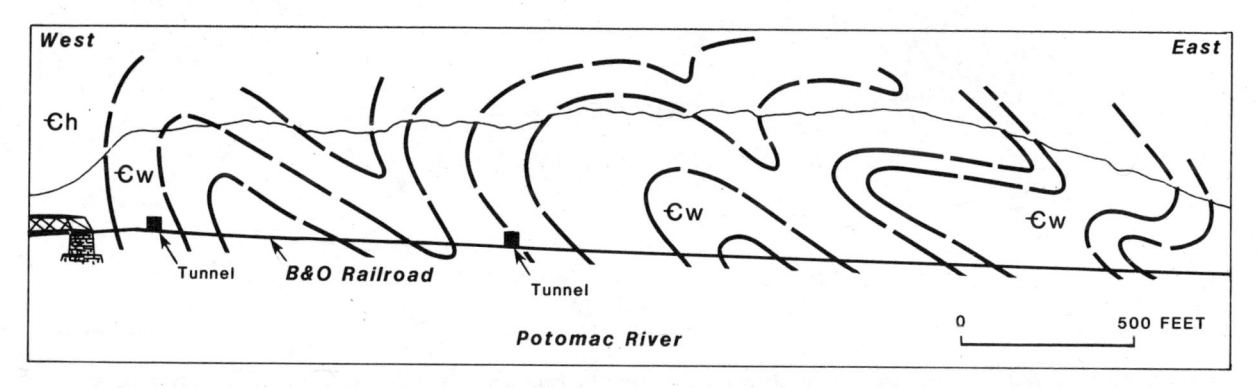

Figure 3. Sketch of geological structures in the Weverton Fm through Elk Ridge, north side of Potomac River, as viewed from Route 340 on the south side of the Potomac River (modified from Cloos, 1951, 1958); Stop 1.

Figure 4. Photograph of Harpers Formation showing cleavage (S_2) dipping 25° right (east), bedding (S_1) dipping 80° left (west), and chevron folds with trace of axial plane parallel to bedding; Stop 2. See Figure 6 for location of photo.

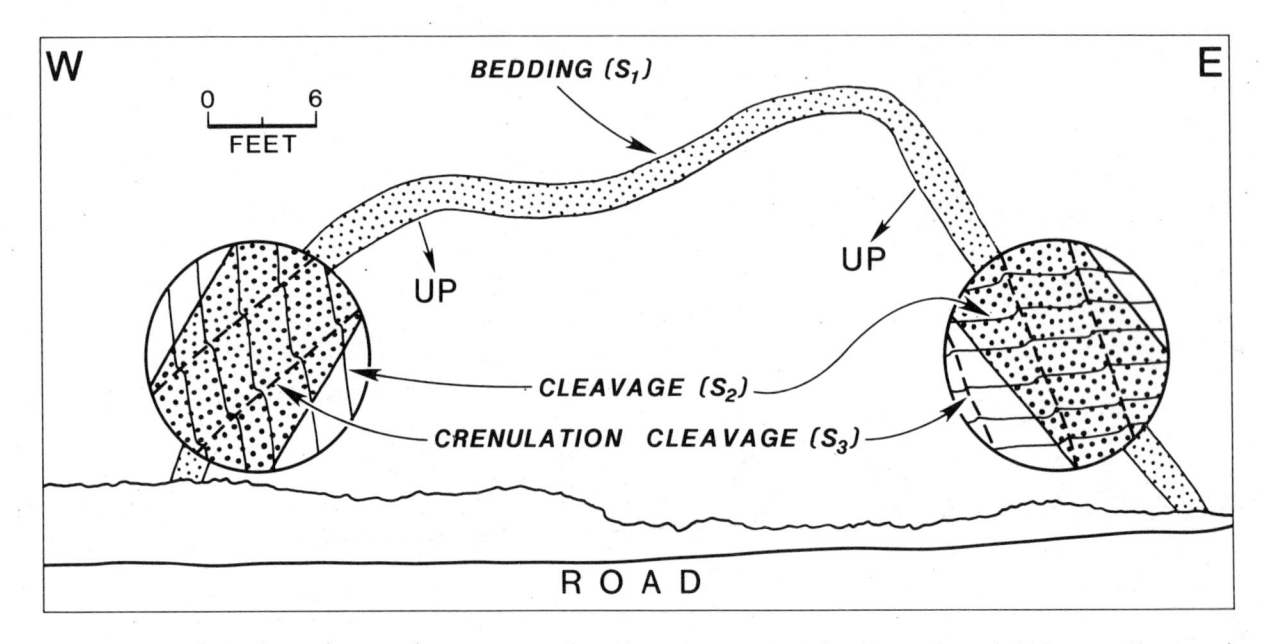

Figure 5. Antiformal syncline in the overturned Harpers Formation after Onasch and O'Conner (in press). See Figure 6 for location.

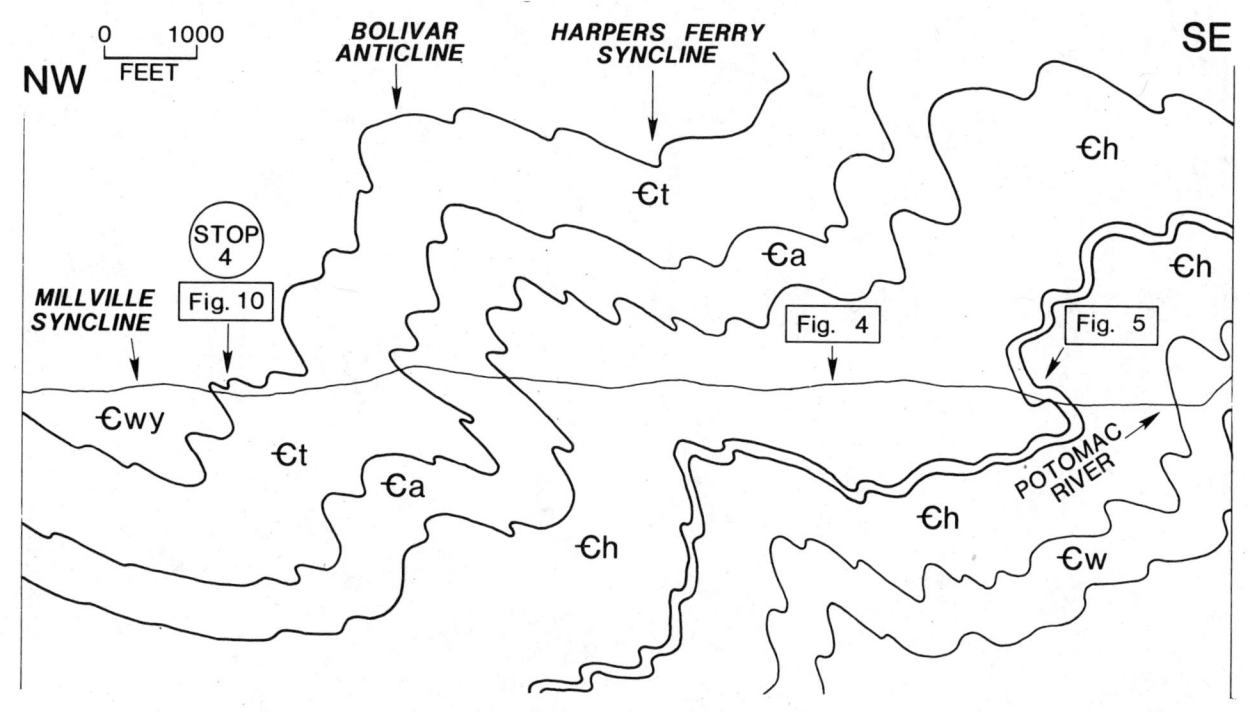

Figure 6. Generalized cross section of Harpers Ferry area; Stops 2 and 4.

3.70 Turn left off Route 340 onto Bloomery Road towards Millville. Bloomery Road follows the Tomstown Fm. in a topographic low, which structurally is the eastern limb of the overturned Millville syncline.
5.40 Scenic downtown Millville; church on left.
5.60 Cross left over railroad tracks.
5.70 Cross over railroad tracks again; Shenandoah River is on the left.
5.90 Turn right, cross railroad tracks again, and proceed to Millville Quarry offices on left.

6.00 **STOP 3 (FIGURE 7)**

Millville Quarry offices. We will stop at three locations around and within the quarry to examine the geology and local structures. First proceed around the quarry to the southeast rim for an overview. This location shows the broad, low-angle dips on the limbs of the Millville syncline. This open fold is responsible for the wide outcrop belt of the Tomstown Dolomite. East of the quarry, the rocks in the Tomstown steepen rapidly and are overturned (Figure 8). This overturned limb of the Millville syncline is continuous with the overturned rocks of the Tomstown Fm. in the abandoned quarry to the north along the Bloomery Road (Dean et al., 1990a).

Continue to the north quarry rim, where, just north of the haulage road, Tomstown Dolomite is exposed in a shallow pit. Here, the Tomstown dolomite is on the west limb of the Millville syncline and dips east at 27 degrees. The dominant planar anisotropy is steeply dipping solution cleavage. The exceptionally thick residue along the solution cleavage and the close spacing indicate that significant layer parallel shortening occurred by tectonic pressure solution. Proceed into the quarry to view the south face at the southeastern end of the quarry. In the hinge zone of the Millville syncline, well-exposed backthrusts offset dark beds in the dolomite.

Lunch. Retrace route back to Route 340 via the Bloomery Road.

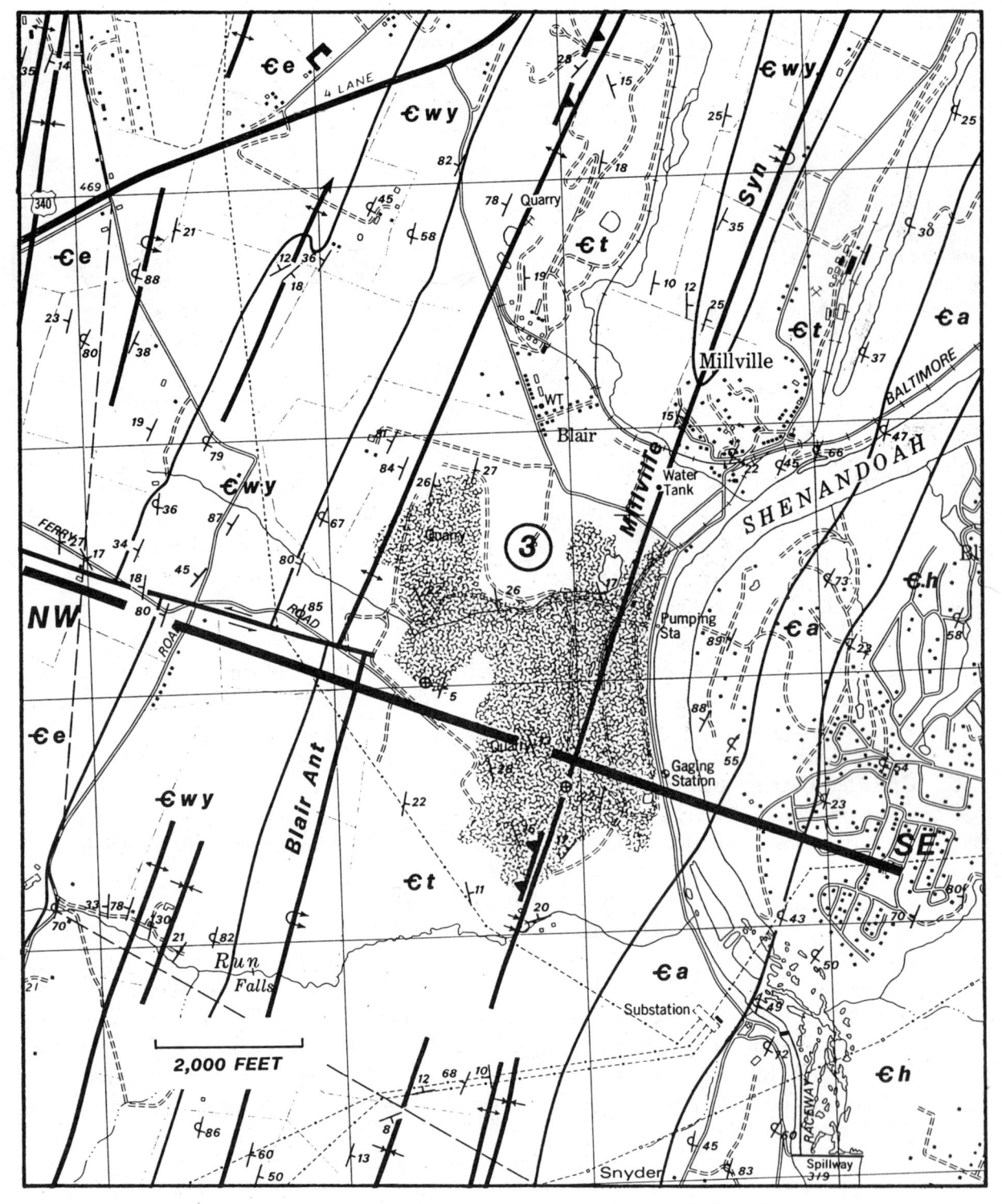

Figure 7. Geologic map of Millville quarry area in Tomstown Dolomite; Stop 3.

8.30 Intersection of Bloomery Road and Route 340. Cross Route 340 and proceed north on Bakerton Road, which lies in a topographic low on the eastern limb of the Millville syncline. The ridge on the right is the Antietam Formation. The ridge on the left is within the Millville syncline, which is cored by the Waynesboro Formation.

9.80 Railroad underpass: STOP AND HONK HORN! Then proceed through, turn left, and stop 500 feet on left.

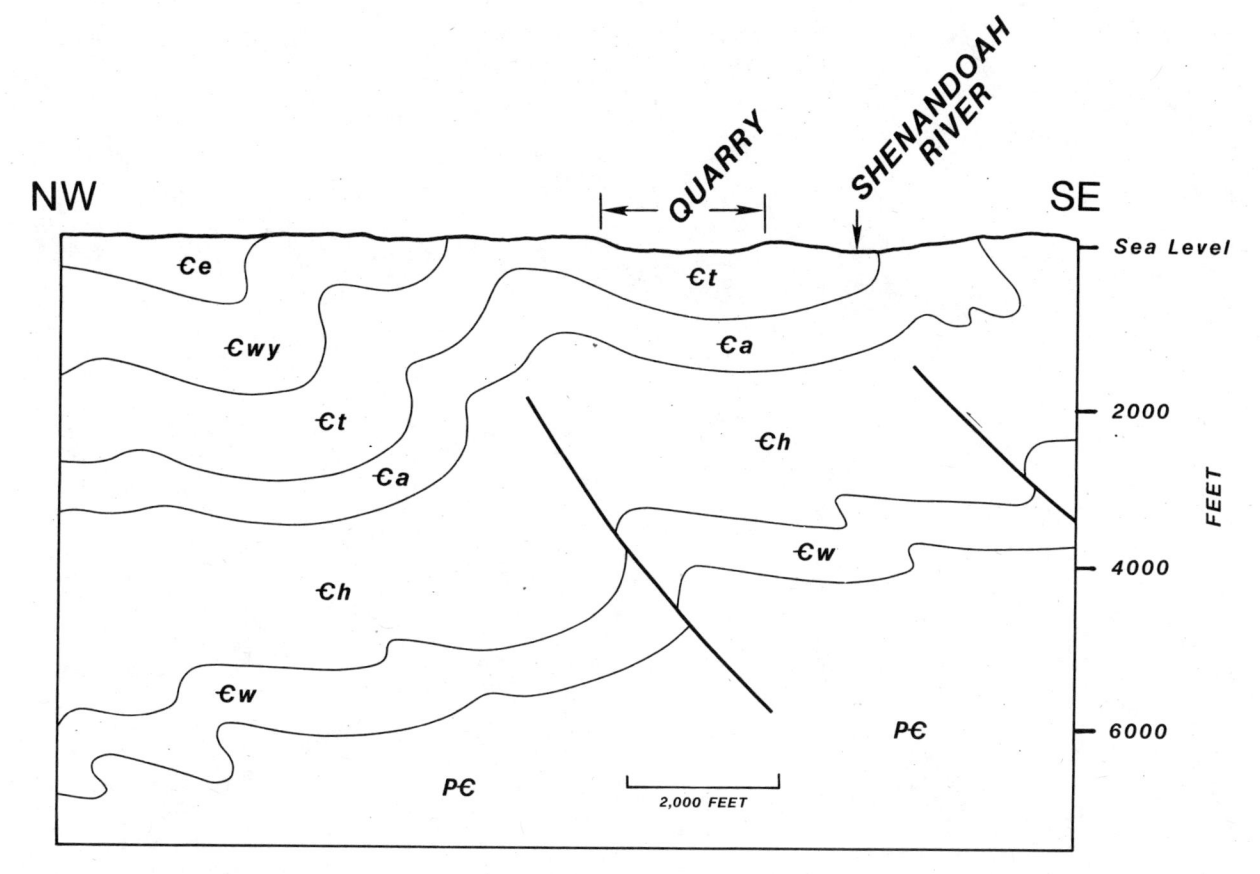

Figure 8. Cross section through Millville quarry, located in Tomstown Dolomite of the Millville syncline; Stop 3.

9.90 STOP 4 (FIGURE 9)

Walk approximately 1,000 feet west along the left side of the railroad tracks.
<u>CAUTION: COMMUTER AND FREIGHT TRAINS TRAVEL FAST ALONG THIS STRETCH OF TRACKS. BE CAREFUL.</u>

Along the western part of this section the lowermost sandstones of the Lower Cambrian Waynesboro Formation (Dean et al., 1990a) are exposed. Presumably, the shales stratigraphically above these sands serve as the deep-seated detachment horizon at the base of the Waynesboro sheet (Kulander and Dean, 1986) some 30,000 ft beneath our present position. Proceeding eastward from the west end of the section, the gentle folds in the Waynesboro Fm. end as the beds steepen at the Waynesboro-Tomstown contact. The remainder of the section to the east is intensely folded Tomstown Dolomite and limestone. The structure through this short section is a small-scale example of the overturned folds and the overall structural style of the Blue Ridge and eastern Great Valley. Figure 10 is a sketch from a panoramic photograph of most of this section, with geology reconstructed above and below the section line. In this section, the variations in rock ductility are clearly evident. Limestone has undergone significant ductile flow, whereas dolomite has deformed by concentric folding and fracturing. Numerous small-scale structures are present, indicating the contrasts in deformation and the facing directions. These features include flow cleavage, deformed oolites, S and Z folds, stromatolites (Figure 11), dolomite rip-up clasts, and fibrous mineralized extension joints.

Proceed north on the Bakerton Road along the Waynesboro-Tomstown outcrop belt.

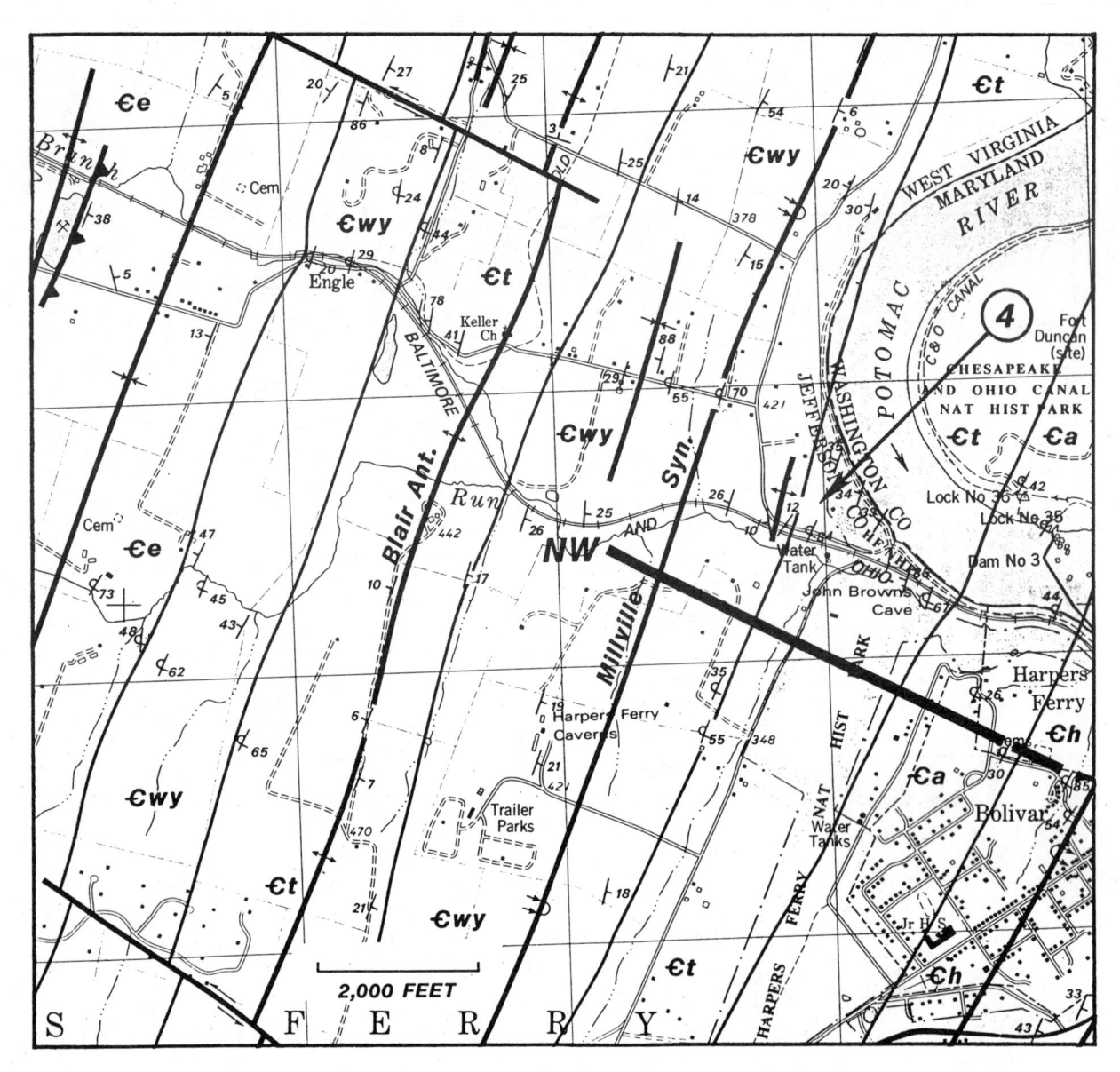

Figure 9. Geologic map of B&O (now CSX) railroad cut northwest of Harpers Ferry; Stop 4.

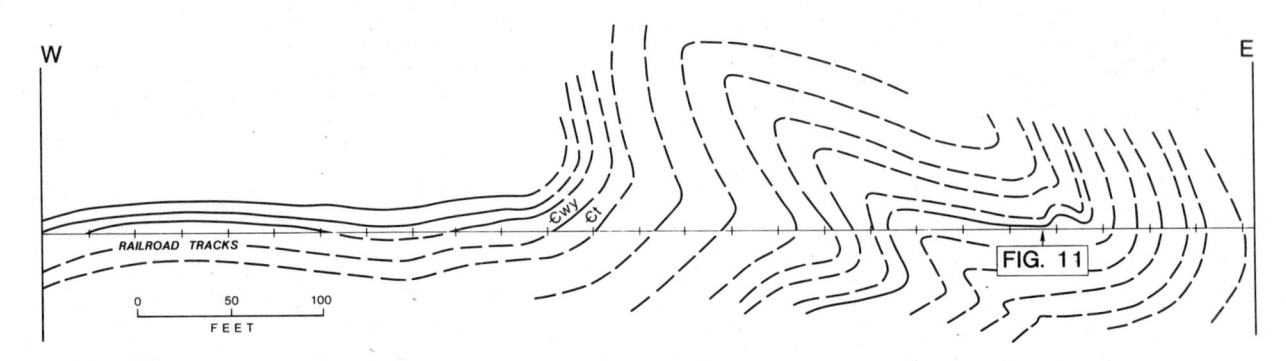

Figure 10. Cross section of folded rocks of the Tomstown and Waynesboro Fms. along B&O (now CSX) Railroad from panoramic photograph; Stop 4. See Figure 6 for location.

38

Figure 11. Photograph of Tomstown Dolomite that shows bedding dipping steeply to left (west, perpendicular to pencil at top center) and stromatolites that indicate bed top is west; Stop 4. See Figure 10 for location.

11.90 Bakerton Road turns left. DO NOT TURN. Proceed northeastward towards Moler Crossroads and Shepherdstown.
14.90 Moler Crossroads: Turn left, then take an immediate right and proceed towards Shepherdstown.
17.70 Turn right towards Shepherdstown on Highway 230. This road follows the strike of the Upper Cambrian Conococheague Formation.
18.50 Turn right at four-way road junction. Remain on Highway 230.
18.60 Again, turn right at four-way road junction. Old Yellow Brick Bank, now a restaurant, is across the junction on the left. Proceed east across a Conococheague Fm. outcrop belt.

19.30   STOP 5 (FIGURE 12)

Walk east along the river road for approximately 1,000 ft. At the west end of this section, the contact between the Conococheague and Elbrook Formations lies in the small stream (Dean et al., 1987). The almost continuous section of Elbrook to the east occupies the western overturned limb of a small anticline. Small-scale structures here include boudinage, deformed oolites, and cleavage. Approximately 0.1 mi east of the beginning of the section is the anticlinal hinge, with a small cave. Note the pronounced cleavage fan around the hinge zone. For the next 0.1 mi east, on the forelimb of the anticline, bedding dips gently 25–55° to the southeast, with a wide variation of cleavage dips. Algal stromatolites are also present along this section. Near the end of the section is a small syncline of the Conococheague Fm. Finally, the Elbrook Fm. reoccurs, dipping 60–70° W.

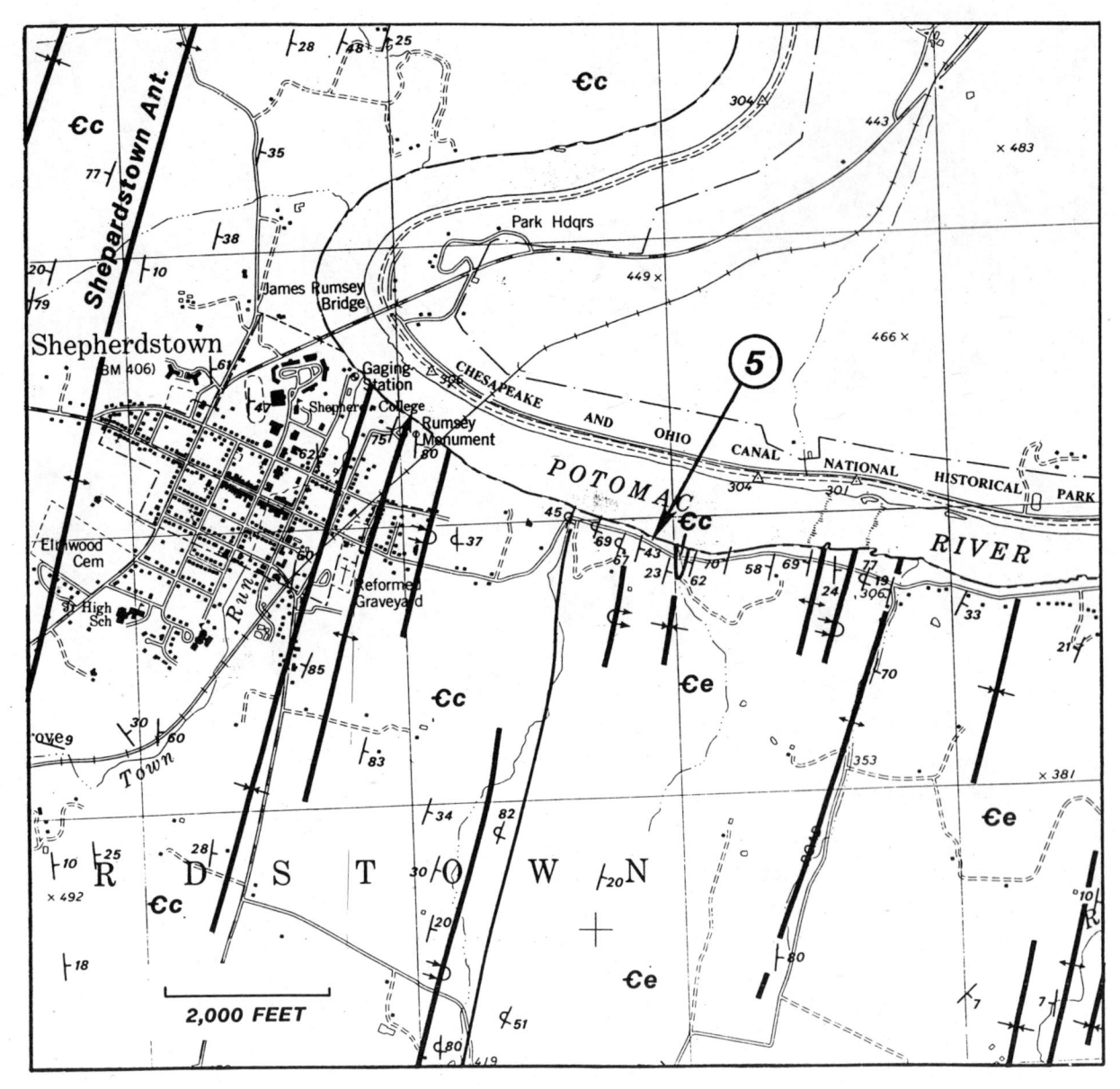

Figure 12. Geologic map of Shepherdstown area; Stop 5.

21.50   At road intersection at Old Yellow Brick Bank in Shepherdstown, proceed west through town.
21.70   At the intersection of Routes 480 and 45, proceed west on Route 45 towards Martinsburg. Road crosses a 2.5-mile expanse of folded rocks in the Conococheague Fm.
27.70   Overturned west limb off the Blairton anticline in the Chambersburg Limestone. The contact of the Chambersburg Limestone and Martinsburg Fm. occurs at the approximate location of Opequon Creek.
29.90   At intersection of Route 45 and Route 9, turn right on Route 9.
36.10   Route 9 takes a bend to the right. North Mountain fault lies just west of Route 9, approximately along the position of the road trending S25W. Some of the homes along this fault trace have near-artesian flow in water wells.
36.40   Route 9 bends left and crosses the trace of the North Mountain fault 0.1 miles before the road intersection to the right. At this point, the North Mountain fault has placed the Elbrook

40

	Formation upon shales of the Martinsburg Fm.
36.60	Turn right at road intersection. Highs Dairy Store is on the northwest corner. Proceed north along a fault slice (horse) of Martinsburg shale within the footwall of North Mountain fault.
37.00	Pass old Hedgesville High School (now Middle School) and turn right on Allensville Road (Route 901).
37.80	Turn left on Allensville Road just before railroad tracks. Proceed towards North Mountain railroad cut.
38.00	Cross North Mountain fault trace in a topographic low just around sharp bend to the left in the road. Proceeding northeast around sharp bend to the right, the road lies just west of the North Mountain fault trace. Overturned Tuscarora Sandstone (Lower Silurian) caps North Mountain to the left.
38.60	Remains of the old Martinsburg shale brick plant on right side of road. Tuscarora Sandstone is exposed in cuts on left side of road.
39.20	Turn right and cross railroad tracks on dirt road.

STOP 6 (FIGURE 13)

Walk east through railroad cut. Discussion will begin at eastern end as we walk west back to the bus.

CAUTION: ONCE AGAIN, TRAINS TRAVEL FAST THROUGH THIS SECTION. BE ESPECIALLY CAREFUL HERE ONCE IN THE RAILROAD CUT.

The low ridge approximately 0.25 mi east of our location is underlain by the Elbrook Formation. The North Mountain fault trace lies 600 feet east of where we stand and runs just east of the abandoned farm house. The North Mountain fault, and related fault-bound horses, is one of the major overthrusts of the Appalachians (Dean et al., 1987). It extends from near the Pennsylvania-Maryland border to the trend change between central and southern Appalachian structures in southwestern Virginia, a distance of approximately 175 miles (Kulander and Dean, 1986). In the Great Valley and Blue Ridge, the North Mountain fault is on a decollement above shales of the Martinsburg Fm. The entire Great Valley and Blue Ridge are allochthonous and comprise the Massanutten–Blue Ridge sheet, which has effectively doubled the thickness of the Cambrian-Ordovician section. At the first location, the Elbrook has been thrust onto Chambersburg Limestone, which in turn is thrust as a small horse against the Martinsburg Formation (Figure 14A, B).

Approximately 5,000–7,000 ft of Cambrian and Ordovician rocks encompassing the Elbrook Fm., Conococheague Fm., Beekmantown Fm., St. Paul Group, and Chambersburg Limestone are omitted by faulting at this locale. Some interpretations suggest that the entire displacement of the Massanutten–Blue Ridge sheet is accommodated here by upward ramping of the North Mountain fault in the western Great Valley. However, our minimum estimates (Kulander and Dean, 1988) indicate that 30–35 mi of displacement are necessary for the emplacement of this overthrust sheet. From our present location, the North Mountain fault complex ends just 15–20 mi northeastward in Pennsylvania. This requires the displacement of the thrust sheet to die out rather abruptly, although the Blue Ridge and Great Valley in southern Pennsylvania still require major overthrusting of the Massanutten–Blue Ridge sheet. A more plausible explanation to accommodate this major overthrusting is to transfer movement along the North Mountain ramp to a buried detachment in the Martinsburg Fm. (Dean et al., 1990b). Thus, as the North Mountain fault displacement decreases northeast, displacement increases westward at the Martinsburg level under the Valley and Ridge. The increase in surface folding and faulting in Maryland and Pennsylvania to the northwest and north of our location are consistent with this interpretation.

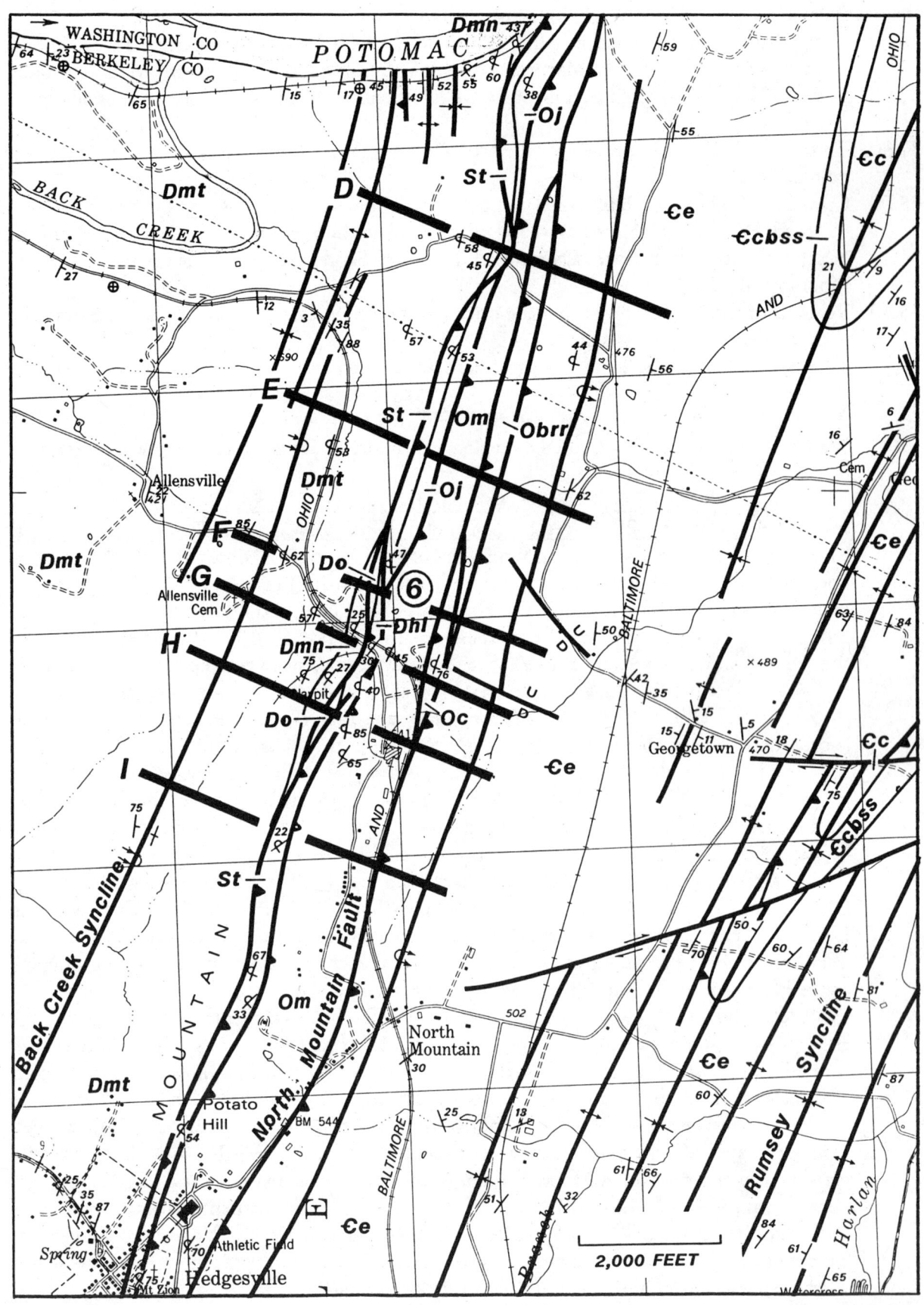

Figure 13. Geologic map of North Mountain area; Stop 6. Cross section lines D-I are illustrated in Figure 14.

42

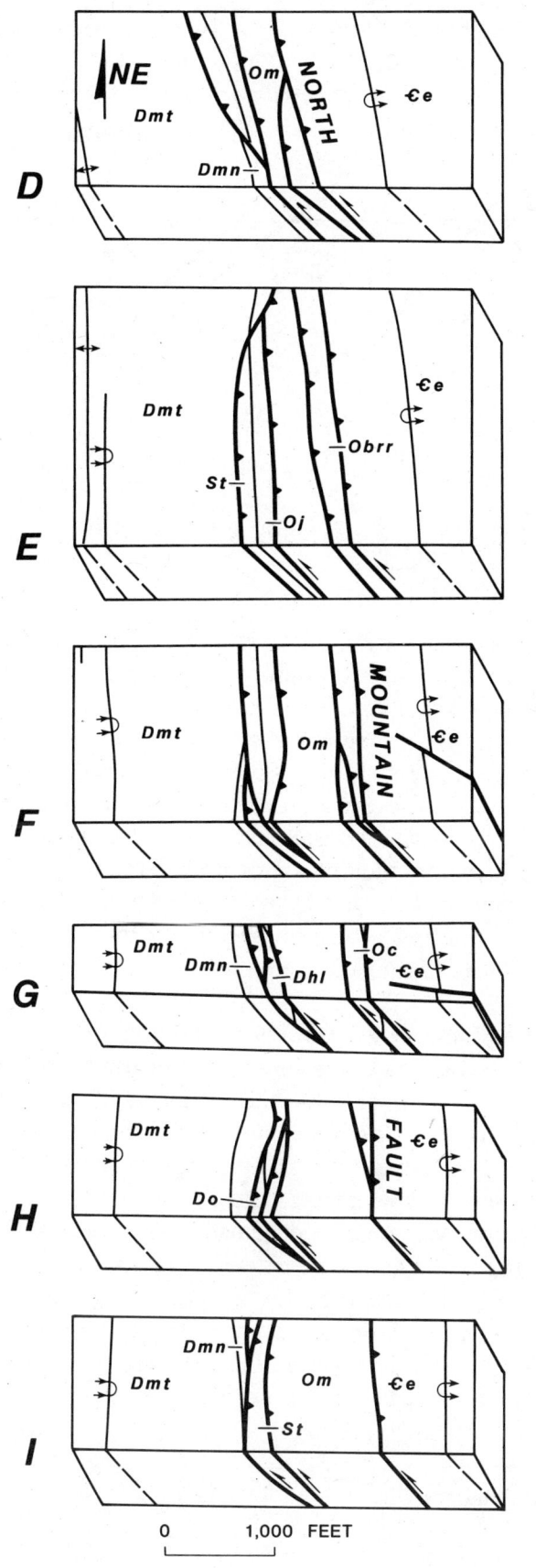

Figure 14A. Block diagrams through North Mountain fault zone; Stop 6.

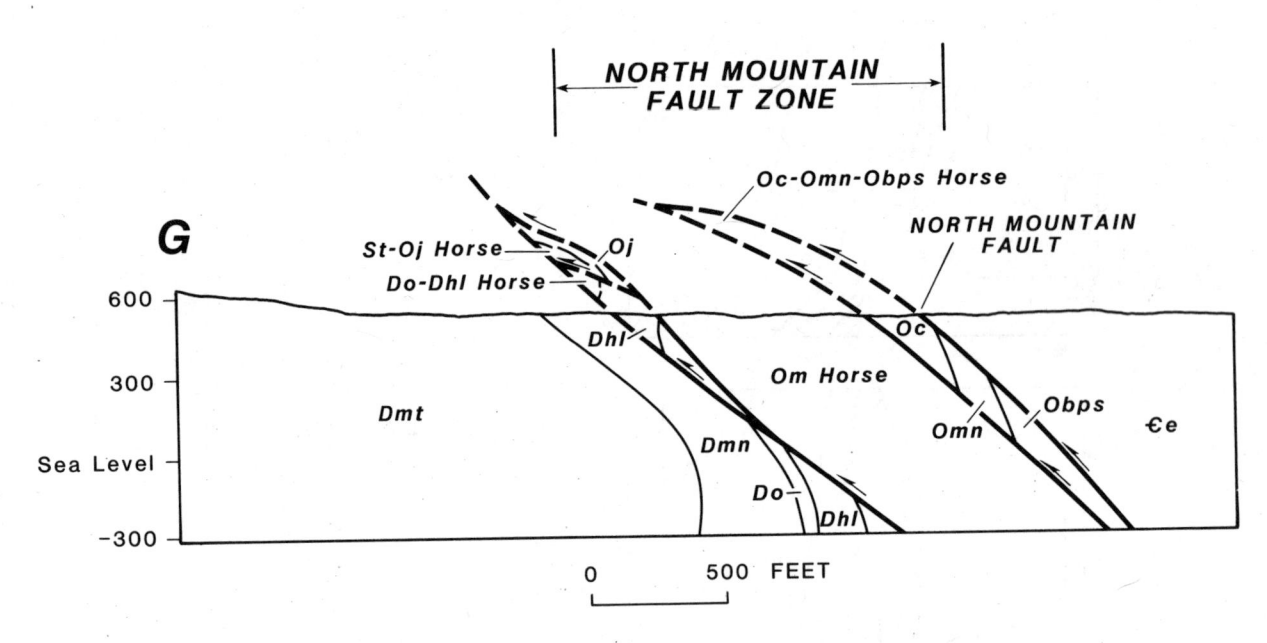

Figure 14B. Cross section through B&O (now CSX) railroad cut at North Mountain fault; Stop 6.

Proceeding northwest along the railroad tracks, we cross approximately 1,200 ft of folded and predominantly overturned rocks of the Martinsburg Fm. The shale is smoke-blackened from decades of coal-burning locomotives. The Martinsburg Fm. is intensely fractured, and has a pervasive cleavage. Before entering the deep railroad cut, the Martinsburg Fm. section ends and a thin outcrop of fossiliferous, cherty, Helderberg Limestone (Lower Devonian–Upper Silurian) can be seen on the left (south). The Martinsburg-Helderberg juxtaposition results from a major footwall splay off the upward ramping North Mountain fault (Figure 14B), or a splay from the westward translated Martinsburg level detachment. This fault has truncated much of the Upper Ordovician section (i.e., Martinsburg to Juniata Formations) and the entire Silurian section, a total stratigraphic omission of approximately 3,000 ft. The offset of the fault-terminated Helderberg Limestone is probably close to this figure.

The western end of the Helderberg Limestone section lies in fault contact with black shales of the Middle Devonian Needmore and Marcellus Fms. Fault displacement on this additional splay is minimal, likely on the order of 200–300 ft since only part of the Helderberg Limestone, all of the Lower Devonian Oriskany Sandstone, and part of the black shale sequence have been deleted. Entering the deep railroad cut, the Needmore-Marcellus shale interval is overturned and cut by several low-dipping thrust faults. A few carbonate lenses and carbonate beds, typical of this interval, are present. Small-scale structural features include well-developed cleavage, boudinage, and late-stage extension faults. Approximately 200 ft into the deep cut, the black shale grades into siltstone and shale of the fossiliferous Mahantango Formation of Middle Devonian age. From this point to the western end of the railroad cut, the Mahantango Fm. is overturned. The section has pervasive cleavage and is intensely fractured. Deformed fossils are present at several zones throughout this section.

Proceed back towards Hedgesville via the same route.

40.50 Turn right on Allensville Road towards Hedgesville.

41.60   Intersection of Allensville Road (Route 901) and Route 9 in Hedgesville. At the stop sign, turn right and head west on Route 9.
42.10   Hinge of Back Creek syncline. From here to the west of Back Creek, the western limb of the syncline is made up of moderate to low eastward-dipping rocks of the Mahantango Fm. This structure is continuous along the entire length of the eastern panhandle of West Virginia and is a major footwall syncline developed in response to the emplacement of the Massanutten–Blue Ridge sheet and upward ramping of the North Mountain fault complex.
45.50   Johnstown
47.50   Turn left off Route 9 towards the Woods Resort.
48.00   Bear right at the stop sign.
50.00   Turn right off hard-top road at Woods Resort sign. Collect keys for double-occupancy rooms from trip leaders. Dinner from the menu (prepaid) is on an individual basis. There will be an informal presentation and discussion downstairs in the restaurant at 8:00 PM on the tectonics of the region and the geology of Meadow Branch syncline.

Wednesday, March 13, 1991

50.00   Breakfast (prepaid) is on an individual basis. Return keys to front desk or trip leaders. Depart the Woods Resort at 8:00 AM and return to Route 9.
52.50   Intersection of Route 9, and turn right on Route 9.
52.60   Turn left on Rustic Tavern Road. The road follows and crosses folded rocks of the Mahantango Fm. with poor exposures.
54.60   Turn left at T intersection and proceed northeast along strike in folded Mahantango Fm.
56.50   Continue to the right at road intersection and stop sign. Railroad tracks are on the right.
56.70   Four-way road intersection in Cherry Run: turn right and cross railroad tracks, then turn left immediately and follow the dirt road parallel to the tracks towards railroad dispatchers building. <u>Our Field Headquarters are on the left.</u>

57.60   STOP 7 (FIGURE 15)

Park at the dispatcher building and walk west along the railroad tracks for 1,000 ft. Helderberg Limestone is exposed on the south side of the tracks. We have proceeded down section from Cherry Run into the core of a small anticline made up mainly of Helderberg Limestone. This feature plunges out just south of this location at the Lower Devonian level. The amplitude of this structure increases northward and a complete Silurian sequence is exposed in the fold core in Maryland. At our location, this gentle anticline is cut by three thrust faults on the western flank (Figure 16). The most obvious structure is a back thrust that places the Helderberg Limestone on top of the Oriskany Sandstone near the west end of the railroad cut. Here the Oriskany Sandstone is only 10–15 ft thick, in contrast to 250 ft of Oriskany Sandstone in the quarries at Berkeley Springs 15 mi to the west, which we will see later this morning. The Helderberg Limestone on the hanging wall is inordinately thin. This suggests a forethrust placing rocks of the Helderberg Fm. on black shales of the Needmore and Marcellus Fms. The folded black shales on the footwall of this forethrust also are thin and are interpreted to be cut by another forethrust, which places Needmore and Marcellus rocks on rocks of the Mahantango Fm. This fault is considered to be a continuation of a fault previously mapped across the Potomac River in Maryland (Edwards, 1978). The eastern limb of the anticline is poorly exposed. However, the Oriskany Sandstone

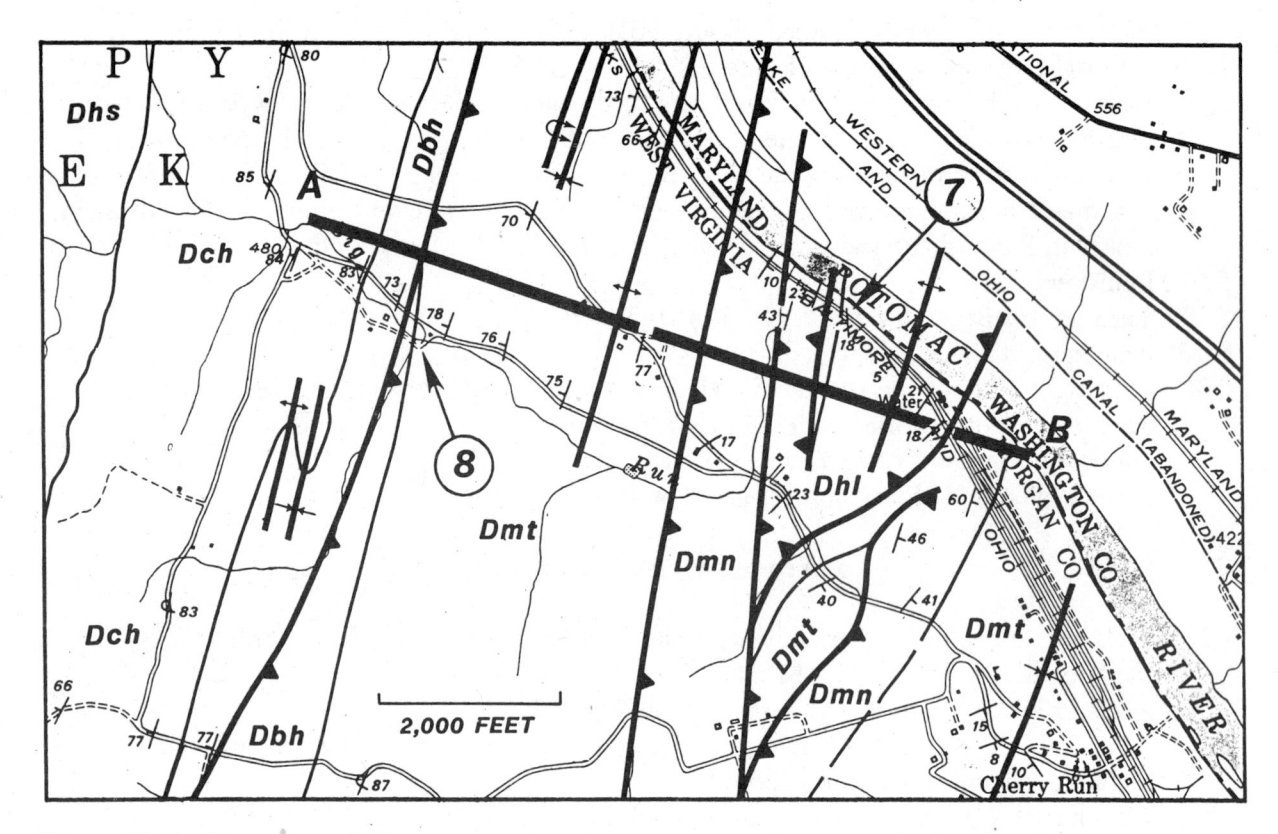

Figure 15. Geologic map of Cherry Run area; Stops 7 and 8. Line A–B illustrated in Figure 16.

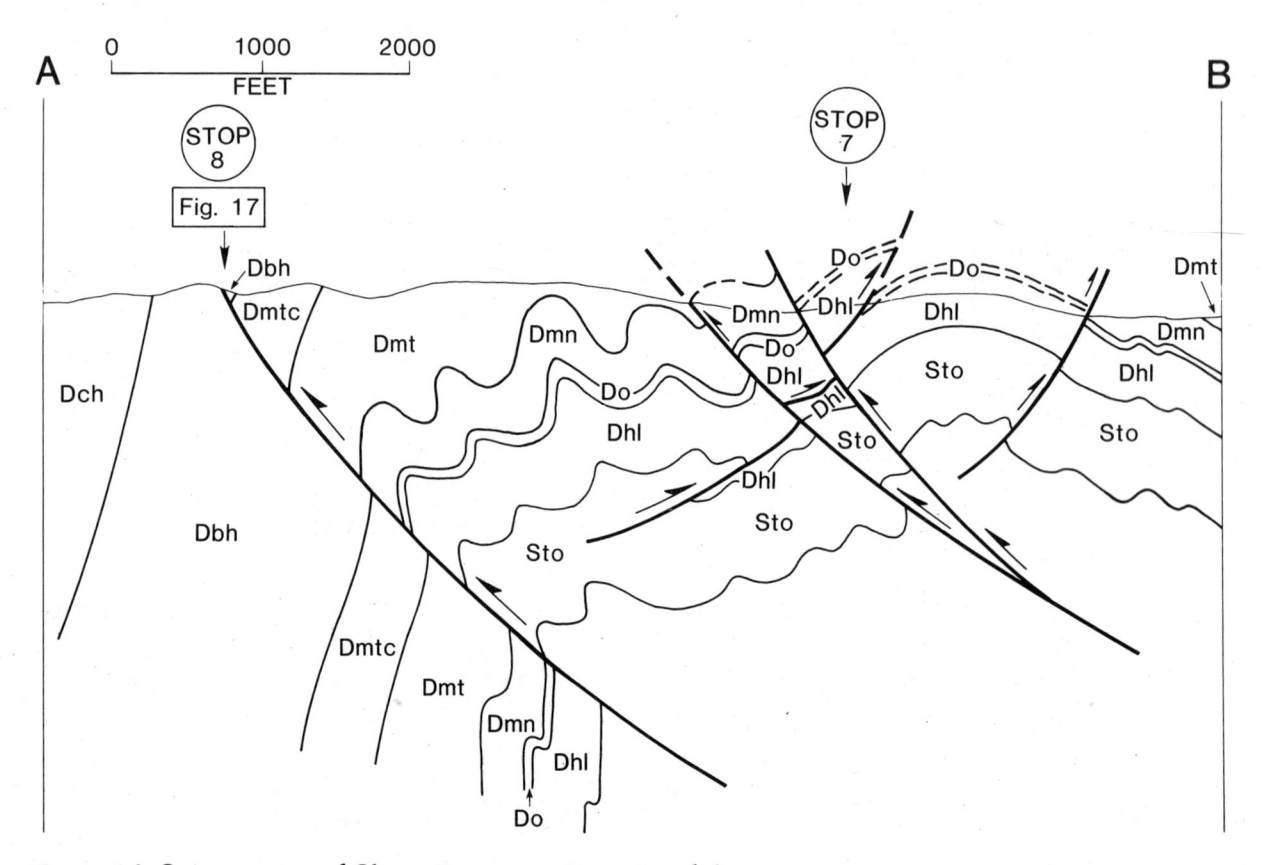

Figure 16. Cross section of Cherry Run area; Stops 7 and 8.

46

Figure 17. Photograph of tight anticlines in Brallier Formation; Stop 8.

is not present and must be faulted out by another backthrust, which places Helderberg Limestone against the black shales of the Needmore and Marcellus Formations.

Other features in the Helderberg Limestone at this location include fossils, chert beds, and solution cleavage. The solution cleavage is perpendicular to bedding, and lack of bedding offsets indicates that cleavage developed primarily from layer-parallel shortening before folding.

Proceed back to Cherry Run.

58.50 Intersection in Cherry Run: proceed straight ahead through the intersection.
59.00 Road intersection at hilltop: continue straight ahead.
59.60 Cross thrust shown in Figure 15 which is present at west end of railroad section. Trace of this fault occurs just before the Y road intersection. Take left fork at intersection and proceed west across folded Mahantango Fm.

60.40 **STOP 8 (FIGURE 15)**

Walk east a short distance. Approximately 100–200 ft east are intensely folded rocks of the Devonian Brallier Fm. (Figure 17), which lie on the footwall of an eastward-dipping thrust fault. The fault places rocks of the upper Mahantango Fm. against rocks of the Brallier Fm. (Figure 16). The fault trace lies in the small gully. Fossiliferous siltstone of the Clearville

Member of the Mahantango Fm. is present some 200–600 ft east of the fault. To the west along the road, the contact of the Brallier and Chemung Fms. lies approximately at the road bend where Big Run crosses the road. Because only 800 ft of the Brallier Fm. is present here, approximately 800–1,000 ft have been truncated by this thrust fault.

Proceed west.

60.60 At intersection of road from the left, continue straight around road bend to the right, across Big Run and up hill.

61.10 At top of the hill road intersection, continue to the left. Exposures are poor for some distance along this upland section of road.

61.70 Road intersection. Turn left at church. For the next 2.5 mi we will travel along strike in outcrops of the Hampshire Fm. just east of the hinge of Meadow Branch syncline.

62.60 Road intersects on right, keep left.

63.00 Road intersects on right, keep straight ahead.

63.70 Intersection with unpaved roads. Follow hardtop road to right, then sharp left in 0.15 mi.

64.60 Intersection with Route 9. Turn right and proceed northwest.

72.50 Eastern limb of Cacapon Mountain anticlinorium ahead with Oriskany Sandstone quarries north (to the right) of Berkeley Springs.

74.20 Junction of Route 9 and Route 522. Turn right at intersection. Resistant Oriskany Sandstone holds up east limb of Cacapon Mountain anticlinorium. Proceed through Berkeley Springs along Routes 9 and 522.

74.70 Turn left and take Route 9 towards Paw Paw. Proceed up east slope of Cacapon Mountain anticlinorium, which is underlain by Oriskany Sandstone. Note: The Castle was for sale for $3 million.

75.30 Top of hill is the contact of Oriskany Sandstone and Helderberg Limestone around sharp right bend in road. Proceed down section into Middle Silurian age rocks as the road descends the hill.

78.00 STOP 9 (FIGURE 18)

Overlook at the Panorama Steak House and photo stop. The overlook is on Lower Silurian age Tuscarora Sandstone on the western flank of the Cacapon Mountain anticlinorium. The geological marker is incorrect; the Tuscarora Sandstone is not part of the Clinton nor does it form Fluted Rocks.

Seen from the overlook, the first low ridge (1.5 mi away) is held up by the Oriskany Sandstone as the stratigraphy proceeds up section from the Cacapon Mountain anticlinorium. The ridge in the distance (4.5 mi away) is Sideling Hill syncline composed by Lower Mississippian "Pocono" sandstone (Rockwell and Purslane Formations). Proceed west as the road winds down the dip-slope of Tuscarora Sandstone.

78.40 Contact between the Tuscarora Sandstone and Rose Hill Formation of Middle Silurian age.

79.60 Cross the Cacapon River and immediately turn left on an unpaved road. Folded rocks of the Rose Hill and Keefer Fms. are on the right.

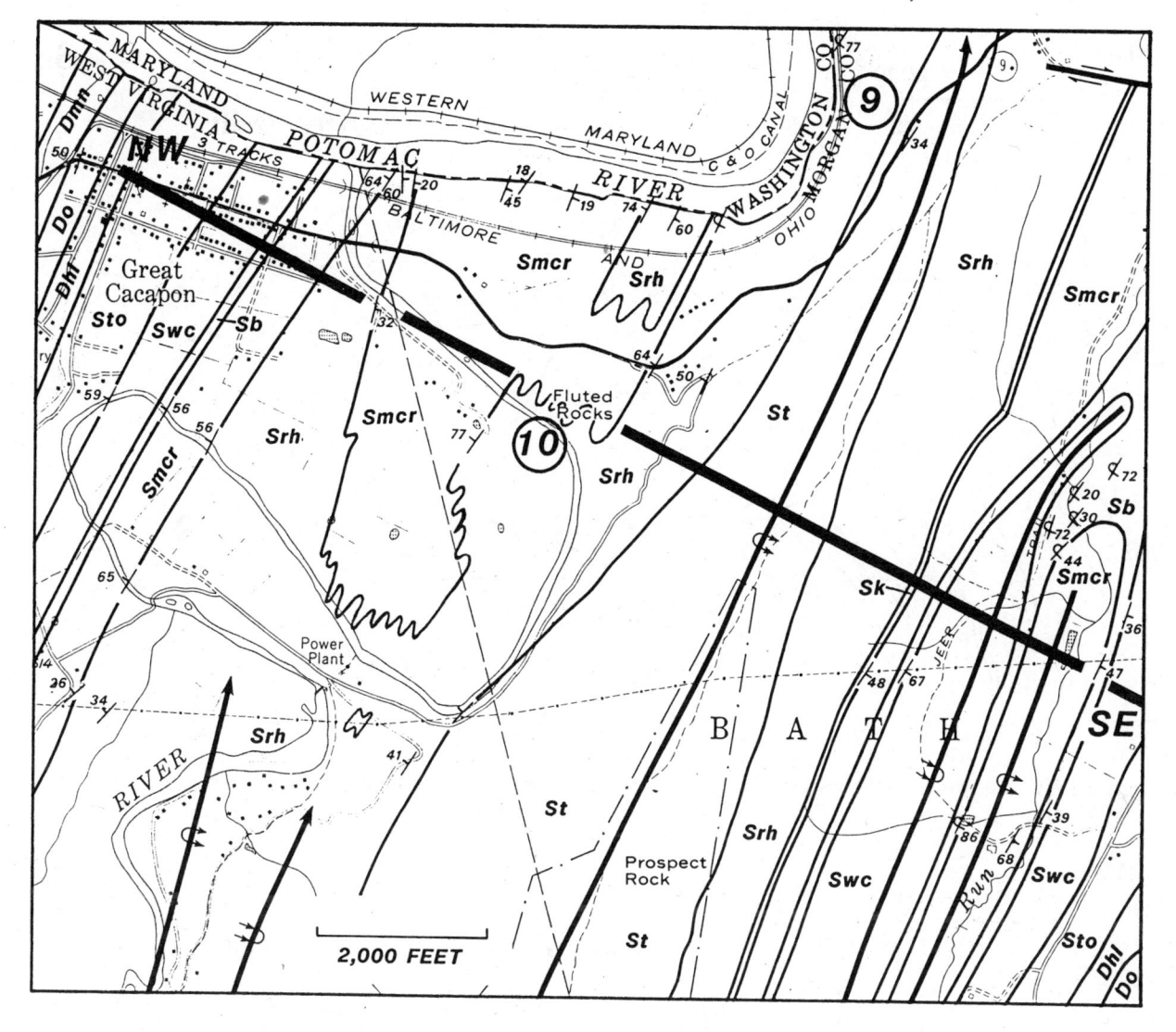

Figure 18. Geologic map of Great Cacapon area; Stops 9 and 10 (geology by Martin, 1964).

80.40 **STOP 10 (FIGURE 18)**

The famous "Fluted Rocks," composed of the intricately folded Keefer Sandstone and underlying Rose Hill Fm. (first described by Stose and Swartz, 1912), form the slope on the north side of Cacapon River (Figures 19 and 20). From a modern-day perspective, it would appear certain that the intricate folding of the Keefer Sandstone and broad expanse of its folded outcrop would necessitate a décollement in the underlying shales of the Rose Hill Fm. Figure 19A depicts the décollement to rise over the anticlinorium. This necessitates gravity sliding off the rising anticlinorium or early detachment of the Keefer Sandstone above the Rose Hill Fm. before development of the anticlinorium. Alternatively, Fluted Rocks may have originated in response to a late forelimb thrust that locally served as a detachment fault in the Rose Hill Fm. (Figure 19B).

Proceed to the north side of the river for a wonderful walk along the base of this folded sequence.

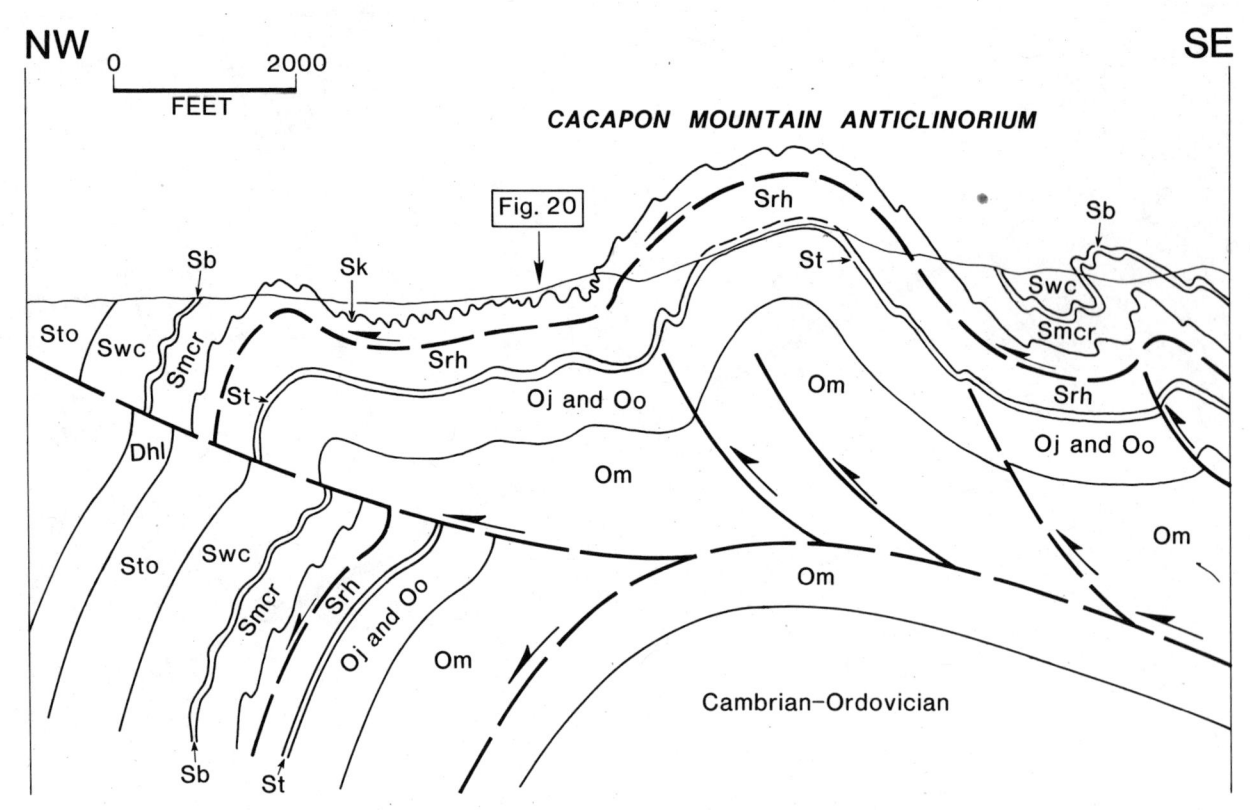

Figure 19A. Interpretive cross section of Fluted Rocks generated by a decollement in the Rose Hill; Stop 10.

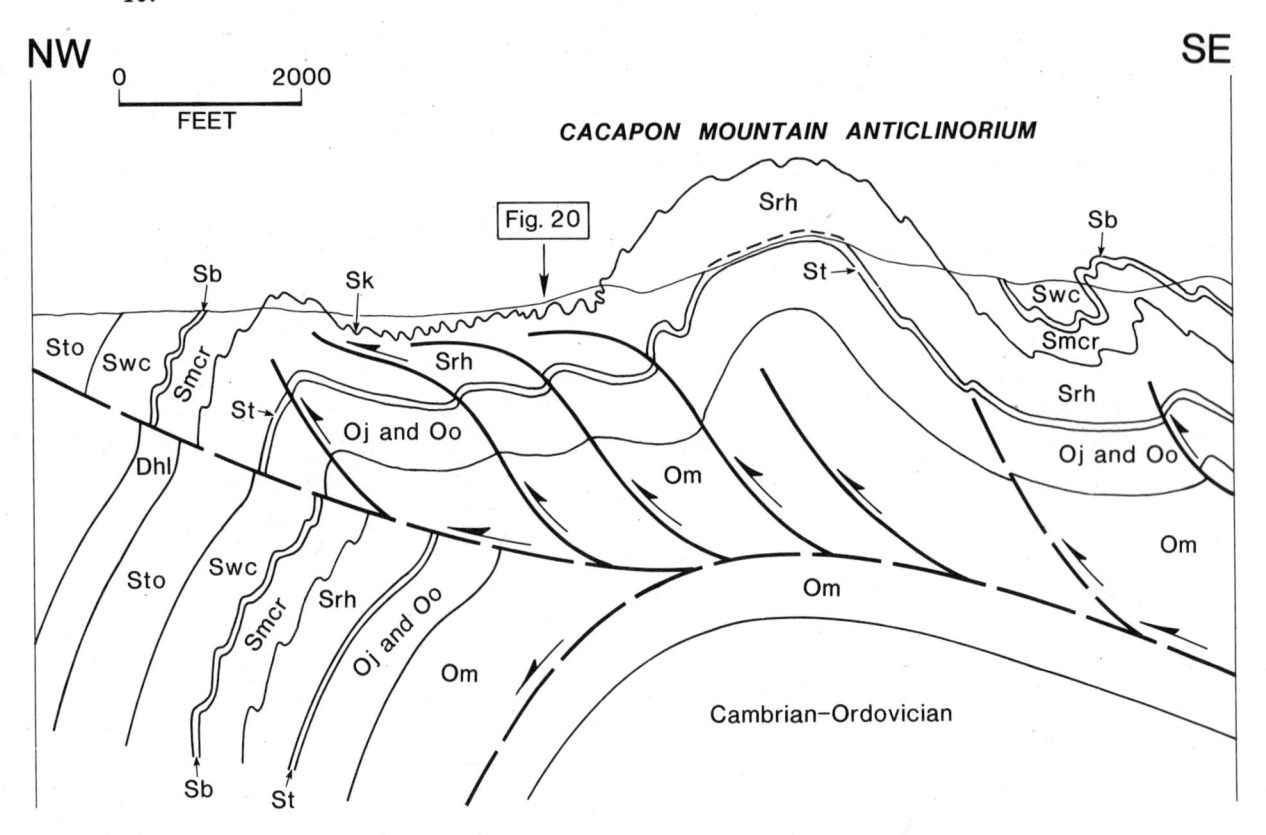

Figure 19B. Interpretive cross section of Fluted Rocks generated by thrust faults; Stop 10.

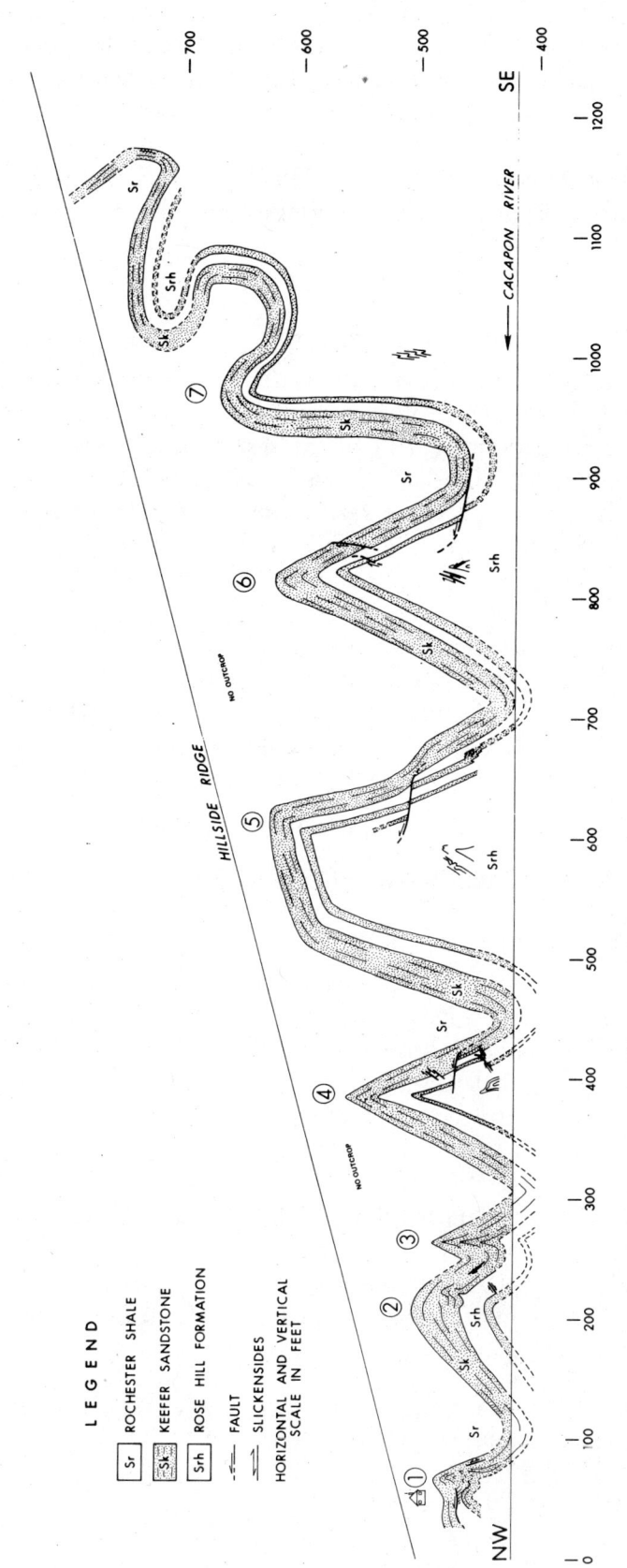

Figure 20. Cross section through Fluted Rocks; Stop 10.

81.20 Turn right at Route 9 and cross Cacapon River.

81.80 STOP 10 CONTINUED. Walk for approximately 1,000 ft along the base of Fluted Rocks using Figure 20 (Lessing, 1987). Numerous small-scale structures are present along this traverse, including bedding plane thrusts, faults, wedges, axial-plane cleavage, S and Z folds, kink bands, quartz-filled joints, and a variety of sedimentary features.

Backtrack to Berkeley Springs.

86.10 Turn right (south) on Routes 9 and 522 in Berkeley Springs at the base of the mountain. Proceed through the stoplight to Berkeley Springs State Park on the right and the Country Inn.

86.40 **STOP 11 (FIGURE 21)**

Country Inn: we will have lunch here (prepaid). This is an area famous for its thermal springs, frequently visited by George Washington, who surveyed the town (and certainly slept here). The elevated temperature (72°F) of the spring water is caused by deep circulation (1,800 ft) down the dip of the Oriskany Sandstone (Figure 22) and the geothermal gradient (1°F/100 ft). After lunch and a brief perusal of the springs site, we will return to Baltimore.

Proceed north on Route 522, directly along the eastern flank of Cacapon Mountain anticlinorium.

88.90 US Silica glass plant is on the right with about 250 ft of Devonian age Oriskany Sandstone quarried on the left. This sand was used to make the 200 inch (20 ton) Mt. Palomar reflecting lens. Devonian age Needmore and Marcellus black shales are sporadically exposed along the left side of the highway. Continue north on Route 522, cross Potomac River to Hancock, Maryland, and follow signs for I-70 east to Baltimore.

ACKNOWLEDGMENTS

The authors appreciate the reviews and editing by Art Schultz, Dave Brezinski, and Katherine Lessing, which have improved this field trip text. Special thanks for permissions and logistics are due Don Campbell, Superintendent, Harpers Ferry National Historical Park (Stop 2); Gary Gray, General Manager, Millville Quarry (Stop 3); Ray Johnston and Andy Micheals (Woods Resort); and Ms. Jane K. Majewski, Fluted Rocks (Stop 10). Cartography by Dan Barker and Ray Strawser and word processing by Betty Schleger have salvaged a host of scribbles, notes, and undecipherable field maps.

REFERENCES CITED

Cloos, E., 1951, History and geography of Washington County and structural geology of Washington County, in Physical features of Washington County, Maryland: Department Geology Mines and Water Resources, County Report 14, p. 1–97, 124–163.

———, 1958, Structural geology of South Mountain and Appalachians in Maryland: Studies in Geology No. 17, Johns Hopkins University Press, Baltimore, 85 p.

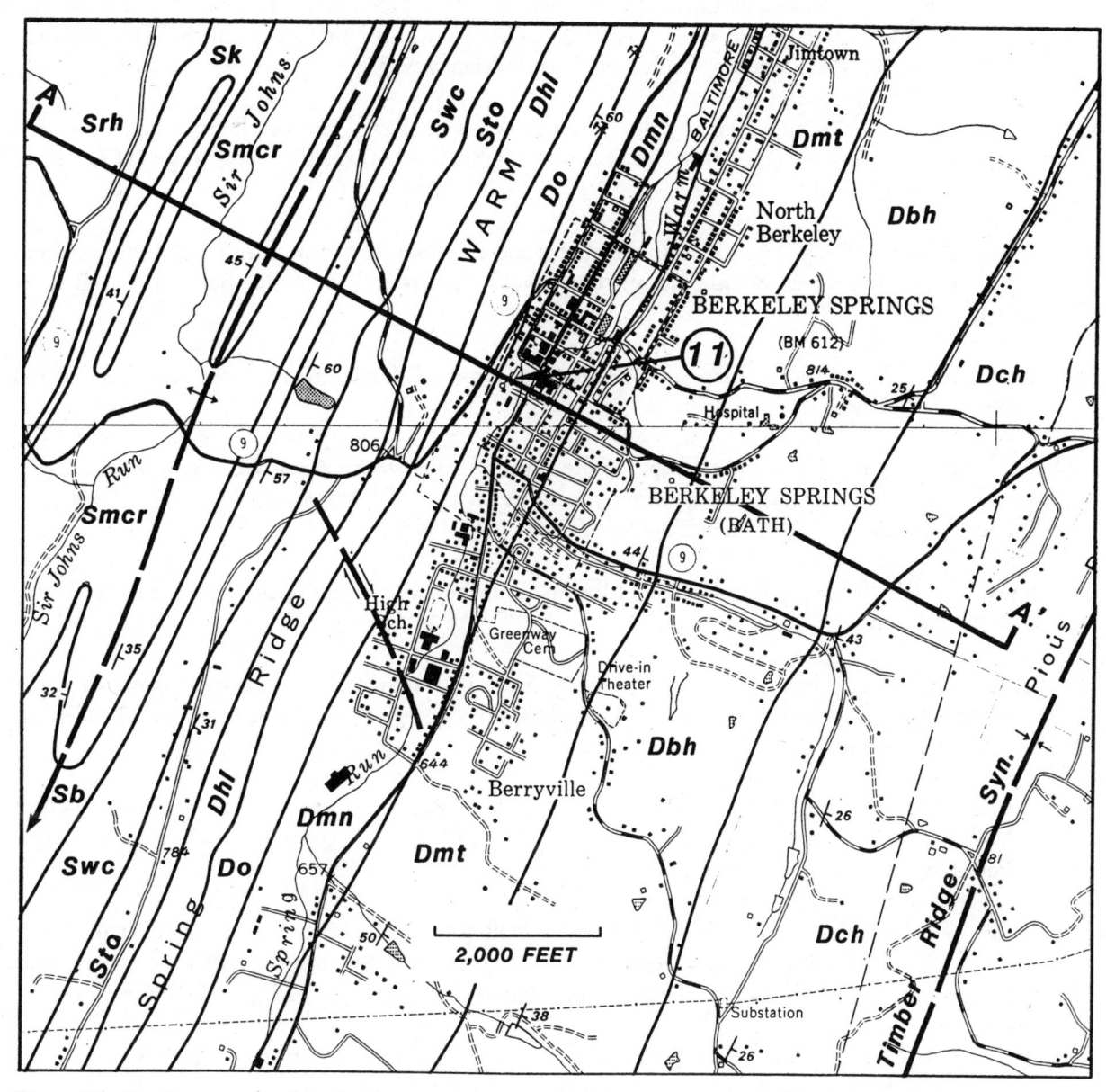

Figure 21. Geologic map of Berkeley Springs area; Stop 11 (geology by Minke, 1964).

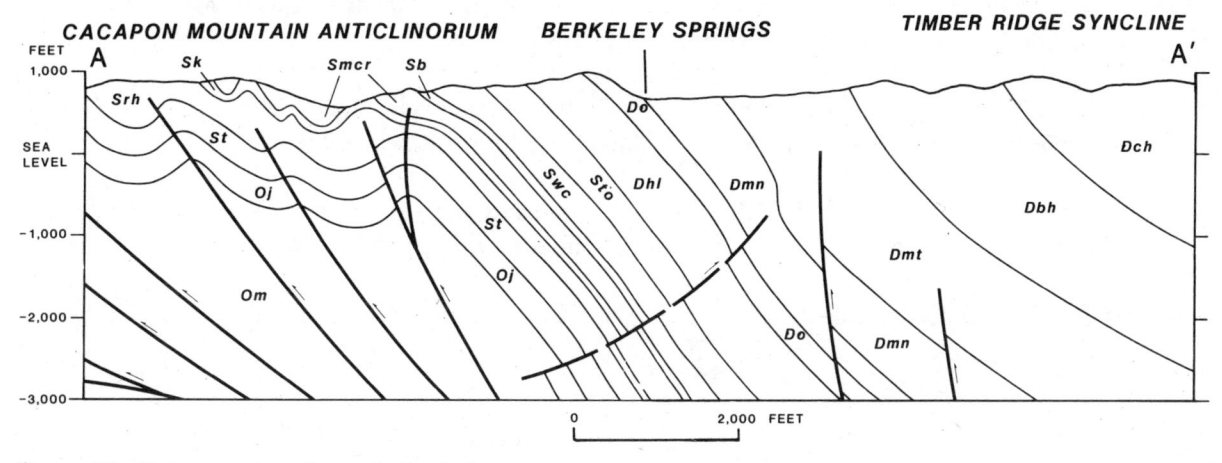

Figure 22. Cross section through Berkeley Springs; Stop 11.

Dean, S. L., Kulander, B. R., and Lessing, P., 1987, Geology of the Hedgesville, Keedysville, Martinsburg, Shepherdstown, and Williamsport quadrangles, Berkeley and Jefferson counties, West Virginia: 1:24,000 scale, West Virginia Geological Survey, Map WV-31.

——, ——, and ——, 1990a, Geology of the Berryville, Charles Town, Harpers Ferry, Middleway, and Round Hill quadrangles, Jefferson County, West Virginia: 1:24,000 scale, West Virginia Geological Survey, Map WV-35.

——, ——, and ——, 1990b, The structural geometry and evolution of foreland thrust systems, northern Virginia—Alternative interpretation: Geological Society of America Bulletin, v. 102, p. 1442–1445.

Edwards, J., Jr., 1978, Geologic Map of Washington County [Maryland]: 1:62,500 scale, Maryland Geological Survey.

Faill, R. T., 1985, Evolving tectonic concepts of the central and southern Appalachians, in Drake, E. T., and Jordan, W. M., eds., Geologists and ideas—A history of North American geology: Geological Society of America, Centennial Special Volume 1, p. 19–46.

Grimsley, G. P., 1916, Jefferson, Berkeley, and Morgan Counties: West Virginia Geological Survey County Report, 644 p.

Hatcher, R. D., Jr., and others, 1989, Alleghanian orogen, in Hatcher, R. D., Jr., and others, eds., The Appalachian-Ouachita Orogen in the United States: Geological Society of America, The Geology of North America, v. F-2, p. 233–318.

Kulander, B. R., and Dean, S. L., 1986, Structure and tectonics of central and southern Appalachian Valley and Ridge and Plateau provinces, West Virginia and Virginia: American Association of Petroleum Geologists Bulletin, v. 70, p. 1674–1684.

——, and ——, 1988, The North Mountain–Pulaski fault system and related thrust sheet structure, in Mitra, G., and Wojtal, S., eds., Geometries and mechanisms of thrusting, with special reference to the Appalachians: Geological Society of America Special Paper 222, p. 107–118.

Lessing, P., 1987, Fluted Rocks—A magnificent exposure to the forces of Mother Nature: West Virginia Geological Survey, Mountain State Geology, p. 20–28.

Martin, R. H., 1964, Geology of the Great Cacapon, West Virginia–Maryland quadrangle, unpublished M.S. Thesis, West Virginia University, Morgantown, 34 p.

Minke, J. G., 1964, Geology of the Hancock and Stotlers Crossroads quadrangles in West Virginia: unpublished M.S. Thesis, West Virginia University, Morgantown, 42 p.

Mitra, S., 1987, Regional variations in deformation mechanisms and structural styles in the Appalachian orogenic belt: Geological Society of America Bulletin, v. 98, p. 569–590.

Nickelsen, R. P., 1956, Geology of the Blue Ridge near Harpers Ferry, West Virginia: Geological Society of America Bulletin, v. 67, p. 239–270.

Onasch, C. M., and O'Conner, J. V., in press, Structure and geomorphology of the Harpers Ferry area: Northeastern Geology.

Stose, G. W., and Swartz, C. K., 1912, Paw Paw–Hancock folio: U.S. Geological Survey, Geologic Atlas 179, 24 p.

Woodward, H. P., 1941, Silurian system of West Virginia: West Virginia Geological Survey, v. 14, 326 p.

———, 1943, Devonian system of West Virginia: West Virginia Geological Survey, v. 15, 655 p.

———, 1949, Cambrian system of West Virginia: West Virginia Geological Survey, v. 20, 317 p.

———, 1951, Ordovician system of West Virginia: West Virginia Geological Survey, v. 21, 627 p.

3
TECTONIC HISTORY OF THE BLUE RIDGE BASEMENT AND ITS COVER, CENTRAL VIRGINIA

Mervin J. Bartholomew, Sharon E. Lewis, Scott S. Hughes
Montana Bureau of Mines and Geology
Montana Tech, Butte, MT 59701

Robert L. Badger
Department of Geology
SUNY College at Potsdam, Potsdam, NY 13676

A. Krishna Sinha
Department of Geological Sciences, Virginia Polytechnic
Institute and State University, Blacksburg, VA 24061

INTRODUCTION

Bloomer and Werner's (1955) synthesis of the central Virginia Blue Ridge has remained the principal published research in that region for several decades. Since the early eighties, however, substantial effort has gone into mapping and analyzing middle Proterozoic rocks of the Virginia Blue Ridge. These were metamorphosed during the Grenville event (ca. 1.0 Ga), and subsequently formed the crystalline basement upon which the late Proterozoic to Paleozoic rift and platform sequence accumulated prior to Taconic metamorphism and Alleghanian deformation of the Appalachian orogen. More recently, principal contributions to the geology of the central Virginia Blue Ridge include: initial characterization of the Pedlar and Lovingston massifs and their respective units (Bartholomew et al., 1981); extensive petrologic and geochemical work on the Roseland Anorthosite and associated titanium deposits primarily in the Lovingston massif (Herz and Force, 1984; 1987); geochronology and geochemistry of both massifs (Sinha and Bartholomew, 1984; Pettingill et al., 1984); and portrayals of regional correlations and possible palinspastic reconstructions of Grenvillian basement rocks of the Appalachians (Bartholomew and Lewis, 1984; 1988; in press).

Recent geochemical data (e.g., Hudson, 1981; Sinha and Bartholomew, 1984; Herz and Force, 1984; Lewis et al., 1986) include analyses for major rock units in both the Pedlar and Lovingston massifs; however, they had little trace element information, and most of these data only reflected major oxide systematics. Some trace element data are presented by Herz and Force (1987), but these were obtained by semi-quantitative methods for the purpose of addressing mineral potentials and general variations among rocks in the Lovingston Massif. Recently, trace element abundances in Pedlar Massif units were obtained using instrumental neutron activation analysis (INAA) in order to approach a more detailed understanding of the charnockites, granulite gneisses, and related rocks in this terrane. Forthcoming chemical data for the Lovingston Massif representatives, as well as other major rock units in the Blue Ridge basement, will be used for comparison with the Pedlar Massif and regional petrogenetic studies. Therefore, our initial petrogenetic interpretations, presented herein, are based largely on the chemical systematics within the Pedlar Massif.

This field trip is intended as an overview of both the crystalline rocks and some of the cover rocks of this part of the Blue Ridge, and thus draws not only on the previously mentioned collection

TABLE 1. *Timing of Events Affecting the Central Virginia Blue Ridge.*

TIMING (Ma)	EVENTS AFFECTING THE CENTRAL VIRGINIA BLUE RIDGE
~200 Ma	-Extensional displacement on Paleozoic thrusts and intrusion of diabase dike swarms associated with Mesozoic opening of the Atlantic Ocean.
~273-303 Ma	-Alleghanian subgreenschist-facies thermal event which post-dated early Alleghanian emplacement of the Blue Ridge and Pulaski thrust sheets. -Alleghanian thrusting produced ramp anticlines which refold Taconic foliation; possibly final movement on Rockfish Valley fault.
~450 Ma	-Taconic greenschist-facies metamorphism and development of dominant fold-style with axial planar foliation in Catoctin Formation.
~540-570 Ma	-Deposition of Early Cambrian rift-to-drift Chilhowee Group.
~570 Ma	-Late Proterozoic Catoctin volcanism. -Deposition of early rift-related sediments of the Late Proterozoic Lynchburg Group.
~650-730 Ma	-Emplacement of early rift-related plutonic rocks (granites) like the Late Proterozoic Rockfish River Pluton. -Major angular unconformity.
~1000 Ma	-Peak of Grenville granulite-facies metamorphism; emplacement of the Roseland Anorthosite diapir; reequilibration of Roses Mill and Shaeffer Hollow U-Pb systems.
~1050 Ma	-Emplacement of Pedlar River Charnockite Suite in Pedlar massif and charnockite plutons in the Lovingston massif.
~1100 Ma	-Emplacement of Archer Mountain Suite (ferrodiorite, quartz monzonite, granitic pegmatite) in Lovingston massif. -Emplacement of Hills Mountain Granulite Gneiss.
~1130-1150 Ma	-Accumulation of volcanic & volcanoclastic units in both massifs (Nellysford and Lady Slipper Granulite Gneisses in Pedlar massif and Stage Road Layered Gneiss in Lovingston massif).
~1490 Ma	-Inherited crustal component in PRCS indicated by Sm-Nd data.
~1870 Ma	-Age of detrital zircons in coarse augen gneisses of the Stage Road Layered Gneiss.

of basement papers but also on recent geochemistry (Badger and Sinha, 1988) of the Catoctin Formation plus sedimentology of the overlying Chilhowee Group (Schwab, 1986; Simpson and Eriksson, 1989; Simpson and Sundberg, 1987) and underlying Lynchburg Group (Wehr, 1985; in press; Wehr and Glover, 1985). All of these units nonconformably overlie the Grenvillian basement. Current work on structural analysis (Bartholomew, 1988a, b; 1989a, b) and other new data, including that from a number of theses, are incorporated herein to provide a basis for inferences regarding many remaining unresolved questions in the central Virginia Blue Ridge. Table 1 represents our chronology of events relevant to this region.

THE GRENVILLE EVENT

In central Virginia, rocks that were subjected to Grenvillian metamorphism (ca. 1.0 Ga) are found in two massifs separated by the Rockfish Valley fault zone (Figure 1). The Pedlar and Lovingston massifs, respectively, lie to the west and east of the east-dipping fault zone. The term "massif," as used in this region, is not a stratigraphic term nor is any stratigraphic connotation implied. Rather, "massif" has consistently been used in the European sense of a distinctive mountainous terrain composed of a distinctive assemblage of igneous/metamorphic rocks. The Pedlar and Lovingston massifs are thus distinguished from each other by their respective (and contrasting) rock associations, which reflect different igneous/metamorphic/structural evolutions, and by their different physiographic aspects, which reflect those contrasting rock assemblages. The contrasts between the Pedlar and Lovingston massifs have been discussed previously in the aforementioned papers, but continuing work in this region justifies a brief discussion of each massif and the rock units visited on this trip.

PEDLAR MASSIF

The Pedlar massif is a terrane containing five igneous charnockite suites and associated layered granulite gneisses, which were all subjected to deep crustal (higher pressure) granulite-facies conditions of metamorphism (ca. 755° C from one orthopyroxene/clinopyroxene pair, Herz, 1984) during the Grenville event (ca. 1.0 Ga). Paleozoic-age retrograde metamorphism under lower greenschist-facies conditions (chlorite-biotite grade) produced relatively minor effects which generally are visible in thin section. Andesine is ubiquitous and commonly occurs with orthopyroxene + garnet in both the charnockite suites and the associated granulite gneisses. Paleozoic thrust faults are distinguished by well-foliated ductile deformation zones of variable widths containing greenschist-facies mineral assemblages. Greenschist-facies metamorphism of the overlying Catoctin Formation was dated (Rb-Sr) at about 450 Ma (Badger and Sinha, 1988; Mose and Nagel, 1984). In the portion of the Pedlar massif crossed by this field trip, the Pedlar River Charnockite Suite and the Nellysford and Lady Slipper granulite gneisses are the principal rock types. One small mafic granulite occurs as well.

Pedlar River Charnockite Suite (PRCS)

Using Streckeisen's (1974) classification, the PRCS compositions for which we currently have trace element data all fall in the farsundite subfield of the charnockite field (hypersthene granite). Vesuvius Megaporphyry compositions fall in the charnockite field and the mafic granulite (pyroxenite) plots near the edge of the jotunite field (with norite). Other PRCS data (Hudson, 1981; Herz and Force, 1984; 1987; Sinha and Bartholomew, 1984) suggest some overlap from the charnockite field into the jotunite (monzonorite), mangerite (hypersthene monzonite), and charno-enderbite (hypersthene granodiorite) fields. The suite (including the Vesuvius Megaporphyry) has a

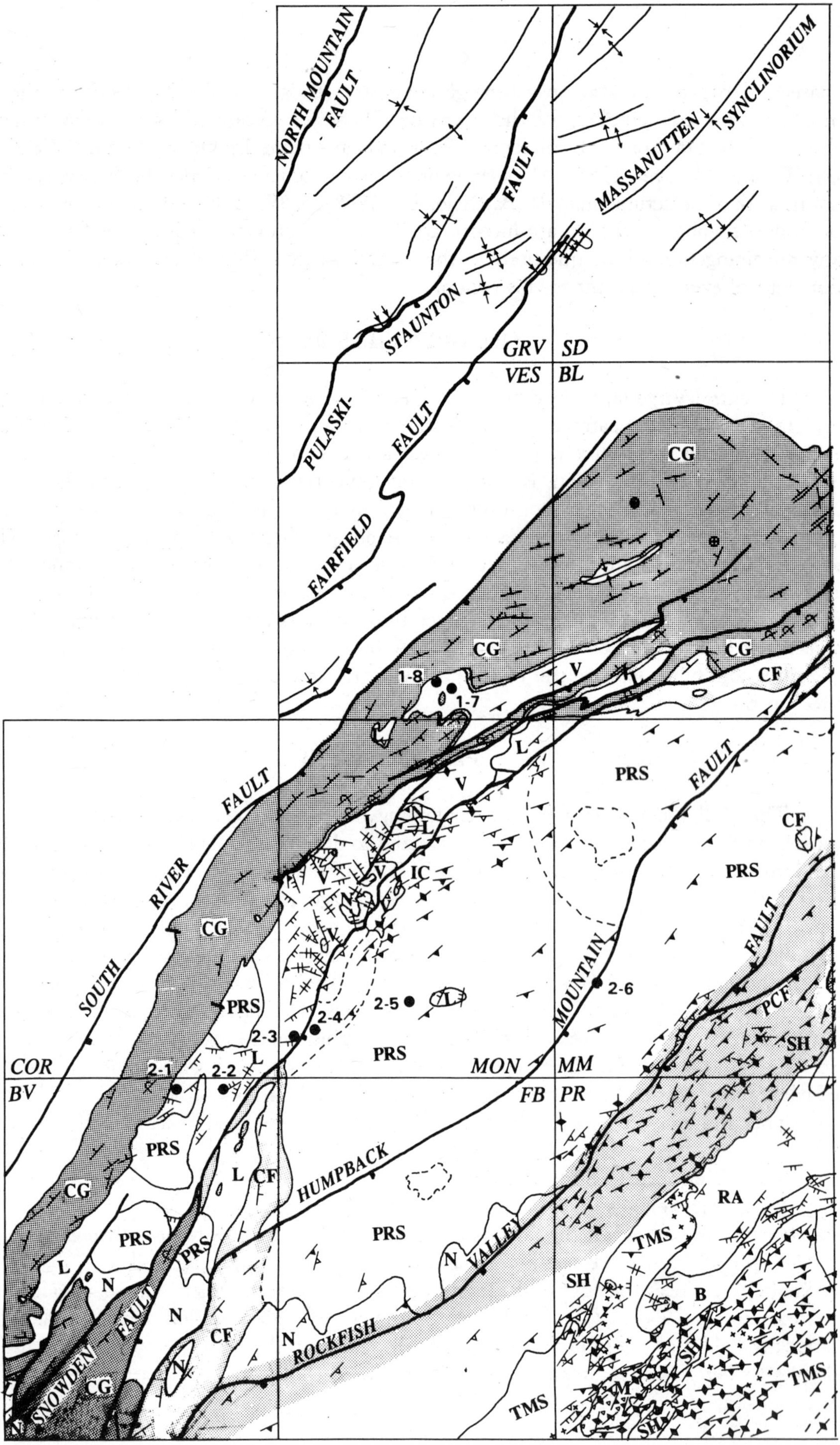

NORTH MOUNTAIN
FAULT
STAUNTON
FAULT
MASSANUTTEN
SYNCLINORIUM
GRV
SD
VES
BL
PULASKI-
FAULT
FAIRFIELD
CG
1-8
1-7
V
CF
L
PRS
FAULT
SOUTH
RIVER
FAULT
IC
2-3
2-4
2-5
2-6
MOUNTAIN
PCF
SH
COR
BV
2-1
2-2
MON
MM
FB
PR
HUMPBACK
RA
TMS
B
N
VALLEY
ROCKFISH
SNOWDEN
FAULT

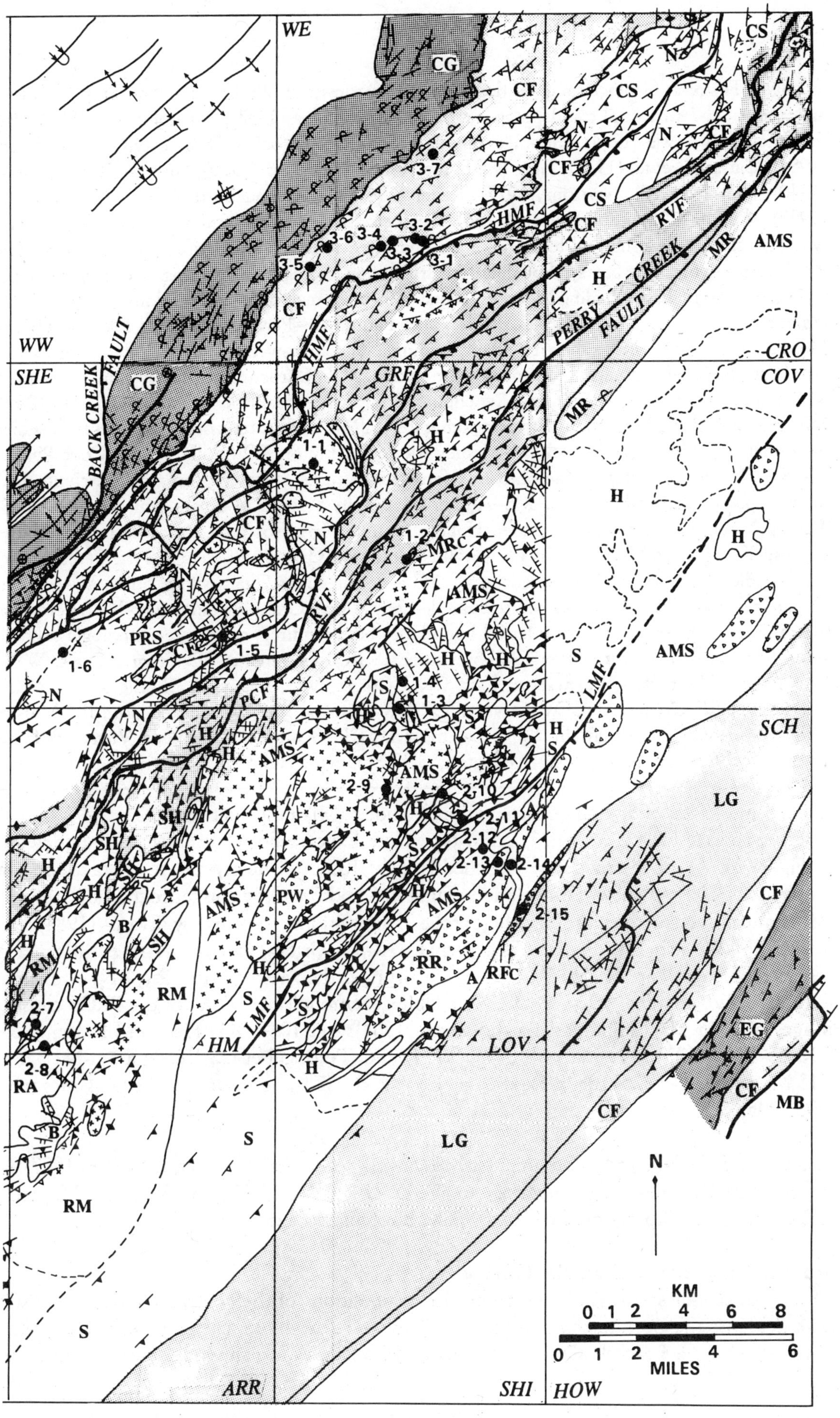
WE
WW
SHE
CRO
COV
SCH
HM
LOV
ARR
SHI
HOW
BACK CREEK FAULT
PERRY CREEK FAULT
HMF
GRF
RVF
PCF
LMF
CG
CF
CS
N
AMS
H
MR
MRC
S
PRS
SH
B
RM
RA
PW
RR
RFC
A
LG
EG
MB
1-1
1-2
1-3
1-4
1-5
1-6
2-7
2-8
2-9
2-10
2-11
2-12
2-13
2-14
2-15
3-1
3-2
3-3
3-4
3-5
3-6
3-7
N
KM
0 1 2 4 6 8
0 1 2 4 6
MILES

Preceding pages: *Figure 1.* Geologic map of the Blue Ridge in central Virginia showing Stop locations, major units, and structural trends; map is modified after Bailey (1983), Bartholomew (1977), Bartholomew and Lewis (1984), Bloomer and Werner (1955), Brock (1981), Davis (1974), Evans (1984), Gathright et al. (1977), Herz and Force (1987), Hillhouse (1960), Hudson (1981), Moore (1940), Nelson (1962), Nystrom (1977), Rader (1967), Sinha and Bartholomew (1984), Wehr (1983), Werner (1966), and Wilson (1986). Geological unit symbols and abbreviations: standard symbols used for bedding and foliation, solid and open foliation symbols denote dips >60° and <60° respectively, two bar symbol denotes Grenvillian segregation layering and compositional layering; intermediate shading denotes broad zones of mylonitic rocks along faults, HMF–Humpback Mountain Fault, RVF–Rockfish Valley Fault, PCF–Perry Creek Fault, LMF–Lawhorne Mill Fault; Cambrian strata shown by dark shading, CG–Chilhowee Group, EG–Evington Group; late Proterozoic rift sequence shown by light shading, CF–Catoctin and Swift Run Formations with coarse conglomerate (CFc), LG–Lynchburg Group with basal Rockfish Conglomerate (RFc), MR–Mechum River Formation with conglomerate (MRc); late Proterozoic A-type plutons shown by triangles, IC–Irish Creek stock, M–Mobley Mountain Pluton, PW–Polly Wright Cove Pluton, RR–Rockfish River Pluton, A–amphibolite; ferrodiorite suites of Lovingston massif include charnockite bodies denoted by x's, AMS–Archer Mountain Suite, RM–Roses Mill Pluton, TMS–Turkey Mountain Suite; SH–Shaeffer Hollow Granite; RA–Roseland Anorthosite; B–Border Gneiss; S–Stage Road Layered Gneiss; H–Hills Mountain Granulite Gneiss; in the Pedlar massif small bodies of charnockite denoted by x's, PRS–Pedlar River Charnockite Suite, V–Vesuvius Megaporphyry, CS–Crozet Charnockite Suite; L–Lady Slipper Granulite Gneiss; N–Nellysford Granulite Gneiss; dotted lines show inferred boundaries between units or between lithologic subdivisions of units. 7.5' min. quadrangle abbreviations: GRV–Greenville, SD–Stuarts Draft, WW–Waynesboro West, WE–Waynesboro East, CRO–Crozet, VES–Vesuvius, BL–Big Levels, SHE–Sherando, GRF–Greenfield, COV–Covesville, COR–Cornwall, MON–Montebello, MM–Massies Mill, HM–Horseshoe Mountain, LOV–Lovingston, SCH–Schuyler, BV–Buena Vista, FB–Forks of Buffalo, PR–Piney River, ARR–Arrington, SHI–Shipman, HOW–Howardsville.

Peacock alkali-lime index of 56 and a restricted mafic index (MI) range between 82 and 92 (Figure 2). The mafic granulite has an MI of 37. AFM-type plots indicate both calcalkaline and tholeiitic affinities (Figure 3). All of the suites are enriched in light REE and nearly all are characterized by a negative Eu anomaly and show Th depletion (Figures 4 and 5). Only the leucocharnockites and associated granites are enriched in Th. Microprobe compositions reported by Evans (1984) are: pyroxene - $En_{21}Fs_{35}Wo_{43}Rh_1$, $En_{28}Fs_{44}Wo_{28}Rh_0$; garnet - $Al_{71}Py_{19}Gr_8Sp_2$; feldspar - $An_{35}Ab_{64}Or_1$. Herz (1984) reported a temperature of 755° C based on coexisting orthopyroxene ($En_{33}Fs_{65}Wo_2$) and clinopyroxene ($En_{25}Fs_{30}Wo_{45}$) and a $^{18}O = +9.3‰$. Available geochronological data for the PRCS consists of: U-Pb ages of 1075 Ma (Sinha and Bartholomew, 1984) and 1040 Ma (Herz and Force, 1984; 1987); a Rb-Sr age of 1021 ±36 Ma with $Sr_i = 0.7047 ±6$ (Pettingill et al., 1984); a Sm-Nd age of 1489 ±118 Ma with $Nd_i = 0.51105 ±9$ and $_{Nd} = +6.7$; and a K-Ar age* for the Vesuvius Megaporphyry of 933 Ma.

*Recalculated (by P. D. Fullagar [1986, unpublished], using constants of Steiger and Jager, 1977) K-Ar age reported by Gulf Research Laboratory to W. D. Lowry (10/5/1961) on Vesuvius Megaporphyry: 4.374 $^{40}Ar \times 10^{-4}$ Stp cc/gm; 9.04 ±0.03 %K; recalculated age = 933 Ma.

Nellysford and Lady Slipper Granulite Gneisses

Using Streckeisen's (1974) classification, the compositions of the gneisses overlap the PRCS charnockite field into the jotunite-charno-enderbite fields and show more scatter of peripheral points. The gneisses also have a Peacock alkali-lime index of 56 but have a mafic index range from 68 to

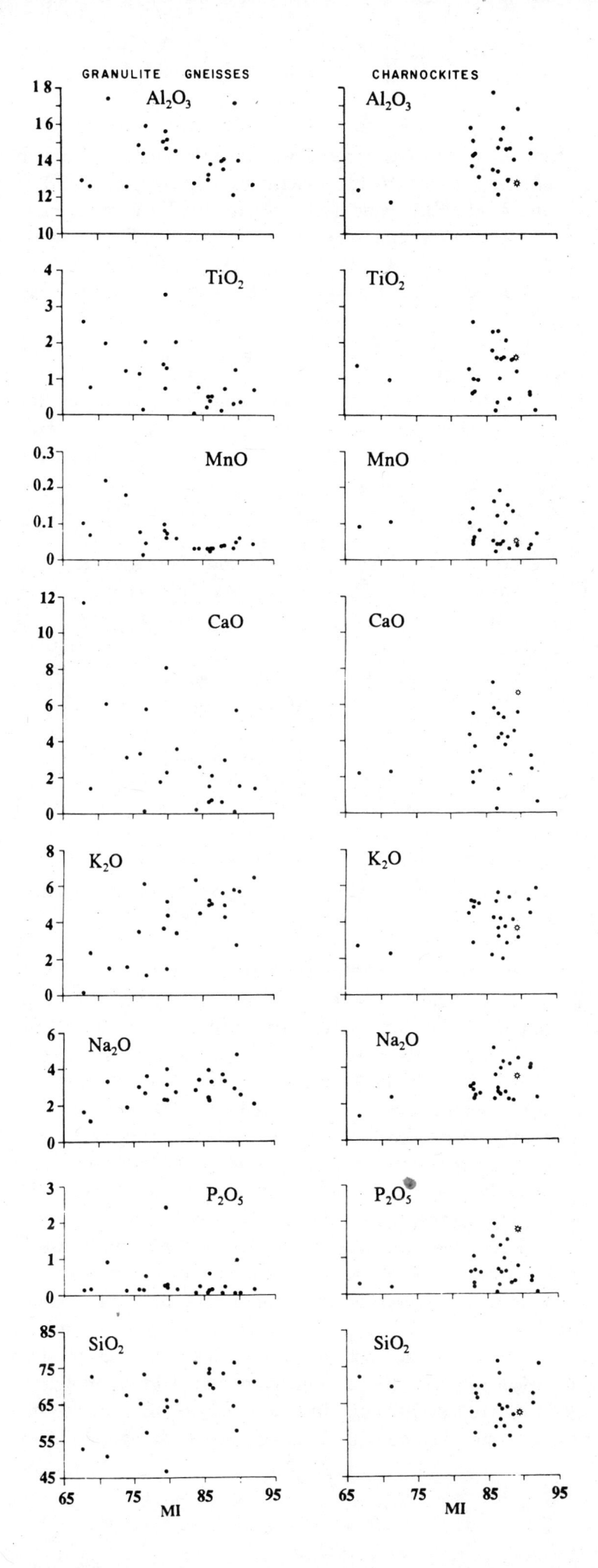

Figure 2. Variations of major oxide compositions in Pedlar Massif granulite gneisses and PRCS charnockites relative to mafic index: $100 \times (FeO + Fe_2O_3)/(MgO + FeO + Fe_2O_3)$. Sources of data: Herz and Force, 1984; Hudson, 1981; Lewis et al. 1986; Sinha and Bartholomew, 1984.

92 (Figure 2). AFM-type plots indicate that the gneisses are predominantly calc-alkaline although some have tholeiitic affinities (Figure 3). The trace-element/REE patterns overlap that of the PRCS (Figure 5). The layered gneisses typically show light REE enrichment, a negative Eu anomaly, and Th depletion. Microprobe compositions for the Nellysford Granulite Gneiss reported by Evans (1984) are: pyroxene - $En_{36}Fs_{38}Wo_{26}Rh_1$, $En_{40}Fs_{57}Wo_2Rh_1$, $En_{39}Fs_{51}Wo_9Rh_1$; garnet - $Al_{65}Py_{16}Gr_{16}Sp_2$; feldspar - $An_{36}Ab_{62}Or_2$, $An_{35}Ab_{63}Or_1$. A U/Pb age of 1130 Ma for the Lady Slipper Granulite Gneiss was reported by Sinha and Bartholomew (1984).

Petrogenetic Interpretations of Geochemical Signatures

The discussion of major oxide variations relies on numerous analyses obtained by the aforementioned authors plus unpublished data. The interpretation of trace element signatures is based solely on the new INAA data obtained for splits from the original samples. Major oxide and trace element abundances in representative units from the Pedlar Massif are listed in Table 2. The chemical data are a compilation from existing XRF analyses for major elements and new INAA determinations for REE and other trace elements.

The charnockites and granulite gneisses in the Pedlar Massif can each be segregated into leucocratic and melanocratic types, thus reflecting a wide range in lithologies and major elements. Variations in major oxides relative to the mafic index (MI, Figure 2) suggest that the granulite gneiss chemistries are prone to overall trends whereas the charnockite chemistries exhibit less variation. Except for two samples having mafic indices less than ~73, the charnockites fall within MI = 82–92 and show no apparent systematic relation to major oxides. The clustering and a definite segregation from units having lower Fe/(Mg + Fe) ratios (including the mafic granulite, not shown, with MI = 37) are possibly related to the separation of charnockitic bodies from parental material during ultrametamorphism or anatexis. Conversely, the granulite gneisses display variations with differentiation (MI), such as the overall increase in K_2O and Na_2O, and overall decrease in TiO_2 and MnO, that likely reflect igneous trends produced within the parental volcanic and volcaniclastic sequences. Although for most major oxides the charnockitic compositions in Figure 2 do not represent an end point of differentiation from the granulite gneisses, the derivation of charnockites from a more mafic granulite gneiss parent can neither be evaluated nor discounted.

Pedlar massif compositions plotted on an A-F-M diagram (Figure 3) indicate similar ranges in leucocratic and melanocratic units (open and closed symbols, respectively), as well as overlapping trends among charnockites (circles) and granulite gneisses (squares). These relations support either a unified process of differentiation among all units or their origins from similar parental compositions. Although some of the mafic units possibly follow an early tholeiitic trend (above the dashed line), a transition between tholeiitic and calc-alkaline compositions is evident in more evolved units. The transitional nature is exemplified in the pair of samples, connected by a tie-line in Figure 3, obtained from adjacent bands of rhythmic layering in a charnockite. Moreover, the open and closed star symbols represent two mafic-appearing granulites that actually display a wide separation in major element chemistry. While the division between leucocratic and melanocratic members is not readily apparent from A-F-M diagrams, a definite segregation is exhibited by these units plotted on a modified A-K-F diagram (Figure 6). It is evident from this diagram that the color index is largely controlled by relative abundances of Na_2O and Al_2O_3, i.e., the normative composition of plagioclase.

Patterns of trace element abundances are shown normalized to primordial earth mantle values of Taylor and McLennon (1985) in Figures 4 and 5. It is readily apparent from Figure 5 that the trace element signatures (Table 2) of the average Nellysford granulite gneiss (NGG) and Pedlar River Charnockite Suite (PRCS) are nearly identical in shape and overall abundances. In addition, the Vesuvius Megaporphyry displays only slight differences, primarily in elevated light REE, Sr, Ba, and Th abundances, from these two major units. The close similarities among these rock types argue for

TABLE 2. Major and Trace Element Geochemistries of Representative Units from the Pedlar Massif.

	NGG (7)	PRCS (5)	PRCS-LG (3)	VMP (B-37)	MG (344A)
SiO2 (%)	63.1	68.5	74.1	65.7	43.9
TiO2	1.18	1.09	0.29	1.00	0.42
Al2O3	14.84	12.99	13.13	14.30	14.50
Fe2O3	2.40	1.42	1.30	1.20	1.70
FeO	4.48	4.03	0.78	3.20	9.30
MnO	0.082	0.089	0.027	0.060	0.160
MgO	1.94	1.47	0.33	0.87	18.8
CaO	3.34	2.62	0.75	3.70	6.60
Na2O	2.77	2.16	2.80	2.40	1.00
K2O	3.67	3.81	5.22	4.80	0.86
P2O5	0.34	0.42	0.11	0.61	0.10
CO2 (%)	0.23	0.10	0.05	0.61	n.d.
Sc (ppm)	18.0	16.1	3.2	15.1	9.0
Cr	47.	47.	5.	12.	118.
Co	14.5	11.4	2.1	9.2	75.
Ni	27.	37.	n.d.	24.	79.
Zn	125.	150.	30.	130.	80.
Rb	132.	91.	215.	129.	72.
Cs	0.36	0.21	0.24	0.61	1.33
Sr	320.	340.	200.	700.	390.
Ba	868.	1090.	620.	2040.	320.
La	52.0	56.2	81.1	89.8	10.6
Ce	99.	109.	144.	207.	17.3
Nd	51.	54.	56.	102.	8.8
Sm	10.8	11.9	9.0	21.7	2.4
Eu	2.61	3.16	1.14	4.78	0.96
Tb	1.46	1.72	0.75	2.50	0.52
Yb	4.07	5.62	1.26	5.06	1.63
Lu	0.60	0.85	0.16	0.67	0.24
Zr	280.	370.	140.	430.	42.
Hf	12.4	17.2	7.5	18.1	1.0
Ta	1.15	1.13	0.53	1.61	0.29
Th	0.8	1.5	33.0	2.9	4.5
U (ppm)	1.1	2.1	3.1	1.5	1.3

NGG, Nellysford Granulite Gneiss; PRCS, Pedlar River Charnockite Suite; PRCS-LG, PRCS- Leucocharnockite/Granite; VMP, Vesuvius Megaporphyry (sample B-37); MG, Mafic Granulite (sample 344A).

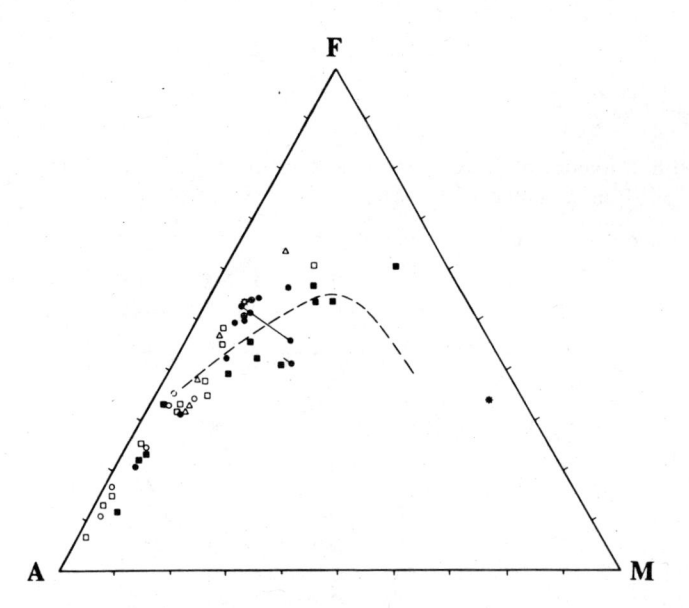

Figure 3. A-F-M (Na$_2$O + K$_2$O − FeO − MgO) diagram of Pedlar Massif rock units. Open and closed symbols are for leucocratic and melanocratic rocks, respectively. Symbols are as follows: circles, charnockites; squares, granulite gneisses; triangles, Vesuvius Megaporphyry; crossed circles, coarse-grained melanocharnockites; and stars, mafic granulites. Data from same sources as Figure 2.

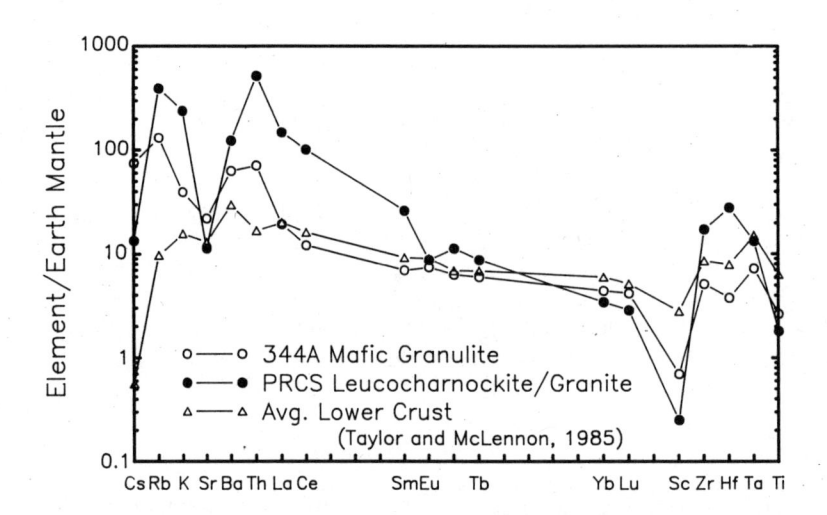

Figure 4. Geochemical patterns of the PRCS leucocharnockite/granite and 344A mafic granulite (Table 2) shown normalized to primordial mantle values of Taylor and McLennon (1985). The affinity of the mafic granulite composition to lower crust is evident in the overall similarity between the two patterns.

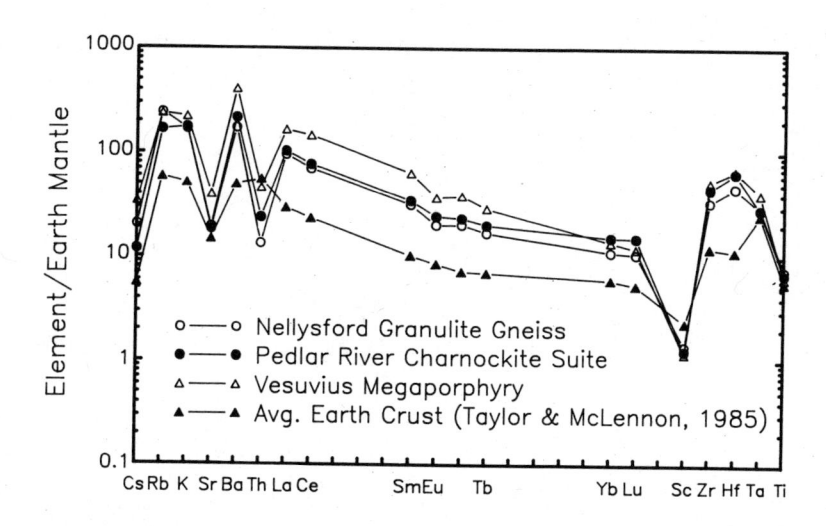

Figure 5. Geochemical patterns of Pedlar Massif rock units NGG, PRCS, and Vesuvius Megaporphyry (Table 2) shown normalized to primordial mantle values. Mantle abundances and the average crust abundances are from Taylor and McLennon (1985). Higher overall abundances of Pedlar Massif units relative to average crust indicates the likelihood of significant evolution from parental compositions prior to Grenville metamorphism.

a common parent (e.g., Sinha and Bartholomew, 1984) regardless of whether these charnockitic rocks are considered to be truly igneous or metamorphic in origin. One possible scenario is the production of charnockitic "magma" from deeper regions within the volcanic/volcaniclastic pile undergoing high-grade metamorphism. We suggest that the charnockite, an intrusive near-solidus body, would be derived as the ultrametamorphic equivalent of overlying granulite gneiss. Magmatic segregations, including the Vesuvius Megaporphyry, would be enabled during emplacement by localized fluid migration.

Alternatively, the NGG could have been produced by retrograde metamorphism of the PRCS or some other unit having the required parental signature. However, the lack of hydrous minerals in both units implies a condition, at the time of metamorphism, of low P_{H2O}, such that mobilization of incompatible elements and large scale segregations of mineral phases would not prevail. The clustering of major oxide systematics in charnockites compared to the trends associated with granulite gneisses (Figure 2) suggests the opposite effect and that differentiation of the NGG occurred prior to metamorphism. Any attempt to show that the NGG suite was either derived by retrograde metamorphism from the PRCS or any other major unit must address the variations among major oxides within the NGG, and the striking similarity of NGG and PRCS trace element patterns. Given that element mobility would be minimized under dry conditions, it is more likely that the chemical signatures in these two major units of the Pedlar Massif reflect primary compositions obtained prior to development of the metamorphic fabric in the NGG.

Average crustal abundances (from Taylor and McLennon, 1985) are also shown on Figure 5 to illustrate the comparison with NGG and PRCS patterns. The average lower crust composition (Taylor and McLennon, 1985) shown in Figure 4 is roughly identical, except for incompatible alkali elements, to the 344A mafic granulite. These relations likely reflect the regional aspects of the NGG and PRCS compositions and their potential association with other Grenville basement rocks. Of particular interest in these comparisons is the depleted Th signature obtained in the NGG and PRCS. The arguments presented above in favor of less mobility of incompatible elements do not seem to

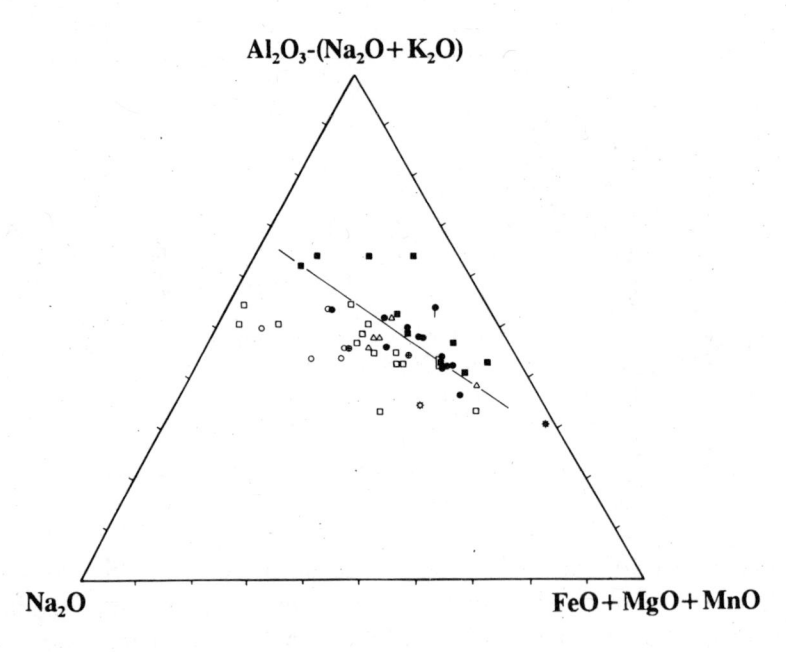

Figure 6. Modified A-K-F diagram of Pedlar Massif units. Symbols are the same as in Figure 3. Data sources same as Figure 2.

hold true for this enigmatic signature although, with regard to mass balance, the PRCS leucocharnockite/granite signature (Figure 4) suggests a viable unit for Th uptake. It may be possible that Th behaved as the least compatible element during lower crustal anatexis and was incorporated preferentially into initial charnockitic compositions and was thereby depleted in the NGG and PRCS units. In order for this mechanism to be operative, an increase in Th mobility relative to all other incompatible elements would have been required.

LOVINGSTON MASSIF

The Lovingston massif contains a more variable assemblage of rock types. Charnockites here are part of the plutonic ferrodioritic suites. These suites intrude several types of layered gneisses. The Roseland Anorthosite, along with its associated Border Gneiss, appears to represent a synmetamorphic (ca. 1.0 Ga) exotic diapir of deep crustal and mantle material (Herz and Force, 1982; Bartholomew and Lewis, 1984). Finally, widespread late Proterozoic-age granites, associated with initial Iapetus rifting, intruded the massif.

Grenvillian metamorphism, within and near the Roseland Anorthosite diapir, occurred under the deep crustal (higher pressure) granulite-facies conditions of about 800° C and 8Kb with $P_T >> P_{H2O}$ (Herz, 1984) with the characteristic mineral assemblage of andesine + orthopyroxene + garnet. Elsewhere in the massif, granulite-facies garnet is absent and metamorphism likely did not exceed the shallower crustal (lower pressure) granulite-facies conditions with a characteristic relict mineral assemblage of andesine + orthopyroxene ± amphibole still preserved in extensively retrograded (Paleozoic) rocks.

Paleozoic metamorphism (ca. 425–450 Ma) reached upper greenschist-facies (garnet grade) conditions at about 400° C (Evans, 1984) in the Lovingston massif, near and southeast of the Lawhorne Mill fault. In many of the rock types in this massif the biotite foliation probably developed during retrograde Paleozoic metamorphism which is dated (K-Ar) at ca. 425–450 Ma in the overlying Lynchburg Group. The Paleozoic garnets have about 20% higher grossular and 10% higher spessartine components than the Grenvillian granulite-facies garnets (Evans, 1984). Although Paleozoic retrograde metamorphism was a key factor with regard to biotite and garnet compositions and foliation development, regionally it did not affect Rb-Sr, U-Pb, or Sm-Nd systematics to any significant degree because these systems all reflect middle Proterozoic-age equilibration.

In the portion of the Lovingston massif crossed by this field trip, the Hills Mountain Granulite Gneiss, the Roseland Anorthosite and Border Gneiss, the Archer Mountain (ferrodiorite + charnockite + granite) Suite, the Stage Road Layered Gneiss, and the (Late Proterozoic) Rockfish River Pluton are the principal rock types found in the basement beneath the latest Proterozoic-age Lynchburg Formation.

Roseland Anorthosite and Border Gneiss

The Roseland Anorthosite is an elongate (14 x 3 km) domal body of megacrystic alkalic andesine antiperthite (Herz, 1969, 1984). Although differing in details, both Hillhouse (1960) and Herz and Force (1987) mapped zones of layered graphitic garnet-granulite gneiss that discontinuously mantle the Roseland Anorthosite dome. Sinha and Bartholomew (1984) formalized the name Border Gneiss (after Hillhouse's 1960 usage) for these distinctive paragneisses which contain some mafic layers, and which have not yet been recognized elsewhere in the Lovingston massif (Bartholomew and Lewis, 1984). The close association of these exotic rocks (anorthosite and paragneisses) in a limited region of the Lovingston massif where deep (higher pressure) granulite-facies mineral assemblages (orthopyroxene + garnet) are preserved suggested their emplacement as a diapir of deeper crustal material into the Lovingston massif (Bartholomew and Lewis, 1984; Sinha and Bartholomew, 1984). The Lovingston massif is generally characterized by shallower (lower pressure) granulite-facies mineral assemblages (orthopyroxene, but no garnet) (Bartholomew et al., 1981; Bartholomew and Lewis, 1984). Herz and Force (1982, 1984, 1987) also concluded that the anorthosite, with its gneissic mantle, was diapirically emplaced at a shallower crustal level based on (1) deformation of the marginal rocks, (2) the structural concordance of internal fabric to the dome margin, and (3) the dome's crosscutting relationship to older Grenvillian contacts.

The Roseland Anorthosite has a Sm-Nd age of 1045 ±44 Ma with Nd_i = 0.51134 ±4 and $_{Nd}$ = +1.0 (Pettingill et al., 1984). A sample of leucocratic neosome from the Border Gneiss has a slightly discordant $^{207}Pb/^{206}Pb$ age of 980 Ma (Herz and Force, 1987). Both ages are consistent with an interpretation that the anorthosite and its gneissic mantle were diapirically emplaced during peak metamorphic conditions of the Grenville event (ca. 1 Ga).

Microprobe compositions of the Roseland Anorthosite reported by Herz and Force (1987) are: antiperthite host- $An_{31}Ab_{68}Or_1$; K-feldspar lamellae- $An_1Ab_6Or_{93}$; orthopyroxene- $En_{55}Fs_{41}Wo_4$ (from chemical analyses, Ross, 1941). Herz and Force (1984) also give feldspar compositions calculated from the whole rock norms ($An_{27-28}Ab_{51-59}Or_{14-21}$) and give $An_{23.5}Ab_{51}Or_{25}Celsian_{0.5}$ (or $An_{31.5}$ calculated for plagioclase only) for the bulk chemical analyses (table 2 of Ross, 1941) of andesine megacrysts. Herz and Force (1984) and Ames (1981) reported ^{18}O values of +7.7 and +8.8 for the Roseland Anorthosite and Border Gneiss, respectively.

Archer Mountain Suite (AMS)

The Archer Mountain Suite of ferrodiorite, charnockite, quartz monzonite, and granite is representative of the four plutonic suites found in the Lovingston massif (Bartholomew and Lewis, 1984). Both the Turkey Mountain Suite and the Roses Mill Pluton, described by Herz and Force (1984, 1987) also lie in the region crossed by the field trip. AMS charnockites typically contain orthopyroxene but lack (granulite-facies) garnet, indicating less deep (lower pressure) granulite-facies conditions during Grenville metamorphism; Roses Mill charnockite contains orthopyroxene + garnet (Herz and Force, 1984, 1987), indicating deep crustal (higher pressure) granulite-facies conditions prevailed in the vicinity of the Roseland Anorthosite. AMS charnockite has a U-Pb concordia age of 1060 Ma (Sinha and Bartholomew, 1984) which is similar to Lukert's (1973) U-Pb concordia age of 1066 ±20 Ma for augen gneiss (presumably derived from a rock similar to AMS charnockite) near Madison, Virginia. U-Pb concordia intercept ages of 1100 Ma and 406 ±35 Ma were reported for both AMS quartz monzonite and AMS granitic pegmatite (Sinha and Bartholomew, 1984; Davis, 1974). The Roses Mill Pluton has a concordant $^{207}Pb/^{206}Pb$ age of 970 Ma (Herz and Force, 1987). Samples from the Archer Mountain and Turkey Mountain suites and from the Roses Mill Pluton yield an Rb-Sr isochron age of 1009 ±26 Ma with Sr_i = 0.7058 ±4 and the Sm-Nd data for the same samples fit the isochron age for the Roseland Anorthosite of 1004 ±36 Ma with Nd_i =0.51139 ±3 and $_{Nd}$ = +1.0 (Pettingill et al., 1984).

Herz and Force (1984), Herz (1984), and Ames (1981) reported microprobe compositions and ^{18}O data for ferrodiorites of both the Roses Mill Pluton (orthopyroxene- $En_{43-60}Fs_{40-56}Wo_{1-2}$; clinopyroxene- $En_{30-36}Fs_{19-26}Wo_{43-46}$; plagioclase- $An_{39}Ab_{58}Or_3$; orthoclase- $An_{32}Ab_{42}Or_{26}$; ^{18}O = +7.6%) and the Turkey Mountain Suite (orthopyroxene- $En_{35-58}Fs_{41-63}Wo_{1-2}$; clinopyroxene- $En_{25-36}Fs_{19-33}Wo_{42-47}$; plagioclase- $An_{35-40}Ab_{58-64}Or_{1-2}$; orthoclase- $An_{24}Ab_{39}Or_{37}$; ^{18}O = +8.2%).

Hills Mountain Granulite Gneiss

This poorly layered gneiss is found over a limited area. It contains poikiloblastic feldspar porphyroblasts and is streaked with crudely aligned mafic mineral segregations (Bartholomew, 1977). Orthopyroxene is present but no (granulite-facies) garnet, indicating less deep (lower pressure) granulite-facies conditions during Grenville metamorphism (Bartholomew et al., 1981). Migmatized zones between the Hills Mountain gneiss and AMS plutonic rocks suggest that the Hills Mountain is the older of the units. The relationship between Hills Mountain Granulite Gneiss and the Stage Road Layered Gneiss, a widespread metavolcanic/metasedimentary sequence (Bartholomew and Lewis, 1984; Sinha and Bartholomew, 1984), is uncertain. But, because the former is of limited areal extent, and the streaked banding in the Hills Mountain is probably of (Grenvillian) metamorphic origin, the Hills Mountain Granulite Gneiss is tentatively interpreted as an early (pre-1100 Ma) plutonic unit intruded into the Stage Road gneisses.

Shaeffer Hollow Granite

The Shaeffer Hollow Granite of Herz and Force (1984; 1987) is found only over a limited area near the Roseland Anorthosite. Herz and Force (1984; 1987) describe it as a coarse-grained granite with 2–5 cm feldspar phenocrysts. It may be similar to the AMS granitic pegmatite of Sinha and Bartholomew (1984). A slightly discordant $^{207}Pb/^{206}Pb$ age of 993 Ma and a discordant $^{207}Pb/^{206}Pb$ age of 1787 Ma were reported for two Shaeffer Hollow samples; the latter sample contained zoned zircons (Herz and Force, 1987).

Stage Road Layered Gneiss (SRLG)

The widespread SRLG is a heterogeneous volcanoclastic sequence of coarsely layered (typically measured in meters) gneisses (Sinha and Bartholomew, 1984). The coarse augen gneiss at the type outcrops yielded two distinct zircon populations. The metamorphically produced, euhedral zircons are nearly concordant at 915 Ma, whereas the rounded detrital zircons yield a $^{207}Pb/^{206}Pb$ age of 1420 Ma with concordia intercepts at 915 Ma and 1870 ±200 Ma (Sinha and Bartholomew, 1984; Davis, 1974). The Rb-Sr isochron age of 1147 ±34 Ma with Sr_i = 0.7047 ±5 (Pettingill et al., 1984) is similar to the U-Pb concordia intercept of 1118 ±10 Ma reported for SRLG augenless gneisses near Madison, Virginia (Lukert, 1973).

LATE PROTEROZOIC-CAMBRIAN IGNEOUS ROCKS AND SYNRIFT DEPOSITS

Rockfish River Pluton

Bartholomew and Lewis (1984) mapped an unnamed supersuite of late Precambrian-age plutonic rocks extending from northwestern North Carolina through the Virginia Blue Ridge, many of which Rankin (1975; 1976) associated with Iapetus rifting. This supersuite includes the plutonic suite of the Crossnore Complex (as used by Odom and Fullagar, 1984) and the Robertson River Suite (Tollo and Arav, in press) as well as numerous plutons not formally included in any suite. The Rockfish River Pluton, described by Davis (1974) and Sinha and Bartholomew (1984) and visited on this field trip, is typical of many of these smaller A-type plutons which commonly contain allanite and fluorite. It has a U-Pb concordia age of 730 Ma (Davis, 1974; Sinha and Bartholomew, 1984) and a Rb-Sr age of 646 ±55 Ma with an Sr_i = 0.7172 ±0.0035 (Mose and Nagel, 1984). Its age of 730 Ma suggests that the Rockfish River Pluton is one of the older plutons associated with the early phase of Iapetus rifting. The more reliable age determinations seem to constrain the plutonism between 730 and 650 Ma (Odom and Fullagar, 1984; Lukert and Banks, 1984; Sinha and Bartholomew, 1984) which is significantly older than the 570 ±36 Ma Rb-Sr age reported by Badger and Sinha (1988) for the Catoctin Formation, which is associated with the later phase of Iapetus rifting.

Lynchburg Group and Mechum River Formation

The Lynchburg Group lies stratigraphically beneath the metavolcanic rocks of the Catoctin Formation exposed along the eastern flank of the Blue Ridge anticlinorium. This thick sequence of clastic rocks is interpreted as a synrift accumulation of clastic sediments derived from Grenvillian rocks (Wehr, 1985; in press; Wehr and Glover, 1985) during the early phase of Iapetus rifting. According to Wehr (1985; in press) the lower Lynchburg Group accumulated near the margin of a rapidly subsiding rift basin and illustrates the transition from alluvial to deltaic to deep-water fan deposits. The Lynchburg unconformably overlaps granitic rocks of the Robertson River Suite (Nelson, 1962; Tollo and Arav, in press), and lithologically similar granite clasts are found in the Mechum River Formation (Lukert and Banks, 1984), a lateral equivalent to the Lynchburg Group (Wehr, 1985). Rocks of the A-type plutonic supersuite in this region have an age range of 650–730 Ma (based on both U-Pb and Rb-Sr data of Lukert and Banks, 1984; Mose and Nagel, 1984; Sinha and Bartholomew, 1984; Rankin, 1976) which places an older limit on the age of the Lynchburg metasedimentary sequence. Fullagar and Dietrich (1976) obtained a Rb-Sr age of 583 ±21 Ma with a Sr_i = 0.7175 ±0.0008 for the Lynchburg in this area and demonstrated the derivation of clasts in the basal Rockfish Conglomerate from Grenvillian-age rocks, which is consistent with dispersal patterns reported by Schwab (1986).

The southern end of the belt of fine-grained metasedimentary rocks of the Mechum River Formation was shown by Nelson (1962) to end near Batesville in the western part of the Covesville quadrangle. Bartholomew (1977) noted an occurrence of phyllite and metaconglomerate on strike with the Mechum River Formation at Batesville, isolated within a zone of mylonitic rock derived from the Archer Mountain Suite. The granitic cobbles in this metaconglomerate are derived from a lithologic type of the Archer Mountain Suite but not that immediately adjacent to the Mechum River metasediments, thus implying fluvial transport of the clasts. Perhaps this coarse conglomerate was part of the fluvial system that supplied similar coarse cobbles deposited as the basal Rockfish Conglomerate of the Lynchburg Group (Figure 1).

Catoctin and Swift Run Formations

The Catoctin Volcanic Province consists of a sequence of metabasalts erupted during Iapetus (Late Proterozoic–Cambrian) rifting between North America and Baltica. On the east limb of the Blue Ridge anticlinorium, the lavas interfinger with and overlie rift-facies metasediments of the Lynchburg Group (Schwab, 1974; Wehr, 1985; in press). On the west limb of the anticlinorium, the lava flows locally interfinger with or overlie the Swift Run Formation, a coarse clastic sequence, of variable thickness, of terrigenous conglomerate and conglomeratic sandstone, derived from the craton (Schwab, 1986), which nonconformably rests on the Pedlar massif. Where Swift Run metasedimentary rocks are absent, lava flows and thin metasedimentary rocks of the Catoctin Formation overlap the crystalline rocks of the Pedlar massif in central Virginia. Rocks of the Catoctin Volcanic Province pinch out both westward and southwestward (Bloomer and Werner, 1955; Werner, 1966; Furcron, 1969; Bartholomew et al., 1981). Despite Taconic greenschist-facies metamorphism, the volcanic stratigraphy in most places is still preserved. Flows are separated by thin discontinuous metasedimentary units of phyllite and siltstone and by zones of epidotized volcanic breccia. Tops and bottoms of flows are frequently vesicular. Graded bedding in the metasedimentary units help determine up-direction.

The age of the Catoctin Formation volcanic rocks is minimally constrained by the overlying Early Cambrian–age Chilhowee Group. A discordant $^{207}Pb/^{206}Pb$ age of 686 Ma (recalculated age done by Lukert and Banks, 1984) was reported by Rankin et al. (1969) for the basal Catoctin rhyolite in Pennsylvania, but an inherited component to these zircons is suspected (D. W. Rankin, personal communication to Badger), thus making the actual age younger than 686 Ma. Later Rb-Sr analysis yielded an unrealistically young age, which actually reflected Taconic metamorphism at 470 ±77 Ma (Mose, 1981; Mose and Nagel, 1984). Mose and Nagel (1984) inferred an age of about 570 Ma for the Catoctin Formation based on a spurious Rb-Sr isochron age of 570 ±15 Ma they used for the Robertson River Pluton. More recently, Badger and Sinha (1988) used a very careful screening process to select only Catoctin Formation samples that contained relict igneous textures and relict igneous mineralogy (clinopyroxene) and samples from more massive portions of flows reflecting constant chemical compositions. Analyzing Rb-Sr from five such samples from the I-64 traverse, plus a clinopyroxene mineral-separate from one sample, yields a line with a slope reflecting an age of 570 Ma with Sr_i = 0.7035 ±1. This age is consistent with stratigraphic and paleontologic evidence, the Rb-Sr age for the basal part of the Lynchburg Group, and is interpreted as the timing of Catoctin magmatism.

Chilhowee Group

Overlying the volcanic rocks of the Catoctin Formation is a sequence of metasandstone and metasiltstone, with some metaconglomerate, of the Chilhowee Group. The lower part of this group reflects a transition from rift to passive-margin deposition during the opening of Iapetus, whereas the upper part of the group marks the inception of passive-margin sedimentation during the Cambrian (Simpson and Eriksson, 1989).

Scolithus tubes present in the Antietam Quartzite, the upper formation of the Chilhowee Group in central Virginia, along with the Balcony Falls fossils (including *Olenellus sp.*) (Walcott, 1891) indicate a Cambrian age for the top of the Chilhowee Group. The recent discovery of the Early Cambrian trace fossils, *Planolites* and *Rusophycus*, in the Unicoi Formation (the basal part of the Chilhowee Group) in southwestern Virginia suggests that the entire Chilhowee Group is Early Cambrian in age (Simpson and Sundberg, 1987), which is consistent with available geochronological data from the underlying Catoctin Formation metavolcanic and Lynchburg Group metasedimentary rocks.

PRECAMBRIAN AND PALEOZOIC OROGENESIS

The geologic database in this region has grown significantly since Griffin (1971) used the geologic map of Bloomer and Werner (1955), plus structural data he collected from this part of Virginia, for his structural fabric analysis. Mitra and Elliott (1980) and Mitra and Lukert (1982) analyzed regional structural data from an even larger portion of the northern Blue Ridge. Bartholomew et al. (1981) and Gryta and Bartholomew (1989) were able to utilize an even more substantial map and fabric database available in the last few years. Although the data coverage is still incomplete, major faults and folds have been delineated and general trends (northwest and east-west) of relict Grenvillian segregation layering are easily discernable, amidst the dominant northeastwardly trending Paleozoic foliation (Figure 1).

Age of Metamorphism and Thrusting

K-Ar, Rb-Sr, and U-Pb data show that the Late Proterozoic–Cambrian–age cover sequence of the Pedlar and Lovingston massifs in central Virginia was metamorphosed ca 425–450 Ma (Table 3). This is consistent with occurrences of metamorphosed Chilhowee Group quartzite clasts in the Ordovician Fincastle Conglomerate (Bartholomew et al., 1981). Earlier work (Dietrich et al., 1969; Furcron, 1969; Robison, 1976) utilizing K-Ar data reported a wide range of values, which showed ages as young as 345 Ma. This was partially the reason that Bartholomew and Lewis (1984) inferred movement on the Rockfish Valley–Fries fault system to have occurred ca 345 Ma. We believe that the scatter of K-Ar data more likely reflects post-metamorphic cooling ages, perhaps in part disturbed by a younger event(s) such as the sub-greenschist thermal event (303–273 Ma, Elliot and Aronson, 1987) which affected Valley and Ridge sediments during Alleghanian thrusting (Lewis and Hower, 1990). Conversely, in the northern part of the Blue Ridge anticlinorium, Mitra and Elliot (1980) argued that the South Mountain cleavage is Alleghanian in age because of the parallelism of metamorphic foliation in the Catoctin-Chilhowee sequence with cleavage in Valley and Ridge rocks in the North Mountain thrust sheet (Cloos, 1947). We suggest that this parallelism must be viewed as coincidental because the greenschist-facies event is clearly of Ordovician age, as established above, and the Valley and Ridge cleavage is much younger. Alternatively, Bartholomew et al. (1981) argued that the Rockfish Valley fault, which is characterized by a zone of ductile deformation under greenschist-facies conditions (the same as ductile deformation zones in northern Virginia described by Mitra, 1977), was synmetamorphic. Attempts by A. K. Sinha to date the Rockfish Valley ductile deformation zone (DDZ) have been unsuccessful due to a lack of disturbance of radiogenic systems during fluid (water only) migration associated with the DDZ. Thus, the age of thrusting can only be inferred from constraining geological arguments.

Bartholomew (1988b; work in progress) has suggested that the Rockfish Valley, Blue Ridge, and Pulaski thrust systems are related and likely moved sequentially, but also nearly synchronously. This linked "mega-system" evolved across both the transition from ductile to brittle behavior and the basement-involvement to décollement-type faulting. This interpretation is based on: 1) all three systems die out northeastward in the same geographic area; 2) all three systems increase markedly in displacement southward, especially southwestward of the Roanoke recess; 3) the high percentage of shortening near the recess of 50%, 60%, and 80% respectively is significantly greater than shortening (Bartholomew, 1987) on the western Valley and Ridge thrust systems; and 4) according to Spencer and Waterbury (1987), the Blue Ridge fault, *per se*, merges into the Rockfish Valley DDZ just south of the James River.

The Humpback Mountain fault (Figure 1) and other faults in the region are characterized by ductile deformation in the basement and brittle behavior higher in the section. Some of these faults truncate folds with axial planar foliation and locally have ramp-flat geometries that refold the

TABLE 3. *Age of Regional Greenschist-Facies Metamorphism in Central Virginia Blue Ridge.*

AGE	METHOD	UNIT	REFERENCE
410 ±25 Ma	K-Ar	Mechums River Fm., Albemarle County	Furcron, 1969
425 ±25 Ma	K-Ar	Lynchburg Group, Fauquier County	Furcron, 1969
>400 Ma	K-Ar biot/hb	Lynchburg Fm., Lovingston massif Amherst County	Robison, 1976
406 ±35 Ma	U-Pb concordia	Stage Road Gneiss, Nelson County	Sinha & Bartholomew, 1984
470 ±77 Ma	Rb-Sr isochron	Catoctin Fm., northern Virginia	Mose and Nagel, 1984
450 Ma	Rb-Sr	Catoctin Fm., Nelson County	Badger & Sinha, 1988

Taconic foliation (Figure 1). Perhaps the following interpretation is most consistent with the above discussion:

   1) Regional greenschist-facies metamorphism occurred ca 425–450 Ma and produced both the chlorite/actinolite/biotite foliation (axial planar to folds in the Lynchburg-Catoctin-Chilhowee cover sequence) and a chlorite/biotite foliation in the Blue Ridge basement.

   2) Early Alleghanian (ca 300 Ma) thrusting in the Blue Ridge was accompanied by a fluid-expulsion event(s) that, locally, in crystalline basement rocks DDZ's, re-oriented micaceous minerals and/or produced retrograde lower greenschist-facies assemblages. The fluid-expulsion front may have migrated westward with Alleghanian thrusting and enhanced cleavage development throughout the Paleozoic section particularly within the North Mountain thrust sheet containing the Massanutten synclinorium.

The Rockfish Valley Fault and the Tye River Imbricate Fan

The Rockfish Valley fault and its associated zone of mylonitic rocks has been mapped and described in numerous publications (Bartholomew, 1977; Bartholomew et al., 1981; Gathright et al., 1977; Herz and Force, 1987). Greenschist-facies mineral assemblages are pervasive within the mylonitic rocks along the zone and suggest temperatures of several hundred degrees centigrade prevailed at the time of ductile deformation associated with thrusting along the Rockfish Valley fault.

The Tye River imbricate fan is a trailing imbricate fan developed northwestward of the Humpback Mountain fault which merges (north of the area) with the Rockfish Valley fault (Bartholomew, 1988b). The fan derives its name from the Tye River fault zone of Werner (1966).

The Humpback Mountain fault, which we will visit on this trip, is the major fault northwestward of the Rockfish Valley fault and was mapped by Bartholomew (1977) in the Sherando quadrangle. The Humpback Mountain fault is more or less parallel to the Blue Ridge Parkway from the western margin of the quadrangle northeastward to Humpback Gap; then it trends more easterly into Glass Hollow in the Greenfield quadrangle. North of Glass Hollow, the fault was incorrectly mapped as the Rockfish Valley fault across the Waynesboro East quadrangle by Gathright et al. (1977). Shear zones along the Humpback Mountain fault were mapped northward across the Crozet quadrangle by Wilson (1986) and extended beyond there by reconnaissance (Gathright, 1976; Bartholomew and Lewis, 1984).

Werner (1966) mapped the trace of the Humpback Mountain fault only 5 km southwestward from the Sherando quadrangle boundary, only to where basement was thrust over Catoctin Formation metavolcanic rocks. Reconnaissance by Lewis et al. (1986) extended the trace of this ductile fault zone from the Big Levels quadrangle across the Massies Mill, Montebello, and Forks of Buffalo quadrangles. Spencer and Waterbury (1987) show this fault as finally merging back into the Rockfish Valley DDZ south of the Buena Vista quadrangle.

Other major faults within the Tye River imbricate fan are the Back Creek and Snowden faults. These and the other lesser faults were mapped by Bloomer and Werner (1955) with modifications from Bartholomew (1977), Werner (1966), Hudson (1981), and Spencer and Waterbury (1987). As discussed below, portions of these faults may have experienced extensional displacement during the Mesozoic (Bartholomew, 1988a) and thus, subsequently, have inverted (younger over older) structural relationships, with rocks of the Catoctin Formation "thrust" over middle Proterozoic–age basement (Bloomer and Werner, 1955; Werner, 1966).

Regional Fabric

Grenvillian segregation layering has been re-oriented, and in some cases probably completely transposed, parallel to axial-surfaces of Paleozoic-age folds (CRO, PR, HM quadrangles, Figure 1) as shown by Gryta and Bartholomew (1989, their Figure 9). However, in other localities (MON, SHE, GRF quadrangles, Figure 1) the strong discordance of Paleozoic foliation with steeply dipping segregation layering (Gryta and Bartholomew, 1989, their Figure 9) suggests that foliation re-orientation was influenced by relict Grenvillian trends as well as subsequent deformational processes.

The principal rock fabric in the area (Figure 1) is the micaceous foliation formed during the Paleozoic. Overall, this foliation is axial planar to folds defined by poles to bedding in the late Proterozoic-Cambrian-age cover sequence (Griffin, 1971; Bartholomew et al., 1981; Gryta and Bartholomew, 1989). This micaceous foliation is part of the "South Mountain cleavage" of Mitra and Elliott (1980), which they attribute to Alleghanian deformation largely because of the coplanar cleavage present in rocks as young as Devonian within the Massanutten Synclinorium (Figure 1). Although these fabrics may be regionally coplanar, the micaceous minerals in the late Proterozoic-Cambrian-age sequence formed during regional greenschist-facies metamorphism dated at about 450 Ma (Table 3). The parallel micaceous fabric in the basement is inferred to have formed primarily at this time as well because almandine (Evans, 1984) garnet formed after biotite in both the Rockfish River Pluton and the older basement gneisses southeast of the Lawhorne Mills Fault (LOV quadrangle, Figure 1) should have occurred during the peak of greenschist-facies metamorphism.

Near many of the major thrusts, foliation orientations are suggestive of mica re-orientation within the more mylonitic zones during thrusting. Bartholomew et al. (1981) demonstrated flattening of foliation dip within the mylonitic zone of the Rockfish Valley fault. Although we now recognize more mylonitic fault zones than did they, the fabric still suggests a marked alignment of foliation and a decrease in dip within the mylonitic zones adjacent to the Humpback Mountain, Rockfish Valley, and Perry Creek faults (northeastern quarter of Figure 1) as well as near the thrust within the Lynchburg Group (SCH quadrangle, Figure 1).

Locally, cleavage orientation is also influenced by both pre-existing fabric anisotropy and inhomogeneity in the regional distribution of lithologies. An example of this influence can be seen in the Archer Mountain Suite (HM, LOV quadrangles, Figure 1) where foliation orientation is locally influenced by the presence of large boudin-like bodies of competent charnockite.

The strong northeast-southwest trend ("South Mountain cleavage") is characteristic of both earlier Taconic metamorphism and later Alleghanian re-orientation of planar fabric as depicted by Mitra and Elliott (1980). Major fold orientations, including those which fold foliation, are approximately N 45° E. Regionally, foliation in the northeastern part of the area, where a wide zone of mylonitic rocks is associated with overturned strata of the late Proterozoic-Cambrian-age cover sequence, is significantly less steep (<< 60°) than elsewhere. Conversely, very steep (> 60°) to vertical foliation characterizes the southeastern flank of the Lovingston massif (LOV, ARR, PR quadrangles), especially near the steeply dipping Lawhorne Mills shear zone. Most likely the steepness of the fault and foliation is due to re-orientation above a ramp on a structurally lower thrust. However, the presence of numerous late Proterozoic A-type plutons near this zone permits the possibility that this zone may have experienced extensional displacement during Iapetus rifting, although kinematic data is not yet available.

Although northeast-striking, southeast-dipping foliation orientations predominate in many localities, variable foliation orientations indicate significate folding by post-formation deformational event(s) (Figure 1). Foliation trend analysis shows that a dome and basin fold-style with northwest-trending axial surfaces produces a regional "crenulation" pattern in regions of high density data (Figure 1). The "crenulation" pattern can be inferred, but not conclusively demonstrated, throughout the northern Blue Ridge from the data of Mitra and Elliott (1980). Locally (along the SHE-GRF quadrangle boundary), a Taconic axial surface is refolded about a northwest-trending axial surface. Such broad warps may be due to oblique ramps formed during Alleghanian thrusting, but the overall "crenulation" pattern does not appear to be consistent with northwest-directed thrusting during the Alleghanian. Within the south-central part of the SHE quadrangle, axial planar foliation within the Catoctin and Swift Run Formations (Bartholomew, 1977) has an S-shaped pattern adjacent to a steeply dipping northeast-trending fault zone inferred (from Mesozoic dike and fault orientations) to have possibly experienced sinistral extensional displacement during Mesozoic opening of the proto-Alantic Ocean (Bartholomew, 1988a). Thus, the "crenulation" pattern may represent the effects of Mesozoic extension on pre-existing Paleozoic foliation.

MESOZOIC EXTENSION

A large portion of the area covered by this field trip was affected by NW-SE extension related to the opening of the Atlantic Ocean during the Mesozoic. The most obvious evidence of extension is the numerous, generally N-NW-trending, diabase dikes which cut obliquely across the Blue Ridge into the Valley and Ridge near Waynesboro, Virginia. *En echelon* sets of dikes in this N-NW-trending swarm suggest a component of sinistral shear parallel to NE-trending faults and dike-sets and a component of dextral shear parallel to NW-trending faults and dike-sets (Bartholomew, 1989b). This pattern of right- and left-lateral dike-sets (Figure 7A) is found from the Waynesboro-Charlottesville area southward to the Danville Basin and into North Carolina, and suggests that gravity (vertical axis) was the intermediate stress throughout this region during the Mesozoic extension. Bartholomew (1989b; work in progress) characterized this strain-field pattern by: (1) Mesozoic basins (like the Danville basin) generally lacking sills; (2) basins bounded on only one flank by major faults; (3) major dike trends that bisect the acute angle between intersecting major fault trends (NE and NW); and (4) local *en echelon* dike patterns indicative of sinistral shear parallel to NE-trending faults and dextral shear parallel to NW-trending faults.

North of the Waynesboro-Charlottesville area Bartholomew (1989b) characterized a different strain-field pattern (Figure 7B) by: (1) Mesozoic basins (like the Culpeper basin) containing sills; (2) NE-trending basins bounded along both flanks by subparallel major faults with opposing dips; and (3) dikes which trend subparallel to the basin-bounding faults. This strain-field pattern suggests that gravity (vertical axis) was the principal stress during the Mesozoic extension.

These different strain-field patterns, north and south of the Waynesboro strain-boundary, allow concomitant NW-SE extension to be accommodated across this entire region without requiring temporally varying stress fields or unexplained periods of basin-related extension and "inactive" dike

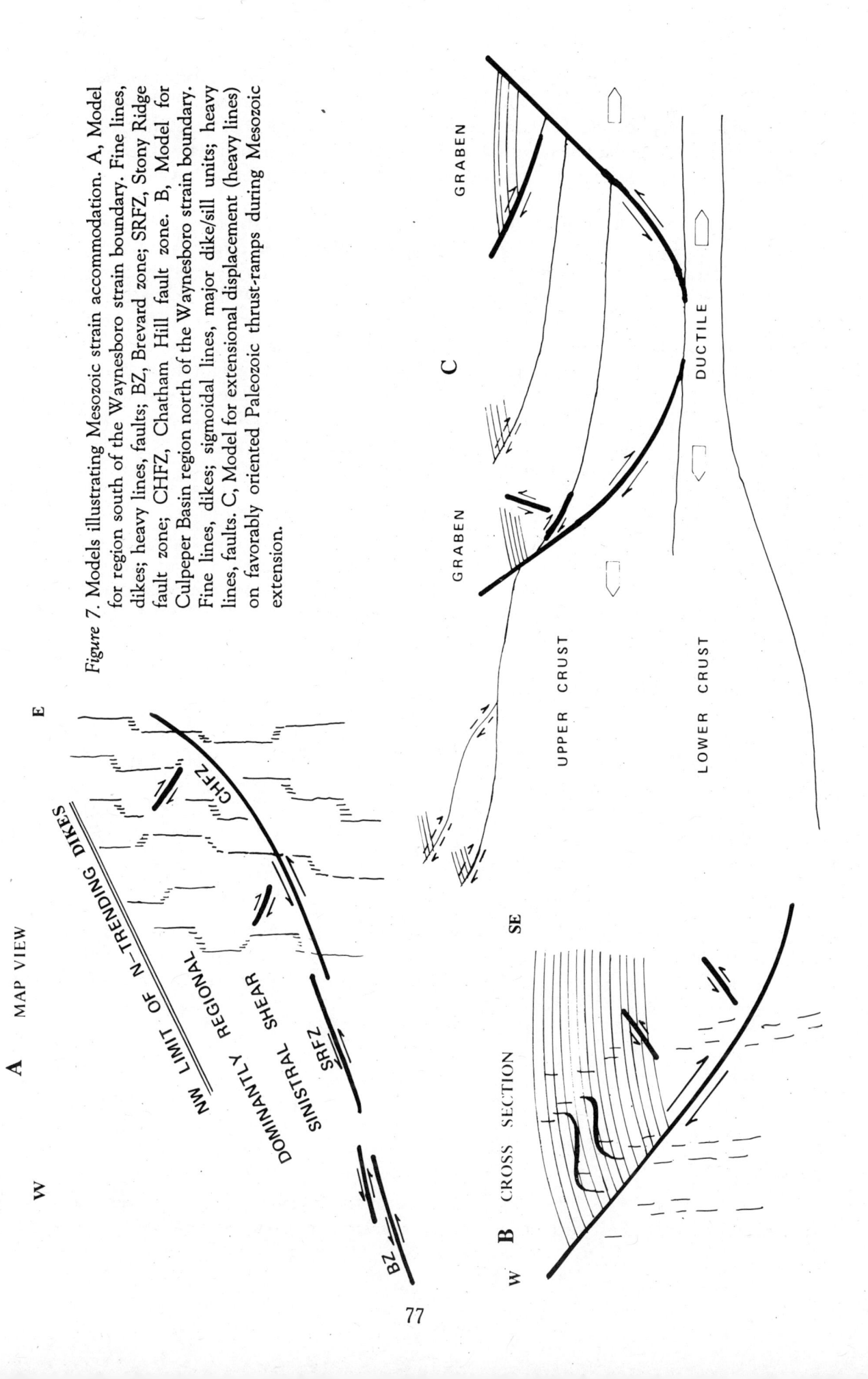

Figure 7. Models illustrating Mesozoic strain accommodation. A, Model for region south of the Waynesboro strain boundary. Fine lines, dikes; heavy lines, faults; BZ, Brevard zone; SRFZ, Stony Ridge fault zone; CHFZ, Chatham Hill fault zone. B, Model for Culpeper Basin region north of the Waynesboro strain boundary. Fine lines, dikes; sigmoidal lines, major dike/sill units; heavy lines, faults. C, Model for extensional displacement (heavy lines) on favorably oriented Paleozoic thrust-ramps during Mesozoic extension.

injection (or vice versa) to account for different dike-swarm orientations relative to the associated basins. The Lewis and Clark zone is a Cordilleran analogue (Bartholomew, 1989b) which acted as a strain-field boundary during a Tertiary extension and the development of the Montana-Idaho Basin and Range (Stickney and Bartholomew, 1987). The Waynesboro strain-boundary likely persisted from Mesozoic time through Tertiary dike emplacement in the Valley and Ridge Province (Bartholomew, 1989b).

Although Mesozoic dikes are the most obvious indicators of the Mesozoic extension in the Blue Ridge, Bartholomew (1988a; 1989a; work in progress) recognized certain fault patterns that are suggestive of Mesozoic extension being accommodated by backslip on favorably oriented Paleozoic thrust faults, generally above a favorably oriented ramp (Figure 7C). Within the region of this field trip, cross sections by both Bartholomew (1977) and Werner (1966) show faults where Catoctin Formation metavolcanic rocks are "thrust" over Grenvillian-age basement. The segments of these thrust faults which exhibit apparent backslip are NE-trending (consistent with the strain-field pattern discussed above) and lie along the southeastern flank of probable Paleozoic-age ramp anticlines deduced from fold and foliation patterns (Figure 1) and balanced cross sections.

STOP LOCATIONS AND DESCRIPTIONS

The latter half of the first day of the trip will be devoted to Grenvillian rocks, primarily of the Pedlar massif, along an E-W traverse from Avon to Vesuvius. The second will be a W-E traverse across both the Pedlar and Lovingston massifs from Buena Vista to near Rockfish Station. Morning of the third day will be devoted to the Catoctin Formation, which overlies the Pedlar massif, exposed along Interstate 64 just east of Rockfish Gap at the junction of the Blue Ridge Parkway and the Skyline Drive (of the Shenandoah National Park) with I-64 and U.S. Highway 250.

DAY 1 (Figures 1 and 8) Leaders: Mervin J. Bartholomew, Sharon E. Lewis, and Scott S. Hughes

STOP 1-1

On State Road 709, 1.35 miles west of its northern junction with State Highway 151, which is 5.35 miles south of the junction of 151 and U.S. Highway 250. Mafic granulite (pyroxenite) from which Virginia Division of Mineral Resources (VDMR) samples R-6787 and R-6788 were collected is exposed along south side of curve. Photomicrograph: Figure 18 of Bartholomew (1977); chemistry: sample 344A in Table 2.

STOP 1-2

On State Road 633, 0.4 miles south of junction with State Road 635; junction is approximately 1.3 miles east (on 635) of the community of Greenfield on State Highway 151. Outcrop along east side of road contains phyllite and metaconglomerate of the Mechum River Formation recognized by Bartholomew (1977, p. 30).

STOP 1-3

On State Highway 6, 2.55 miles north of junction with U.S. Highway 29, at southern junction with State Road 810. Outcrop in turnout on east side of road is charnockite of Archer Mountain Suite.

STOP 1-4

On State Road 810, 0.85 miles north of its southern junction with State Highway 6. Hills Mountain Granulite Gneiss is exposed in roadcut on east side of road. Although weathered, both the

Figure 8. Portion of the Roanoke (1:250,000) topographic map showing field trip locations for day 1 and day 2.

segregation layering and feldspar porphyroblasts (Plate 3A of Bartholomew et al., 1981) that characterize this unit are visible. Unfortunately the type locality, which is less than 0.25 miles SE in a badly overgrown stream valley where VDMR sample R-6780 was collected, has deteriorated from the fresh exposure uncovered in a scoured channel after 1969 Hurricane Camille. Type locality and migmatite zone are illustrated in Figures 6–11 of Bartholomew (1977) and Plate 2A, B of Bartholomew et al. (1981).

STOP 1-5

Note: **Permission** is needed to enter Stoney Creek at Wintergreen; turn west off of State Highway 151 (2.4 miles south of its junction with State Highway 6 near Greenfield) onto State Road 634 (Monocan Drive), then left onto Stoney Creek West to the Creekside Close (cul-de-sac); park in site F-4 and walk about 150 feet due west to Stoney Creek. Exposure in creek bed, from which VDMR sample R-6494 was collected, is the type locality of the Nellysford Granulite Gneiss; outcrop illustrated -Plate 5A of Bartholomew et al., 1981, and Figures 2, 3 (photomicrograph), and 32 (fold) of Bartholomew (1977); sample included in average for Nellysford (Table 2).

STOP 1-6

On State Road 664, 1.7 miles east of junction with Blue Ridge Parkway and 0.45 miles east of the entrance to Wintergreen. Typical PRCS charnockite is exposed for 500 feet up and down road from stream junction; both mafic and felsic dikes cut the charnockite; Paleozoic retrograde effects are visible as locally abundant biotite flakes and as chlorite-coated surfaces; xenoliths are uncommon. Locality illustrated in Plate 5B of Bartholomew et al. (1981); sample included in average for PRCS (Table 2).

STOP 1-7

On State Highway 56 W, 1.6 miles northwest of junction with Blue Ridge Parkway, limited parking on left (SE) side of road at driveway to Golden Ridge Farm. The type locality (illustrated in Plate 3B of Bartholomew et al., 1981) of the Vesuvius Megaporphyry is the exposure along Highway 56, just downhill from this driveway, where the U.S.G.S. bench mark is located. Sample B-37 (Table 2) is from this location.

STOP 1-8

On State Highway 56 W, 2.2 miles northwest of junction with Blue Ridge Parkway, park on right (NW) side at old quarry. This was the unakite quarry in the Vesuvius Megaporphyry where the sample that yielded a 933 Ma K-Ar age was collected.

DAY 2 (Figures 1 and 8) Leaders: A. Krishna Sinha and Mervin J. Bartholomew

STOP 2-1

On U.S. Highway 60 E 0.25 miles east of junction with Blue Ridge Parkway. Type outcrop of Lady Slipper Granulite Gneiss; illustrated in Figure 2A of Sinha and Bartholomew (1984); zircon sample location.

STOP 2-2

On U.S Highway 60 E, 2.7 miles east of junction with Blue Ridge Parkway. Pull off on right by old motel and walk 0.1 mile back along U.S. 60; outcrop shows intrusive contact between PRCS charnockite and Lady Slipper Granulite Gniess.

STOP 2-3

On State Road 633, 1.2 miles east of its junction with State Road 605 (junction is 1.8 miles north on 605 from junction of 605 with U.S. 60 E at Oronoco); park at turnout on right just past creek and walk back along 633 to falls on the Pedlar River; good exposure of streaked layering in Lady Slipper Granulite Gneiss cut by Paleozoic-age shear zones.

STOP 2-4

Park at intersection of Virginia State Roads 633 and 634 (0.4 miles from Stop 2-4); walk back about 100 feet along 633 to last outcrop; outcrop is typical Pedlar River charnockite (illustrated in Figure 2B of Sinha and Bartholomew, 1984), cut by 2–3 cm wide mafic dike.

STOP 2-5

On State Road 634 (which becomes Forest Service Road 63 after passing through Alto), 2.45 miles from junction of 633 and 634; park at turnout on right just past stream; walk back to outcrop on other side of creek. The charnockite here is lineated and contains blue quartz; feldspars show an increase in fracturing.

STOP 2-6

On State Road 634, 10.3 miles from junction of 633 and 634; park on right; walk back to sharp curve; exposure of mylonitic gneiss along the Humpback Mountain fault.

STOP 2-7

On State Highway 56 W along the Tye River, 0.3 miles west of junction with State Highway 151 just south of Roseland. Type locality of Border Gneiss in Tye River (illustrated in Figure 3A, B of Sinha and Bartholomew, 1984).

STOP 2-8

On State Highway 56 E along the Tye River, 0.3 miles east of junction with State Highway 151 just south of Roseland. Roseland Anorthosite cut by charnockite is exposed in roadcut.

STOP 2-9

On U.S. Highway 29 S at junction with State Highway 6 at Woods Mill. Outcrops southwest of intersection are type locality of the Archer Mountain Suite (AMS) (illustrated in Figure 4A, B of Sinha and Bartholomew, 1984). Biotite is the dominant mafic mineral in the AMS ferrodiorite and occurs both in aggregates (probably after pyroxene or hornblende) and along foliation surfaces developed during Paleozoic orogenesis. Mafic dikes (probably of Proterozoic age) cut the AMS and have a biotite foliation as well. Note that the AMS ferrodiorite typically contains abundant leucocratic gneiss xenoliths unlike the PRCS.

STOP 2-10

On U.S. Highway 29 N at junction with State Road 617 (1.55 miles northeast of Stop 2-10); walk about 200 feet back (SW) along U.S. 29 N to last outcrop. Type locality of Stage Road Layered Gneiss (illustrated in Figure 3C, D of Sinha and Bartholomew, 1984) with thick compositional layering. These augen gneisses (also well exposed just NE of intersection) yielded the two distinct zircon populations.

STOP 2-11

On State Road 617, 0.8 miles east of junction with U.S. Highway 29 N; park in pullout on right and walk 100 feet up creek on left (north). Pegmatitic granitoid phase (dated at 1100 Ma) of

Archer Mountain Suite typically has abundant felsic and mafic gneiss xenoliths in it (illustrated in Figure 4C, D of Sinha and Bartholomew, 1984).

STOP 2-12

On State Road 617, 1.75 miles east of junction with U.S. 29 N; outcrop forms bank of Rockfish River. Typical AMS ferrrodiorite containing abundant xenoliths and cut by felsic dikes which are folded about Paleozoic axial surfaces (illustrated in Plate 4A, B of Bartholomew et al., 1981).

STOP 2-13

On State Road 617, 2.55 miles east of junction with U.S. 29 N. Type locality of Late Proterozoic-age Rockfish River Pluton (RPP) (illustrated in Figure 5 of Sinha and Bartholomew, 1984) and mapped and described by Davis (1974). The RRP contains abundant xenoliths and is cut by foliated mafic dikes of probable Late Proterozoic age. Although poorly foliated, the RRP does contain both mica and garnet which developed during Paleozoic retrograde metamorphism.

STOP 2-14

On State Road 617, 2.95 miles east of junction with U.S. 29 N. Mafic rock exposed in roadcut and in Rockfish River is part of a large mafic pluton which cuts the Late Proterozoic-age RRP. This unit is probably related to the abundant small mafic dikes found throughout this region.

STOP 2-15

On State Road 617 at junction with State Road 714, 4.35 miles east of junction with U.S. 29 N; walk across bridge on 714 and follow short road down under bridge on south bank of Rockfish River. This is the classic locality of the Rockfish Conglomerate of the Lynchburg Group (see Nelson, 1962; Wehr, 1983; 1985). Some of the clasts are lithologically similar to various types of Grenvillian rocks exposed in this region, and Rb-Sr data on clasts collected 0.2 miles farther east on S.R. 617 indicate their likely derivation from Grenvillian rocks (Fullagar and Dietrich, 1976).
NOTE: For those people interested in local history, the historic Rockfish store and railroad station, popularized on "The Waltons" TV show, is located about 0.5 miles farther east on State Road 617.

Day 3 (Figures 1 and 9) Leader: Robert L. Badger

Interstate 64 provides one of the best exposures of Catoctin greenstones on the west limb of the Blue Ridge anticlinorium. The stratigraphic section is presented in Figure 10. Graded bedding in the underlying Swift Run Formation and in metaconglomerates in the upper portion of the Catoctin Formation indicate that along the I-64 traverse, the entire section is overturned (Figure 1). Planned stops will look at the basal contact with the Swift Run Formation, along with examinations of one of the metasedimentary interbeds, an amygdaloidal zone, and some interbedded sandstones and conglomerates near the top of the section. Along the Skyline Drive in the Shenandoah National Park north of the I-64 traverse, optional stops provide an opportunity to look at one of the few porphyritic units in the formation and at one of the volcanic breccias. **NOTE: No rock hammers at outcrops and no sample collecting are allowed in the Shenandoah National Park.**

STOP 3-1

On I-64 W at mile marker 103.2, just beyond RR tracks overpass. The overturned contact between gray and tan phyllite of the Swift Run Formation and greenstones of the Catoctin Formation.

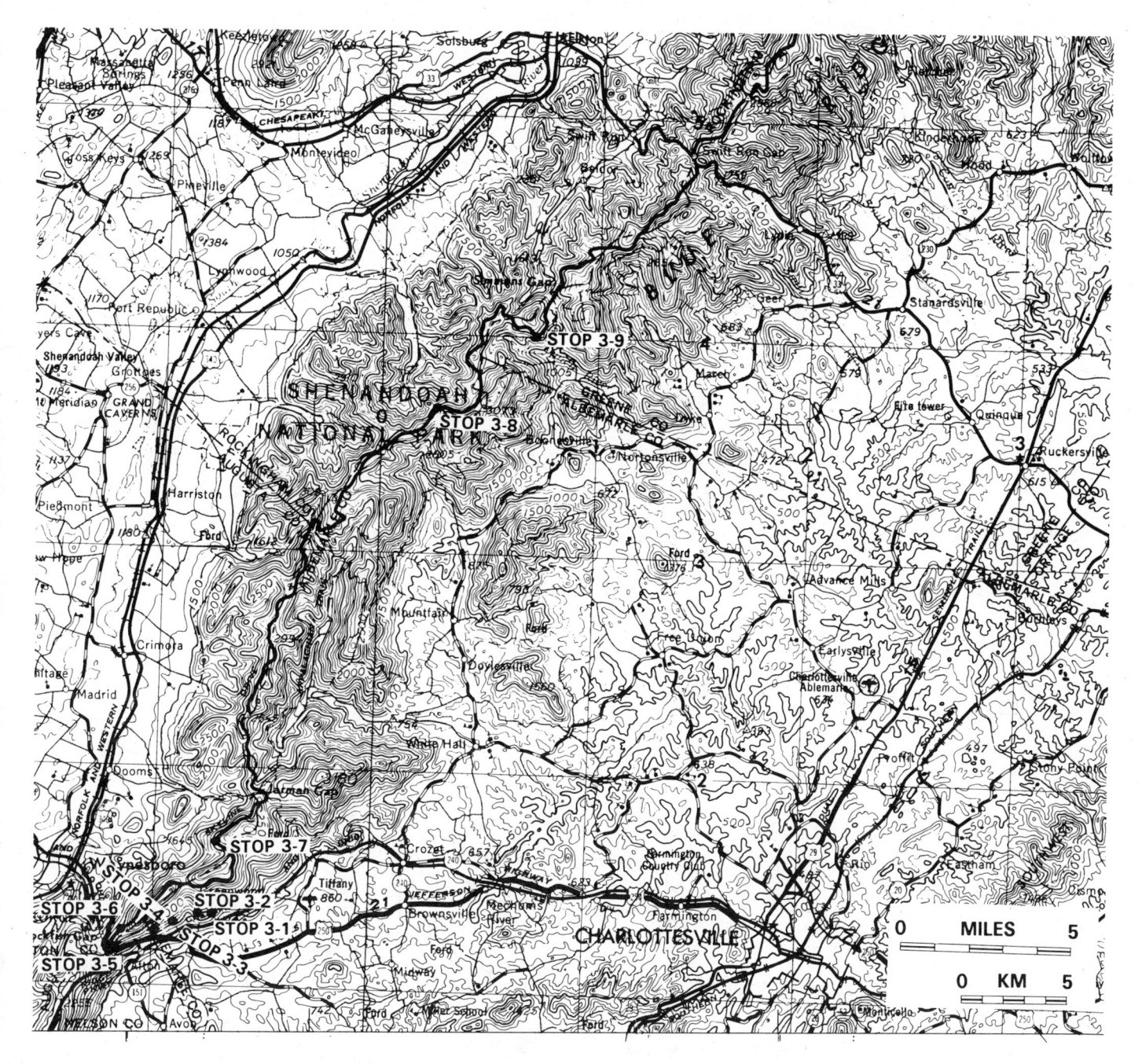

Figure 9. Portion of the Charlottesville (1:250,000) topographic map showing field trip locations for day 3.

STOP 3-2

On I-64 W at mile marker 102.7, at end of chain link fence. Metasedimentary interbed between Catoctin Formation flows here consists of red metasandstone and metasiltstone with dark gray phyllite. Note the strung-out vesicles at base of next flow in sequence and the epidote pods in nearby flows.

STOP 3-3

On I-64 W about 300 feet before mile marker 102. Mesozoic dike cutting Catoctin Formation flows.

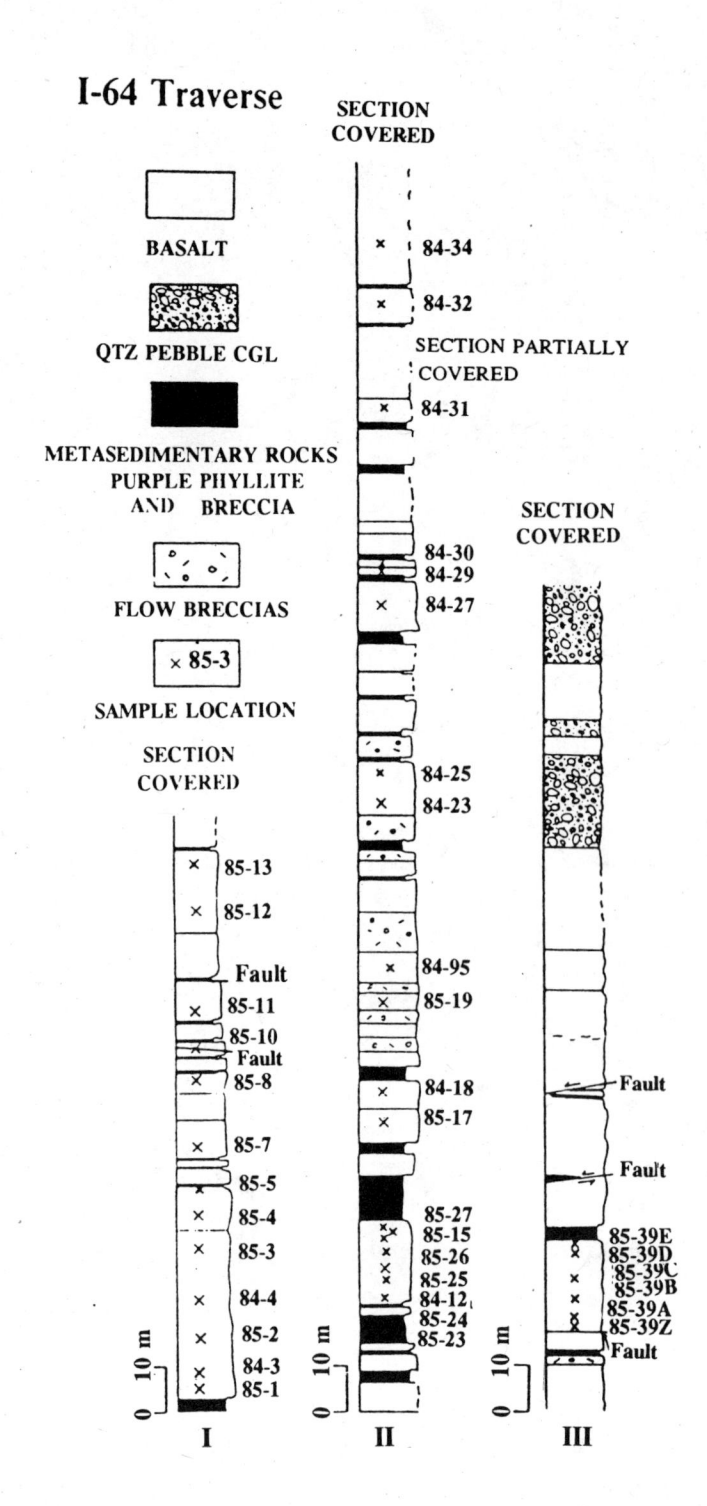

Figure 10. Stratigraphic column of Catoctin Formation along I-64 traverse. Column I is from bottom of the section; column II is from middle to upper part; column III is from near the top of the exposed section. Approximately 50–80 m of section is missing between columns I and II; 30–50 m is missing between columns II and III.

STOP 3-4

On I-64 W at mile marker 101.3. Amygdaloidal zone at the top of flow. Note contrast of fabric in different portions of the flow. At mile 100.4, note landslide which occurred here in 1986; the slip surface was along a thin (5–15 cm) phyllite unit between flows. The major landslide that occurred during construction of this portion of I-64 was also along a phyllitic interbed slip-surface.

STOP 3-5

Take exit 19, intersection of offramp and U.S. Highway 250; park on grass straight ahead; **watch out for POISON IVY vines.** Upper flows of Catoctin Formation metabasalts here are interbedded with metasandstone and metaconglomerate which contains pink feldspar (probably derived from unakite) and blue quartz indicating a Grenvillian crystalline source for the detritus.

STOP 3-6

On Skyline Drive North (SDN), 0.5 miles north of southern entrance (bridge over I-64) to Shenandoah National Park. Excellent outcrop of quartz-pebble metaconglomerate in the upper portion of the Catoctin Formation. Again note the pink feldspar clasts suggesting the source was uplifted basement.

STOP 3-7

Calf Mountain Overlook, 6.8 miles on SDN north of entrance. Refer to Gathright (1976) for description of the regional geology and the view. West of the overlook is an outcrop of metaconglomerate (still within the Catoctin Formation) containing clasts of metabasalt.

STOP 3-8

Big Run Overlook, 24.5 miles on SDN north of entrance; park here and walk 0.3 miles along road to excellent exposure of a porphyritic unit (Catoctin Formation) containing large plagioclase phenocrysts. Such porphyritic flows serve as mappable marker horizons.

STOP 3-9

Loft Mountain Overlook, 31.2 miles on SDN north of entrance. A purplish unit of Catoctin Formation metabasalt and an amygdaloidal zone with epidosite pods and some volcanic breccia are all exposed here. Some slickensides are present as well. Look at the stone wall at the overlook. It is composed of the Erwin Quartzite, an upper unit of the Chilhowee Group, and contains *Scolithus* tubes which are used, in part, to infer a Cambrian age for the Chilhowee farther east on State Road 617.

REFERENCES CITED

Ames, R. M., 1981, Geochemistry of the Grenville Basement rocks from the Roseland District, Virginia: unpublished M.S. thesis, University of Georgia, Athens, 91 p.

Badger, R. L., and Sinha, A. K., 1988, Age and Sr isotopic signature of the Catoctin volcanic province: Implications for subcrustal mantle evolution: Geology, v. 16, p. 692–695.

Bailey, W. M., 1983, Geology of the northern half of the Horseshoe Mountain quadrangle, Nelson County, Virginia: unpublished M.S. thesis, University of Georgia, Athens, 100 p.

Bartholomew, M. J., 1977, Geology of the Greenfield and Sherando quadrangles, Virginia: Virginia Division of Mineral Resources, Publication 4, 43 p.

——, 1987, Structural evolution of the Pulaski thrust system, southwestern Virginia: Geological Society of America, Bulletin, v. 99, p. 491–510.

——, 1988a, Northwestern extent of Mesozoic extensional faulting in the Virginia Blue Ridge: Geological Society of America, Abstracts with Programs, v. 20, n. 7, p. 253.

——, 1988b, The Alleghanian basement–no basement–tectonic transition: Lateral imbrication and the change from ductile to brittle to decollement faulting in the Appalachians: Geological Society of America, Abstracts with Programs, v. 20, no. 7, p. 395.

——, 1989a, Tectonic evolution of Proterozoic margin of ancestral North American craton (ANAC) in the Appalachians from Middle Proterozoic through Mesozoic: A synthesis: 28th International Geological Congress, Washington, D.C., USA, July 9–19, 1989, Abstracts, v. 1, p. 93.

——, 1989b, Appalachian Mesozoic fault/dike patterns: Indicators of lateral Mesozoic stress-field variations similar to a Cenozoic Cordilleran analogue: Geological Society of America, Abstracts with Programs, v. 21, no. 6, p. A64.

——, and Lewis, S. E., 1984, Evolution of Grenville massifs in the Blue Ridge geologic province, southern and central Appalachians, in Bartholomew, M. J., Force, E. R., Sinha, A. K., and Herz, N., eds., The Grenville Event in the Appalachians and Related Topics: Geological Society of America Special Paper 194, p. 229–254.

——, and ——, 1988, Peregrination of middle Proterozoic massifs and terranes within the Appalachian orogen, eastern U.S.A.: Trabajos de Geologia, University of Oviedo, Spain, v. 17, p. 153–163.

——, and ——, in press, Appalachian Grenville Massifs: Pre-Appalachian translational tectonics in Mason, R., ed., Proceedings of the 7th International Conference on Basement Tectonics at Kingston, Ontario: D. Kluwer Academic Publishing, Dordrecht, Holland.

——, Gathright, T. M., II, and Henika, W. S., 1981, A tectonic model for the Blue Ridge in central Virginia: American Journal of Science, v. 281, p. 1164–1183.

Bloomer, R. O., and Werner, H. J., 1955, Geology of the Blue Ridge region in central Virginia: Geological Society of America, Bulletin, v. 66, p. 579–606.

Brock, J. C., 1981, Petrology of the Mobley Mountain Granite, Amherst County, Virginia: unpublished M.S. thesis, University of Georgia, Athens, 130 p.

Cloos, E., 1947, Oolite deformation in the South Mountain fold, Maryland: Geological Society America Bulletin, v. 58, p. 843–918.

Davis, R. G., 1974, Pre-Grenville ages of basement rocks in central Virginia: A model for the interpretation of zircon ages: unpublished M.S. thesis, Virginia Polytechnic University and State University, Blacksburg, 47 p.

Dietrich, R. V., Fullagar, P. D., and Bottino, M. L., 1969, K/Ar and Rb/Sr dating of tectonic events in the Appalachians of southwestern Virginia: Geological Society of America Bulletin, v. 80, p. 307–314.

Elliot, W. C., and Aronson, J. L., 1987, Alleghanian episode of K-bentonite illitization in the southern Appalachian basin: Geology, v. 15, p. 735–739.

Evans, N. H., 1984, Late Precambrian to Ordovician metamorphism and orogenesis in the Blue Ridge and western Piedmont, Virginia Appalachians: unpublished Ph. D. dissertation, Virginia Polytechnic Institute and State University, Blacksburg, 313 p.

Fullagar, P. D., and Dietrich, R. V., 1976, Rb-Sr isotopic study of the Lynchburg and probably correlative formations of the Blue Ridge and western Piedmont of Virginia and North Carolina: American Journal of Science, v. 276, p. 347–365.

Furcron, A. S., 1969, Late Precambrian and early Paleozoic erosional and depositional sequences of northern and central Virginia: Georgia Geological Survey, Bulletin 80, p. 57–88.

Gathright, T. M., II, 1976, Geology of the Shenandoah National Park, Virginia: Virginia Division of Mineral Resources, Bulletin 86, 93 p.

——, Henika, W. S., and Sullivan, J. L., 1977, Geology of the Waynesboro East and Waynesboro West quadrangles, Virginia: Virginia Division of Mineral Resources, Publication 3, 53 p.

Griffin, V. S., Jr., 1971, Fabric relations across the Catoctin Mountain–Blue Ridge anticlinorium in central Virginia: Geological Society of America Bulletin, v. 82, p. 417–432.

Gryta, J. J., and Bartholomew, M. J., 1989, Factors influencing the distribution of debris avalanches associated with the 1969 Hurricane Camille in Nelson County, Virginia, in Schultz, A. P., and Jibson, R. W., eds., Landslides of Eastern North America: Geological Society of America, Special Paper 236, p. 15–28.

Herz, N., 1969, The Roseland alkalic anorthosite massif, Virginia, in Isachsen, Y. W., ed., Origin of Anorthosite and Related Rocks: New York State Museum and Science Service Memoir 18, p. 357–367.

——, 1984, Rock suites in the Grenvillian terrane of the Roseland district, Virginia: Part 2. Igneous and metamorphic petrology, in Bartholomew, M. J., Force, E. R., Sinha, A. K., and Herz, N., eds., The Grenville Event in the Appalachians and Related Topics: Geological Society of America Special Paper 194, p. 200–214.

——, and Force, E. R., 1982, Anorthosite, ferrodiorite, and titanium deposits in Grenville terrane of the Roseland district, central Virginia, in P. T. Lyttle, ed., Central Appalachian Geology, NE-SE GSA 1982 Field Trip Guidebooks, p. 109–119.

——, and ——, 1984, Rock suites in Grenvillian terrane of the Roseland district, Virginia, in Bartholomew, M. J., Force, E. R., Sinha, A. K., and Herz, N., eds., The Grenville Event in the Appalachians and Related Topics: Geological Society of America Special Paper 194, p. 187–214.

——, and ——, 1987, Geology and mineral deposits of the Roseland district of central Virginia: U. S. Geological Survey Professional Paper 1371, 56 p.

Hillhouse, D. N., 1960, Geology of the Piney River–Roseland titanium area, Nelson and Amherst counties, Virginia: unpublished Ph.D. dissertation, Virginia Polytechnic Institute and State University, Blacksburg, 129 p.

Hudson, T. A., 1981, Geology of the Irish Creek tin district, Virginia Blue Ridge: unpublished M.S. thesis, University of Georgia, Athens, 144 p.

Lewis, S. E., and Hower, J. C., 1990, Implications of thermal events on thrust emplacement sequence in the Appalachian fold and thrust belt: Some new vitrinite reflectance data: Journal of Geology, v. 98, p. 927–942.

——, Bartholomew, M. J., Abercrombie, F. N., Herz, N., and Hudson, T. A., 1986, Comparative geochemistry and petrology of the Pedlar River Charnockite Suite and associated granulite gneisses of the northern Blue Ridge Province: Geological Society of America, Abstracts with Programs, v. 18, n. 3, p. 251.

Lukert, M. T., 1973, The petrology and geochronology of the Madison area, Virginia: unpublished Ph.D. dissertation, Case Western Reserve University, Cleveland, Ohio, 218 p.

——, and Banks, P. O., 1984, Geology and age of the Robertson River Pluton, in Bartholomew, M. J., Force, E. R., Sinha, A. K., and Herz, N., eds., The Grenville Event in the Appalachians and Related Topics: Geological Society of America, Special Paper 194, p. 161–166.

Mitra, G., 1977, The mechanical processes of deformation of granitic basement and the role of ductile deformation zones in the deformation of Blue Ridge Basement in Northern Virginia: unpublished Ph.D. dissertation, The Johns Hopkins University, Baltimore, Maryland, 219 p.

——, and Elliott, D., 1980, Deformation of basement in the Blue Ridge and the development of the South Mountain cleavage, in Wones, D. R., ed., Proceedings "Caledonides in the USA," I.G.C.P. Project 27: Caledonide Orogen: Virginia Polytechnic Institute and State University, Department of Geological Sciences Memoir 2, p. 307–312.

——, and Lukert, M. T., 1982, Geology of the Catoctin–Blue Ridge anticlinorium in northern Virginia, in Lyttle, P. T., ed., Central Appalachian Geology, NE-SE GSA 1982 Field Trip Guidebooks: American Geological Institute, p. 83–108.

Moore, C. H., Jr., 1940, Geology and Mineral Resources of the Amherst quadrangle, Virginia: unpublished Ph.D. dissertation, Cornell University, Ithaca, New York, 92p.

Mose, D. G., 1981, Cambrian age for the Catoctin and Chopawamsic Formations in Virginia: Geological Society of America, Abstracts with Programs, v. 13, p. 31.

——, and Nagel, S., 1984, Rb-Sr age for the Robertson River pluton in Virginia and its implication on the age of the Catoctin Formation, in Bartholomew, M. J., Force, E. R., Sinha, A. K., and Herz, N., eds., The Grenville Event in the Appalachians and Related Topics: Geological Society of America, Special Paper 194, p. 167–173.

Nelson, W. A., 1962, Geology and mineral resources of Albemarle County, Virginia: Virginia Division Mineral Resources, Bulletin 77, 92 p.

Nystrom, P. G., Jr., 1977, Geologic map of the Waynesboro West quadrangle, Virginia: plate 2 in Virginia Division of Mineral Resources Publication 3, 53 p.

Odom, A. L., and Fullagar, P. D., 1984, Rb-Sr whole-rock and inherited zircon ages of the plutonic suite of the Crossnore Complex, southern Appalachians, and their implications regarding the time of opening the Iapetus Ocean, in Bartholomew, M. J., Force, E. R., Sinha, A. K., and Herz, N., eds., The Grenville Event in the Appalachians and Related Topics: Geological Society of America, Special Paper 194, p. 255–261.

Pettingill, H. S., Sinha, A. K., and Tatsumoto, M., 1984, Age and origin of anorthosites, charnockites, and granulites in the Central Virginia Blue Ridge: Nd and Sr isotopic evidence: Contributions to Mineralogy and Petrology, v. 85, p. 279–291.

Rader, E. K., 1967, Geology of the Staunton, Churchville, Greenville, and Stuarts Draft quadrangles, Virginia: Virginia Division of Mineral Resources, Report of Investigations 12, 43 p.

Rankin, D. W., 1975, The continental margin of eastern North America in the Southern Appalachians: The opening and closing of the Proto-Atlantic Ocean: American Journal of Science, v. 275-A, p. 298–336.

———, 1976, Appalachian salients and recesses: Late Precambrian continental breakup and the opening of the Iapetus Ocean: Journal of Geophysical Research, v. 81, p. 5605–5619.

———, Stern, T. W., Reed, J. C., Jr., and Newell, M. F., 1969, Zircon ages of felsic volcanic rocks in the upper Precambrian of the Blue Ridge, Appalachian Mountains: Science, v. 166, p. 741–744.

Robison, M. S., 1976, Paleozoic metamorphism of the Piedmont–Blue Ridge boundary region in west-central Virginia: Evidence from K-Ar dating: Geological Society of America, Abstracts with Programs, v. 8, n. 2, p. 257–258.

Ross, C. S., 1941, Occurrence and origin of the titanium deposits of Nelson and Amherst Counties, Virginia: U.S. Geological Survey Professional Paper 198, 59 p.

Schwab, F. L., 1974, Mechum River Formation: Late Precambrian(?) alluvial fill in the Blue Ridge of Virginia: Geological Society of America, Abstracts with Programs, v. 6, p. 69–70.

———, 1986, Latest Precambrian–earliest Paleozoic sedimentation, Appalachian Blue Ridge and adjacent areas: Review and speculation, in McDowell, R. C., and Glover, L., III, eds., The Lowry Volume: Studies in Appalachian Geology: Virginia Tech Department of Geological Sciences Memoir 3, p. 115–137.

Simpson, E. L., and Eriksson, K. A., 1989, Sedimentology of the Unicoi Formation in southern and central Virginia: Evidence for late Proterozoic to early Cambrian rift-to-passive margin transition: Geological Society of America Bulletin, v. 101, p. 42–54.

———, and Sundberg, F. A., 1987, Early Cambrian age for synrift deposits of the Chilhowee Group of southwestern Virginia: Geology, v. 15, p. 123–126.

Sinha, A. K., and Bartholomew, M. J., 1984, Evolution of the Grenville terrane in the central Virginia Appalachians, in Bartholomew, M. J., Force, E. R., Sinha, A. K., and Herz, N., eds., The Grenville Event in the Appalachians and Related Topics: Geological Society of America, Special Paper 194, p. 175–186.

Spencer, E. W., and Waterbury, M. J., 1987, Basement-cover interaction on the northwestern flank of the Blue Ridge in the James River gap area: Geological Society of America, Abstracts with Programs, v. 19, n. 2, p. 130–131.

Steiger, R. H., and Jager, E., 1977, Subcommission on geochronology: Convention on the use of decay constants in geo- and cosmochronology: Earth Planetary Science Letters, v. 36, p. 359–363.

Stickney, M. C., and Bartholomew, M. J., 1987, Seismicity and Quaternary faulting of the northern Basin and Range Province, Montana and Idaho: Bulletin Seismological Society America, v. 77, p. 1602–1625.

Streckeisen, A., 1974, How should charnockitic rocks be named?: Centenaire de la Societe Geologique de Belgique, Geologie des Domaines Cristallins, Liege, 1974, p. 349–360.

Taylor, S. R., and McLennon, S., 1985, The Continental Crust: Its Composition and Evolution: Blackwell Scientific Publications, London, 312 p.

Tollo, R. P., and Arav, S., in press, The Robertson River Suite: Late Proterozoic anorogenic (A-type) granitoids of unique petrochemical affinity, in Bartholomew, M. J., Hyndman, D. M., Mogk, D. W., and Mason, R., eds., Characterization and Comparison of Ancient (Precambrian-Mesozoic) Continental Margins, Proceedings of the 8th International Conference on Basement Tectonics at Butte, Montana, USA: Kluwer Academic Publishers, Dordrecht, Holland.

Walcott, C. D., 1891, Correlation papers; Cambrian: U.S. Geological Survey Bulletin 81, 447 p.

Wehr, F., 1983, Geology of the Lynchburg Group in the Culpeper and Rockfish River areas, Virginia: unpublished Ph.D. dissertation, Virginia Polytechnic Institute and State University, Blacksburg, 254 p.

———, 1985, Stratigraphy of the Lynchburg Group and Swift Run Formation, Late Proterozoic (730–570 Ma), Central Virginia: Southeastern Geology, v. 25, n. 4, p. 225–239.

———, in press, Transition from alluvial to deep-water sedimentation in the lower Lynchburg Group, upper Proterozoic, Virginia, in Bartholomew, M. J., Hyndman, D. M., Mogk, D. W., and Mason, R., eds., Characterization and Comparison of Ancient (Precambrian-Mesozoic) Continental Margins, Proceedings of the 8th International Conference on Basement Tectonics at Butte, Montana, USA: Kluwer Academic Publishers, Dordrecht, Holland.

———, and Glover, L., III, 1985, Stratigraphy and tectonics of the Virginia–North Carolina Blue Ridge: Evolution of a late Proterozoic–early Paleozoic hinge zone: Geological Society of America Bulletin, v. 96, p. 285–295.

Werner, H. J., 1966, Geology of the Vesuvius quadrangle, Virginia: Virginia Division of Mineral Resources, Report of Investigations 7, 53 p.

Wilson, J. K., 1986, Geology of the northern half of the Crozet quadrangle Albemarle Co., Virginia: unpublished M.S. thesis, Unversity of Georgia, Athens, 115 p.

4
TACONIC COLLISION IN THE DELAWARE-PENNSYLVANIA PIEDMONT AND IMPLICATIONS FOR SUBSEQUENT GEOLOGIC HISTORY

Mary Emma Wagner
University of Pennsylvania
Philadelphia, PA 19104

LeeAnn Srogi
Ohio Wesleyan University, Delaware, OH 43015

C. Gil Wiswall
West Chester University, West Chester, PA 19383

James Alcock
Pennsylvania State University at Ogontz, Abington, PA 19001

INTRODUCTION

During the Ordovician Taconic orogeny, the Piedmont of northern Delaware and southeastern Pennsylvania was the site of collision between the ancient North American continent and a magmatic arc that had developed above an eastward-dipping subduction zone (Wagner, 1982; Wagner and Srogi, 1987). The purpose of this field trip is to inspect exposures across the orogen of both the overriding and subducting plates, and to discuss their implications in the framework of the Wagner and Srogi model. In addition, we shall visit sites where the rocks have been affected by post-collision deformation, responsible in part for the present outcrop pattern, in which units from different crustal levels occupy a narrow region of the Pennsylvania-Delaware Piedmont.

Wagner and Srogi (1987) and Srogi (1988) suggested that granulite-facies mafic and quartzofeldspathic gneiss and deep-seated intrusive rocks of the Wilmington Complex (Figure 1) are the infrastructure of a magmatic arc that grew during Cambrian(?) time above an eastward-dipping subduction zone. Collision occurred during the Ordovician when crust of the North American continent was subducted, carrying Grenville gneiss to depths of at least 35 km. Currently the Wilmington Complex occupies the highest structural level, and the West Chester nappe (Figure 1) the lowest. Other basement-cored nappes (Woodville, Avondale, Mill Creek) and accretionary prism and/or forearc basin metasediments (various parts of the Wissahickon Formation) are at intermediate levels (Figure 1). The model is based in part on the metamorphic histories of the Wilmington Complex, the West Chester prong, and parts of the Wissahickon Formation. Both the Wilmington Complex gneiss and the Grenville gneiss are intruded by undeformed igneous rocks that constrain the timing of events.

Coarse-grained undeformed gabbroids and a norite-charnockite suite form deep-seated plutons whose emplacement deformed the metamorphic foliation in the granulite-facies gneiss of the Wilmington Complex. The norite-charnockite suite is 502 ± 20 Ma (Foland and Muessig, 1978), indicating that the granulite-facies metamorphism must be older than Late Cambrian. Peak metamor-

Figure 1. Geologic map of part of the Pennsylvania-Delaware-Maryland Piedmont.

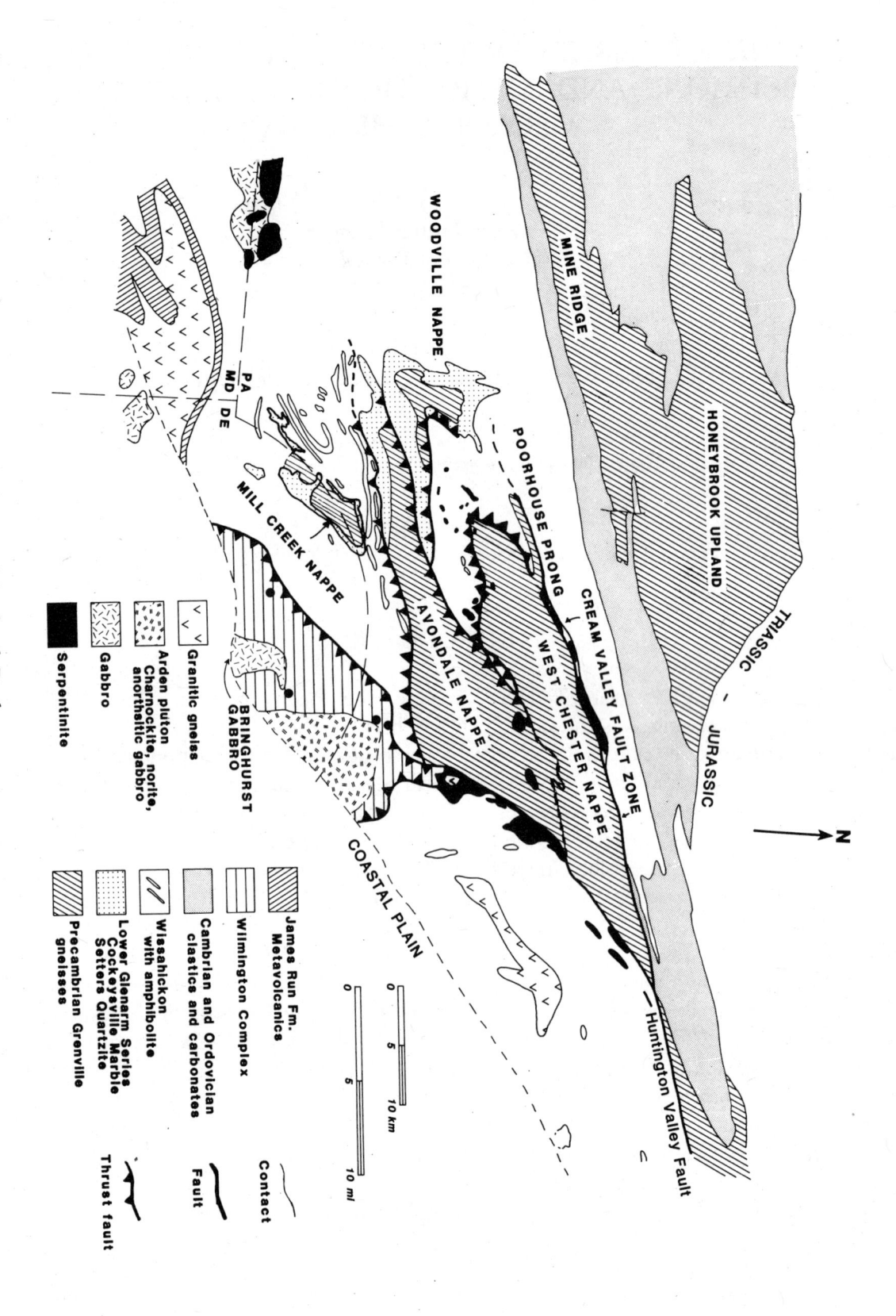

phic temperatures associated with this metamorphism are estimated to be about 800°C at pressures of 0.6–0.7 GPa (Wagner and Srogi, 1987; Srogi, 1988). Spinel-bearing coronas around olivine grew during isobaric cooling of the gabbroids from igneous to granulite-facies temperatures.

Gneiss of the West Chester prong contains granulite-facies assemblages of Grenville age, overprinted by garnet-clinopyroxene-quartz coronas that grew at high pressures, 0.9–1.1 GPa, and amphibolite-facies temperatures, 650–700°C (Wagner and Srogi, 1987). Diabase dikes, intruded after the Grenville rocks were at a shallow crustal level and had cooled significantly from Grenville temperatures of ≈ 800°C, have garnet coronas similar to those in the gneiss. This metamorphism in the dikes is the best evidence that corona formation was caused by later metamorphism rather than isobaric cooling from the Grenville event (Wagner and Crawford, 1975; Wagner and Srogi, 1987). Isotopic data suggest the later metamorphism was associated with the Taconic orogeny (Lapham and Bassett, 1964; Grauert et al., 1973, 1974). The gneiss and dikes least deformed during this prograde metamorphism are in the center of the West Chester prong. We interpret the garnet coronas to have formed at high lithostatic pressure when the continental crust was subducted beneath the Wilmington Complex arc. Coronas are not found in penetratively deformed rocks, suggesting either that they grew before thrusting and nappe formation, or that they grew within the nappes only in rocks not undergoing penetrative deformation.

The Wissahickon Formation reached higher metamorphic temperatures adjacent to the Wilmington Complex than elsewhere in the PA-DE Piedmont. The Wissahickon gneiss is above the second sillimanite isograd both east and west of the Wilmington Complex, with scattered occurrences of cordierite-orthoclase- and spinel-bearing assemblages close to the contact (Crawford and Mark, 1982; Calem, 1987). Peak temperatures are estimated to be 700–750°C (Crawford and Mark, 1982; Calem, 1987; Wagner and Srogi, 1987; Plank, 1988). We interpret these conditions to have resulted from the tectonic emplacement of the hot Wilmington Complex over rocks of the Wissahickon Formation that were already at amphibolite-facies grade (Wagner and Srogi, 1987). Pressure estimates vary widely; most are in the range from 0.3 to 0.7 GPa (Calem, 1987; Plank 1988) with pressures increasing to the north toward the Avondale nappe (Plank, 1988). Alcock (1989a, b) suggested that zoning in garnets from staurolite-grade Wissahickon schist around the margins of the Woodville nappe (Figure 1) indicates a possible pressure increase from ≈ 0.4 to ≈ 0.8 GPa for that part of the Wissahickon.

The West Chester nappe is bordered on the north for most of its length by the Cream Valley fault zone (CVFZ). The fault has had a long and complex history. Where it forms the margin of the West Chester nappe, it separates Grenville gneiss to the south from greenschist-facies metasediments in a tight syncline to the north. There, the fault is nearly vertical, with extensive mylonitization of the Grenville rocks.

Farther west, along strike of the CVFZ, structures dip gently to the south. A structural discontinuity and steep metamorphic gradient separate Wissahickon schist on the south from lower-grade pelitic rocks on the north. The discontinuity is interpreted as the westward extension of the CVFZ (Wiswall, 1990b). We suggest that initially the fault zone formed early in the Paleozoic as a thrust that carried rocks caught in the Taconic suture zone over allochthonous and parautochthonous rocks of the early Paleozoic North American slope/shelf sequence. Later deformation associated with compression produced shearing that was localized in part by the earlier thrusts.

There are fundamental differences in the rocks overlying Grenville gneiss in the various nappes. The rocks of the West Chester nappe and the eastern end of the Avondale nappe are overlain by high-grade schist of the Wissahickon Formation that, in many places close to the contact, encloses pods and lenses of serpentinite and pyroxenite. The gneisses of the Woodville, Poorhouse, and Mill Creek nappes, as well as the western end of the Avondale nappe, are unconformably overlain by the lowest members of the Glenarm Series: the Setters Formation (quartzite and mica schist) and the Cockeysville Marble, which in turn are overlain by Wissahickon schist.

We postulate that the Wilmington Complex was thrust over schist of the Wissahickon Formation. The plate bearing the Wissahickon and Wilmington Complex was in turn thrust above the Mill Creek and Avondale nappes. There is disagreement among the various field trip leaders on the relationships between the Wissahickon Formation and the Woodville nappe. Alcock (1989a) believes, on the basis of discordant relationships between the Wissahickon and the lower Glenarm metasediments, that all of the Wissahickon Formation lies above a single thrust that was emplaced after the Avondale and Woodville nappes were folded. In this interpretation a thrust cuts up-section from the West Chester nappe to the overlying Woodville and Avondale nappes, and the Wissahickon Formation is very thin between the West Chester and Woodville nappes (Figure 2A). Wagner and Srogi (1987) believe there are two or more packages of rock presently called "Wissahickon" that differ in their structural positions. One, with slivers and lenses of serpentinite at its base, was thrust over the West Chester nappe (Figure 2B). Another was thrust above the Mill Creek nappe and the upper limb of the Avondale nappe, where Alcock (1989a) has demonstrated that a metamorphic discontinuity exists between the Wissahickon Fm. and the Cockeysville Marble. This part of the Wissahickon Formation, as well as the Avondale and Mill Creek nappes, do not appear in Figure 2B because they are structurally above the Woodville nappe. In this interpretation, at least some of the Wissahickon around the Woodville nappe, especially on its eastern and northern margins, *underlies* the Grenville gneiss and lower Glenarm metasediments (Figure 2B).

In summary, the sequence in the overriding arc massif (Wilmington Complex) is:

(1) Granulite-facies metamorphism of unknown age, probably latest Precambrian or Cambrian.
(2) Intrusion of gabbro and a norite-charnockite suite ($\approx$500 Ma).
(3) Slow cooling of the igneous rocks at depth to granulite-facies temperatures under a high geothermal gradient.
(4) Rapid cooling associated with thrusting of the Wilmington Complex onto metasedimentary rocks formed in a forearc basin and/or accretionary prism.
(5) Thrusting of the plate with these metasediments at the base over the basement-cored nappes of the subducting plate ($\approx$440 Ma).

The sequence in the subducting plate is:

(1) Granulite-facies metamorphism during the Grenville orogeny ($\approx$1100 Ma).
(2) Erosion and uplift.
(3) Intrusion of diabase dikes during late Precambrian rifting.
(4) High-pressure metamorphism of the Grenville gneiss and the diabase dikes when the continental crust was subducted beneath the magmatic arc; emplacement of accretionary wedge sediments (part of what is presently mapped as Wissahickon Formation) and serpentinites onto the West Chester prong may have occurred at this time.
(5) Thrusting and nappe formation as the Grenville gneiss was overridden by the magmatic arc with pelitic metasediments at its base (a *different* part of the present Wissahickon Formation); recrystallization of many of the granulite-facies gneisses to amphibolite-facies assemblages.
(6) Folding caused by additional compression (the number of these folding events and their timing is unclear).

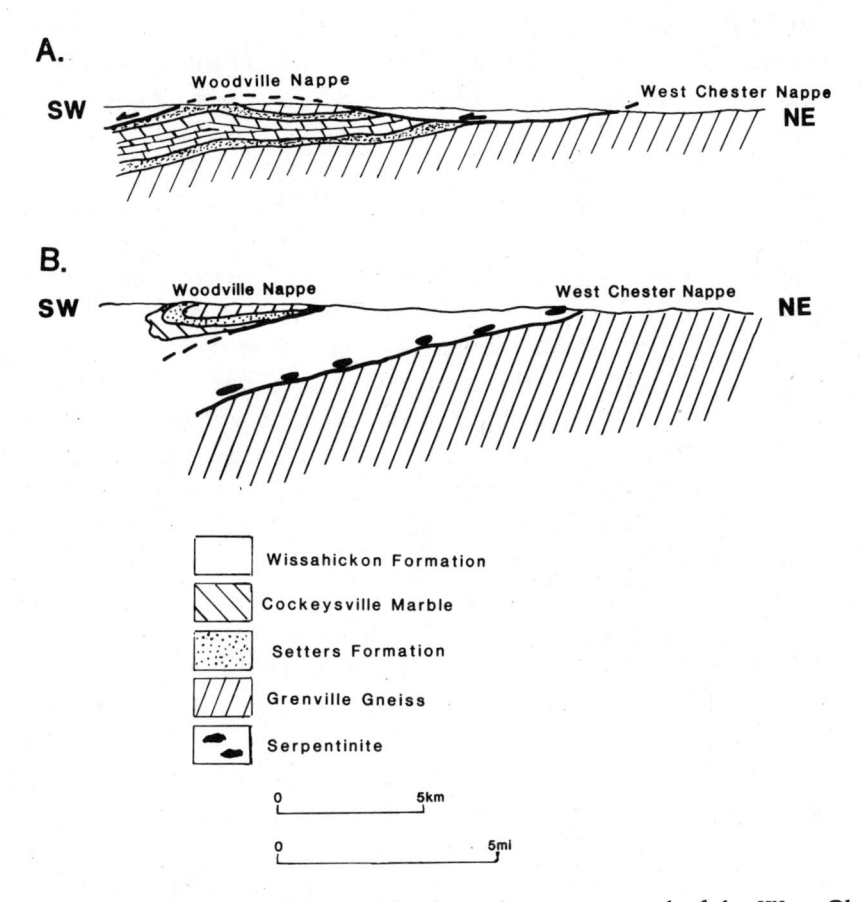

Figure 2. Interpretive cross sections parallel to strike from the western end of the West Chester nappe to west of the Woodville nappe. A, Interpretation in which all of the Wissahickon Formation lies above a single thrust. B, Interpretation in which there are two or more thrust slices of "Wissahickon." One overlies the West Chester nappe and plunges under the Woodville nappe. Another is above the Avondale nappe, which projects above the surface in this section.

(7) Shearing, in part localized by earlier structures associated with thrusting. The later shearing has components of both vertical and strike-slip motion on the northern margin of the Grenville gneiss (Howard, 1988), but has gentle dips farther west (Wiswall, 1990b).

Events (4), (5), and possibly parts of (6) are considered to be Taconic in age.

The sites for this field trip were chosen to evaluate the Wagner and Srogi model and to stimulate discussion on some of the more controversial aspects. The first day will be spent in the arc massif and examining the contact between it and the underlying Wissahickon Formation. The second day we will visit the West Chester nappe, the deepest structural level during the Taconic orogeny, looking first at gneiss that was metamorphosed but not deformed during the Taconic, and then at the effect of deformation on these rocks during thrusting and nappe formation. Rocks occurring in nappes and in thrust slices of structural levels between the arc massif and the West Chester nappe will be visited on the third day. Exposures of pelitic rocks near the westward extension of the CVFZ will illustrate the structural and metamorphic discontinuities across the fault zone. The final stops will be at quarries in Cockeysville Marble along the north side of the Woodville nappe, where the relationship between marble and Wissahickon schist is the source of disagreement among the field trip leaders.

THE WILMINGTON COMPLEX: THE OVERRIDING MAGMATIC ARC

The Wilmington Complex (Figure 1) has been mapped by Bascom and Miller (1920), Bascom and Stose (1932), Ward (1959), Woodruff and Thompson (1975), Mark (1977), and Srogi (1982, 1988). It is adjacent to the Wissahickon Formation along most of its margin, and is covered by Coastal Plain sediments to the southeast (Figure 1). We will examine the contact relationships with the Wissahickon Formation at Stop 3, in Brandywine Creek State Park (Figure 3).

The Wilmington Complex is composed of pyroxene-bearing felsic and mafic gneiss, rare garnet-bearing gneiss, and numerous small plutons (Figure 1). The major plutons have been mapped and described by Ward (1959), Woodruff and Thompson (1975), Thompson (1975), Mark (1977), Brick (1980), Wagner et al. (1987), Wagner and Srogi (1987), and Srogi (1988). The largest of the intrusions are the Bringhurst and Arden plutons; the Arden pluton covers about 20 km^2. Most of the igneous intrusions consist of gabbronorite, gabbro, olivine gabbro, and olivine gabbronorite. The Arden pluton contains gabbroids as well as a suite of pyroxene-bearing quartz diorites, granodiorites, and granites. Contacts of the igneous rocks with the surrounding gneiss are intrusive and *not* tectonic in nature. Most of the igneous rocks contain undeformed igneous textures and mineralogy. We will look at the igneous lithologies and their contact relationships at Stops 1 and 2 (Figure 3).

Foland and Muessig (1978) determined a Rb-Sr whole-rock isochron age of 502 ± 20 Ma for some of the felsic rocks of the Arden pluton, which they interpreted as the age of igneous crystallization. Near the plutons, the fabric of the gneiss is disrupted by plastic flow and local migmatization. This suggests that the gneissic foliation, defined by the layering of granulite-facies minerals, predates the igneous intrusions and is at least 500 Ma old. Zircons from one sample of felsic gneiss were analyzed for U-Pb isotopic composition (Grauert and Wagner, 1975), and the nearly concordant lower intercept of 441 Ma was interpreted as the age of granulite-facies metamorphism during the Taconic orogeny. However, the field evidence for the timing of high-temperature metamorphism suggests that the U-Pb age may represent cooling of the Wilmington Complex following tectonic emplacement onto the North American continent (Wagner and Srogi, 1987).

Quantitative estimates of P-T conditions in the gneiss (Wagner and Srogi, 1987; Srogi, 1988) indicate that metamorphism took place at moderate depths in the crust (P ≈ 0.6–0.7 GPa; ca. 21–28 km), and at high peak temperatures (T ≈ 800°C ± 50°C). The gabbroids contain coronas of orthopyroxene, pargasitic amphibole, and spinel surrounding olivine. We interpret the minerals of the coronas to have formed by reaction between olivine and plagioclase during subsolidus cooling at pressures of spinel-gabbro stability. Field and petrographic evidence in the gneiss and igneous rocks suggests that the magmatism produced a thermal overprint on the country rocks. These rocks were previously metamorphosed to granulite-facies conditions by ≈ 500 Ma and subsequently cooled at depth in the crust. Neither the igneous rocks nor the gneiss show evidence for pervasive deformation or recrystallization following cooling.

The Wilmington Complex differs from high-grade terranes of the Grenville Province in North America. The gneiss does not have the lithologic diversity of rocks in the Grenville province (e.g., Adirondacks, Wiener et al., 1984; West Chester prong, Wagner and Crawford, 1975; Reading Prong, Drake, 1984). The anorthosites and associated igneous rocks in the Grenville Province are mostly Proterozoic in age and have an alkaline affinity (e.g., Morse, 1982). By contrast, the Cambrian-age felsic suite of the Arden pluton is calc-alkaline (Wagner and Srogi, 1987).

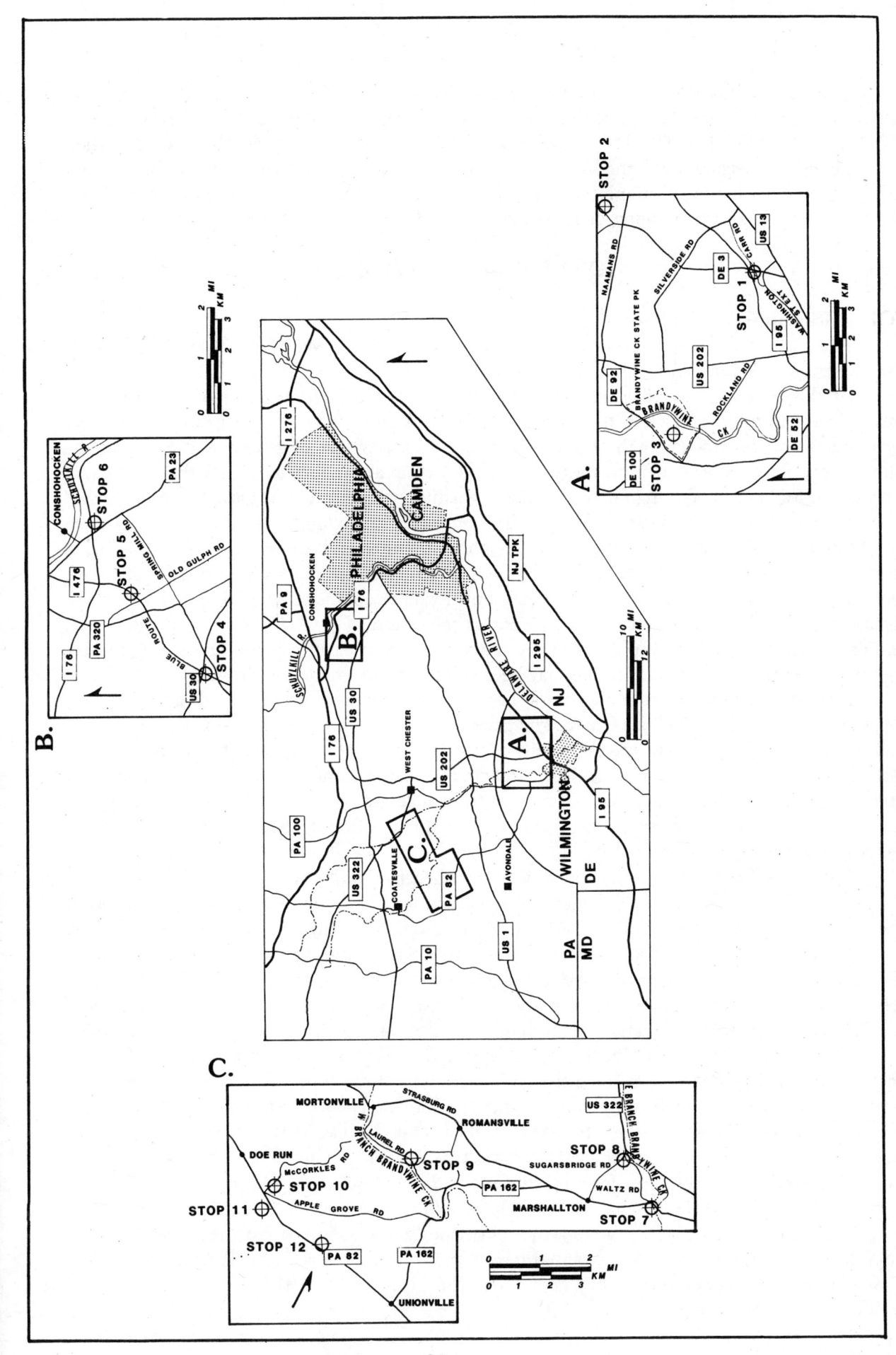

Figure 3. Road map showing stop locations. A, Wilmington Complex; B, West Chester nappe; and C, CVFZ and Woodville nappe.

Igneous rock compositions and metamorphic history of the Wilmington Complex suggest an origin in the lower portions of a microcontinent or mature island arc system (Crawford and Crawford, 1980; Wagner and Srogi, 1987). The Wilmington Complex may be a more deeply exposed portion of the Cambro-Ordovician arc of the Virginia and Maryland Piedmont (Pavlides, 1981; Sinha et al., 1980). Tectonic emplacement onto North America in the Late Ordovician Taconic orogeny did not produce widespread recrystallization or retrograde metamorphism in the Wilmington Complex.

STOP DESCRIPTIONS (1-3)

STOP 1 Bringhurst Woods Park

The entrance to the park is on Carr Road between DE Rte 3 (Marsh Road) and Washington Street Extension (Figure 3A).

At this stop, also described in Wagner et al. (1987), the gabbroic rocks of the Bringhurst pluton and the intrusive contact of the gabbroids into the Wilmington Complex gneiss are exposed. The textures and mineralogy of the Bringhurst gabbroids are typical of gabbroids throughout the Wilmington Complex. Because of the rarity of some rock types at this locality we ask that you refrain from hammering on or collecting samples of the pegmatitic gabbroids or the garnet-bearing gneiss.

Several types of gabbro can be found in the boulders in and along the sides of Shellpot Creek. The gabbroids typically have coarse-grained subophitic textures and are not deformed. The most common type is gabbronorite, containing both clinopyroxene and orthopyroxene. Samples of olivine gabbro and gabbronorite, some displaying igneous flow textures, are found about 100 m downstream. Pegmatitic gabbroids are found in the woods across the stream. Upstream from the parking area, there are several examples of very fine-grained mafic rock enclosed in gabbro. These are probably xenoliths of mafic gneiss, although some may be autoliths of fine-grained gabbroids.

About 45 m upstream from the power line, the contact between the gabbroids and the gneiss is exposed in the bank of Shellpot Creek. The gabbroids truncate the foliation in the gneisses, and small offshoots of the gabbroid into the gneiss occur. In some places adjacent to the contact, the gabbroids have a flow foliation. Near the contact, the metamorphic banding in the gneisses is disrupted and some felsic gneisses appear migmatitic. Similar features occur elsewhere around the margins of the Bringhurst pluton (Brick, 1980), the Arden pluton (Rdesinski, 1983), and some smaller intrusive plugs (Luborsky, 1983).

Typical Wilmington Complex gneiss was described by Ward (1959) as banded gneiss with alternating mafic and felsic layers. However, the banding is not always present, and large areas consist of only felsic gneiss or mafic gneiss. The gneiss is dark-colored on fresh surfaces, and the foliation is visible only on weathered surfaces. Rare garnet-bearing gneiss occurs as lenses or pods less than a meter in any dimension interlayered with felsic and mafic pyroxene gneiss. Garnet-bearing gneiss has been found several meters upstream from the contact and farther upstream at the easternmost extremity of Bringhurst Woods Park, just south of the bridge carrying Carr Road over Shellpot Creek.

The textural features and field relationships in the rocks at this stop are among the most important pieces of evidence in defining the geologic history of the Wilmington Complex. Foliated, granulite-facies gneiss is intruded by gabbroids with undeformed igneous textures. This indicates that penetrative deformation that formed the gneiss is older than the magmatism. While the age of the Bringhurst pluton is not known, similar gabbroids occur in the Arden pluton, where they are intruded by roughly contemporaneous, more felsic igneous rocks, dated at 502 ± 20 Ma (Foland and Muessig, 1978). It is reasonable to assume that all the gabbroic and noritic rocks in the Wilmington Complex are similar in age, and that the granulite-facies mineralogy and foliation in the gneiss are older than ≈ 500 Ma. Further isotopic work is required to constrain the ages of the igneous rocks, the timing of granulite-facies metamorphism in the gneiss, and the effects of the 440 Ma Taconic orogeny on the Wilmington Complex.

Disruption of the gneissic fabric near the contact suggests that intrusion of the gabbroic magmas produced a thermal overprint on the previously metamorphosed gneiss. Mineral assemblages and compositions in the gneiss constrain peak metamorphic conditions to about 800°C ± 50°C and 0.6–0.7 GPa (Wagner and Srogi, 1987; Srogi, 1988). Both felsic and mafic gneisses contain andesine to labradorite plagioclase, orthopyroxene, clinopyroxene, and magnetite, with variable amounts of quartz, brown-green hornblende, and biotite. The garnet-bearing gneiss near the Carr Road overpass contains abundant biotite and sillimanite with a strong preferred orientation, typical of gneiss that occurs some distance from pluton contacts. Garnet-bearing gneiss located closer to the plutons, including the sample found within several meters of the contact in Bringhurst Woods Park, typically contains garnet, Mg-rich cordierite, hercynitic spinel, Al-rich hypersthene, and corundum. These minerals lack preferred orientation and are interpreted to have formed as a restite from incongruent melting of biotite during peak thermal metamorphism associated with igneous intrusion (Srogi, 1988; Srogi et al., 1985).

The gabbroic rocks have petrographic evidence for intrusion and cooling at depth in the crust, consistent with pressures of ≈ 0.65 GPa estimated from the garnet-bearing gneiss (Wagner and Srogi, 1987). Coronas surrounding olivine and separating it from plagioclase are composed of low-Ca orthopyroxene next to olivine, with an outer corona of pargasitic amphibole next to plagioclase. Spinel is commonly found in symplectic intergrowth with the orthopyroxene and/or the amphibole, but garnet is not observed in the coronas. Pargasitic hornblende also rims some pyroxene grains. Formation of these corona assemblages requires a fluid phase. The lack of penetrative deformation of the gabbroids suggests that the coronas formed during subsolidus cooling at mid- to lower-crustal pressures, rather than during a separate metamorphic event. We interpret the tectonic setting of the Bringhurst pluton to be a convergent plate boundary, consistent with the intrusion of gabbroids into continental or arc-related crust. Moreover, the mineral chemistry of the Bringhurst gabbroids is similar to that in gabbros from layered mafic complexes in Alaska, such as the La Perouse gabbro (Loney and Himmelberg, 1983) and the Border Ranges complex (Burns, 1985).

STOP 2 Afton-Timbers Park, Arden Pluton

From the intersection of DE Rtes. 3 (Marsh Road) and 92 (Naamans Road) (Figure 3A), follow Marsh Road east to the entrance of the Afton Park subdivision on the right. At the T-intersection, turn left and park in the cul-de-sac at the end of the road.

A variety of igneous lithologies typical of the Arden pluton are exposed at this stop. The compositions of the igneous rocks and the relationships between different igneous units provide constraints on the timing and tectonic setting of magmatism in the Wilmington Complex.

The Arden pluton is a composite pluton consisting of gabbroic rocks and a suite of pyroxene-bearing quartz diorites (quartz norites), granodiorites (opdalites), and granites (charnockites), (Wagner and Srogi, 1987; Srogi, 1988). Rocks of the quartz norite-charnockite suite occur in the streambed and in the park east of the parking area. Euhedral to subhedral phenocrysts of orthoclase define an igneous flow lineation. Plagioclase phenocrysts are somewhat smaller than the orthoclase phenocrysts and some have zoning that is visible in the outcrop. There is no evidence for penetrative deformation in the igneous rocks. Mafic enclaves are present, some of which may be xenoliths of Wilmington Complex gneiss. Most of the mafic enclaves, however, are recrystallized gabbros, suggesting that the quartz norite-charnockite suite is the younger magmatic unit in this composite pluton. Gabbroic rocks of the Arden pluton are exposed in the woods north-northeast of the parking area and playground. A variety of mafic lithologies is present, ranging from gabbronorites with subophitic to intergranular textures, to amphibolitic rocks with more lineated fabrics.

Careful mapping has demonstrated that gabbroids with pristine igneous textures and mineralogy occur away from contacts with rocks in the quartz norite-charnockite suite. Recrystallized, hornblende-rich gabbroids with more gneissic fabrics are found next to the quartz norite-charnockite suite. Field relationships suggest that the gabbroid was below its solidus in some places and above its solidus in other places at the time the granitic magma was intruded. This resulted in contamination and recrystallization of the gabbroids. The gabbroic rocks are probably the same age as, or slightly older than, the 502 Ma felsic suite.

The rocks of the quartz norite-charnockite suite contain plagioclase ($\approx An_{35}$), orthopyroxene, clinopyroxene, microperthitic orthoclase, biotite, and accessory Fe-Ti oxides, apatite, and zircon. Orthoclase phenocrysts typically have myrmekite along their margins. Zoning in plagioclase phenocrysts is due to variable amounts of K-feldspar blebs in a compositionally uniform plagioclase host. The pyroxenes have textures and compositions indicating that they crystallized from a magma at temperatures of about 850°–950°C, and were not formed as a product of subsolidus recrystallization (Srogi, 1988). The quartz norite-charnockite suite crystallized from a relatively dry granitic magma, and some geochemical variation in the rocks was produced by the migration of late-stage liquids near the end of crystallization (Srogi, 1988; Srogi and Lutz, 1987a, b). A similar model was proposed for the Meatiq Dome, Egypt, by Sultan et al. (1986).

The quartz norite-charnockite suite of the Arden pluton is calc-alkaline (Wagner and Srogi, 1987), and it is similar in whole-rock chemistry to plutonic rocks from convergent tectonic regimes, such as the Coast Batholith of Alaska and British Columbia (Barker et al., 1986). It is also similar, in both mineralogy and chemistry, to portions of the Barrington Tops batholith, an orogenic complex in the New England Province of Australia (Eggins and Hensen, 1987). The granitic rocks of the Arden pluton are not anorthosites, nor are they chemically related to granitic rocks associated with anorthosite massifs (Wagner and Srogi, 1987).

Many of the Arden gabbroids are similar in texture and mineralogy to the Bringhurst gabbroids, and also have spinel-bearing coronas around olivine (Srogi, 1988). However, the Arden gabbroids typically contain more hornblende than the Bringhurst gabbroids, and they have greater abundances of both compatible and incompatible elements. These features suggest that the Bringhurst and Arden plutons may have been contemporaneous but were not co-magmatic.

STOP 3 Brandywine Creek State Park, Wilmington-Wissahickon contact

The entrance to the park is from Rockland Road $\approx$ .3 mi E of the intersection of DE Rtes 92 and 100 (Figure 3A). Park in the southern of the two parking areas in the park.

In Brandywine Creek State Park and the Nature Preserve on the east side of Brandywine Creek, felsic and mafic gneiss of the Wilmington Complex and pelitic to semi-pelitic schist and gneiss of the Wissahickon Formation are exposed. The contact between the Wilmington Complex and the Wissahickon Formation is not exposed in any single outcrop, but can be inferred from the distribution of the rock units in this area (Srogi, 1982; Wagner and Srogi, 1987). The orientation and nature of the contact are controversial and have important implications for the tectonic history of the area.

Wilmington Complex gneiss can be examined in boulders next to the parking area and is similar to the gneiss seen at Bringhurst Woods Park (Stop 1). At Marsh Overlook (Figure 4), located downhill to the southeast from the parking area, the pelitic and semi-pelitic schist and gneiss of the Wissahickon Formation occur in outcrops. In the Wissahickon Formation, a strong foliation strikes northeast and dips steeply to the northwest, and the rocks are tightly folded. Rocks of the Wissahickon Formation are above the second sillimanite isograd and contain garnet, biotite, sillimanite, plagioclase, quartz, K-feldspar, cordierite, hercynitic spinel, and ilmenite. Many of the rocks are extremely garnet-rich, and in some places they are migmatitic.

Downstream from Marsh Overlook, the Wilmington Complex gneiss outcrops in an abandoned quarry just northeast of Rockland Road. The foliation in the gneiss also strikes northeast and dips steeply northwest. A variety of lithologies is present, including hornblende-rich gneiss. If time permits, we will cross Brandywine Creek on Rockland Road and follow the trail northeast. The hillside contains numerous boulders of Wilmington Complex gneiss. Ward (1959) first described the change from felsic gneiss to more abundant mafic gneiss that occurs here.

The contact between the Wilmington Complex and the Wissahickon Formation was previously interpreted as a metamorphosed concordant stratigraphic contact (Ward, 1959) because of the similar orientations of the foliation in both the Wissahickon and Wilmington gneiss. Hager and Thompson (1975) described the contact as a fault; Woodruff and Thompson (1975) interpreted the contact as a steep normal fault dipping to the southeast, nearly perpendicular to the dip of the foliation in the two units. The Wilmington Complex was interpreted to be on the hanging wall.

Srogi (1982) and Hamre (1983) found, however, that the Wilmington Complex structurally overlies the Wissahickon Formation (Figure 4). While the foliations may suggest a steep contact between these units, the overall pattern of outcrop is more consistent with a contact that dips at a shallow angle to the southeast. The attitude of the contact is about N46°E, 5°SE, based on a planar regression through nine contact points, and remains unchanged for at least 2–3 km along strike. The occurrence of the Wissahickon Formation at Marsh Overlook and of the Wilmington Complex in the quarry northeast of Rockland Road (Figure 4) are predicted by the planar regression model (Srogi, 1982). Southwest of Brandywine Creek State Park, Hamre (1983) found that the contact surface is about 100 feet higher in elevation, although the strike and dip of the surface is the same (Figure 4). Field evidence also suggests a shallowly dipping contact to the northeast, in the Marcus Hook and Media Quadrangles (Mark, 1977; M. L. Hill and D. Valentino, pers. comm., 1985; D. Valentino, pers. comm., 1990).

The relationship of the contact between the Wilmington Complex and the Wissahickon Formation to the foliations and deformation histories of the two units is not clear. There is petrographic evidence for high-temperature ductile deformation in both the Wissahickon Formation (Valentino, 1986; 1988) and the Wilmington Complex. Wagner and Srogi (1987) interpreted the contact to be a thrust fault. They showed that the Wissahickon Formation reaches its highest metamorphic grade adjacent to the Wilmington Complex. Their model explains this by emplacement of hot Wilmington Complex over amphibolite-facies rocks of the Wissahickon Formation along a mid-crustal level thrust fault. This heated the Wissahickon rocks and caused incongruent melting of biotite to form cordierite and spinel (Calem, 1987). If this model is correct, the absence of deformation fabrics parallel to the contact is puzzling. An alternative model is that the steep foliations developed after tectonic juxtaposition of the two units. Some field evidence for offset of the contact by later shearing has been recognized (D. Valentino, pers. comm., 1990).

METAMORPHISM AND DEFORMATION IN THE SUBDUCTED PLATE

Introduction

This portion of the trip examines the metamorphic and structural history of a section of the North American continental margin which was overridden by the Wilmington Complex magmatic arc during the Taconic orogeny. The area is bounded on the south by the Wilmington Complex, on the north by the CVFZ, and on the east by the Rosemont fault (Figure 1). Within this terrane are several fault-bounded blocks cored by gneiss of Grenville age ($\approx$ 1100 Ma). These are the Poorhouse, West Chester, Woodville, Avondale, and Mill Creek nappes (Figure 1). With the exception of the West Chester nappe, each massif is associated with the Setters Quartzite and/or Cockeysville Marble of the lower Glenarm Series in varying structural relations. Wissahickon schist occurs throughout the area. These rocks were complexly deformed during the Taconic orogeny, and perhaps again later in the Paleozoic. We interpret each of the basement-cored structural blocks as a nappe within an imbricate stack. We will visit locations in the West Chester nappe and examine its relationship to overridden slope/rise rocks to the north across the CVFZ (Figure 3B, C).

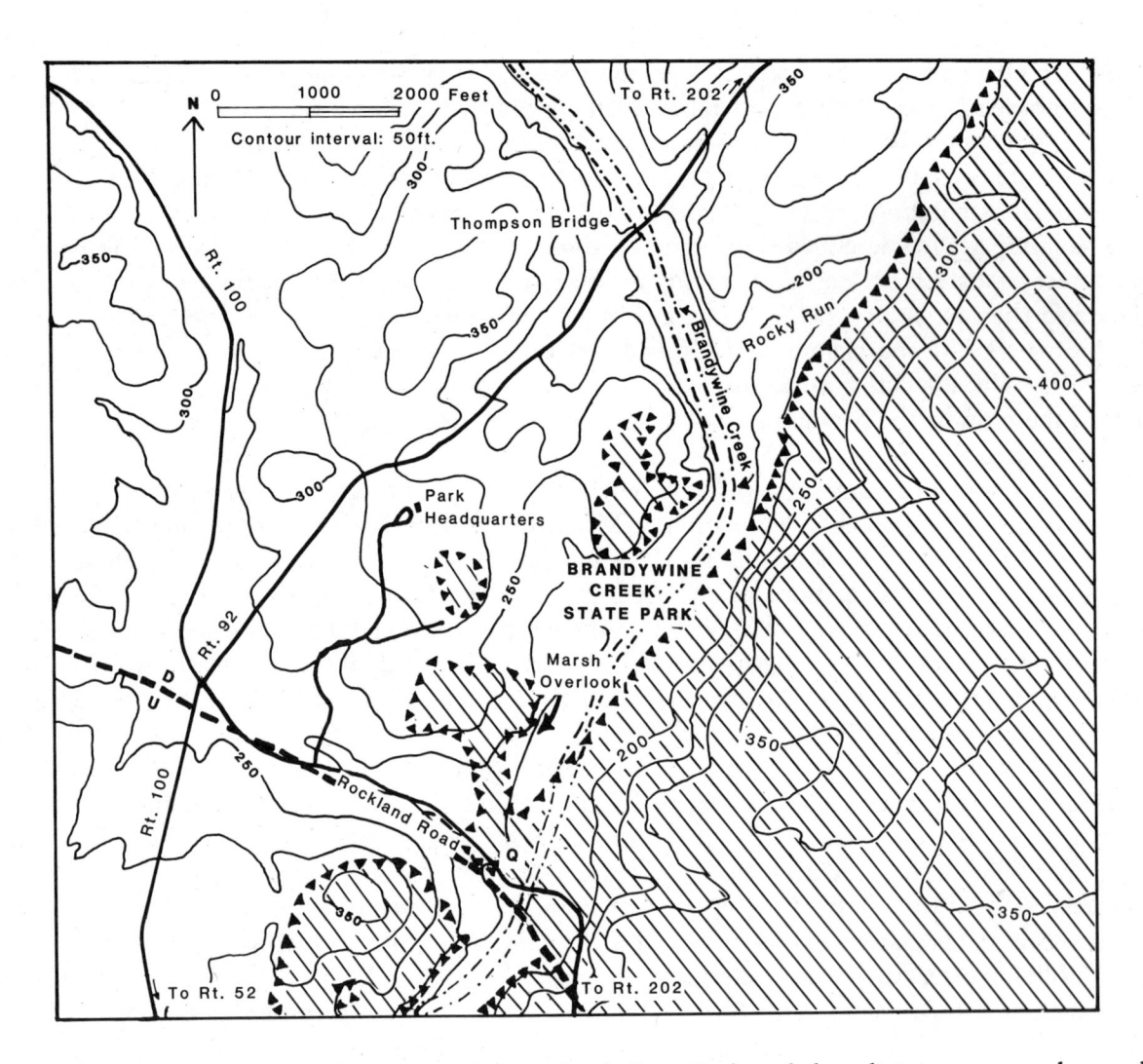

Figure 4. Interpretive geologic map of Brandywine Creek State Park and the adjoining area to the south. The barbed line shown is the intersection with the topography of a plane striking N45°E and dipping 5° southeast and corresponds very closely to the contact between the Wilmington Complex and the Wissahickon Formation. The heavy dashed line is a possible fault that offsets the contact vertically by about 100 feet, as suggested by Hamre (1983). Diagonal lines represent Wilmington Complex; unpatterned, Wissahickon Formation. Contour interval: 50 ft.

Metamorphic History

In the West Chester nappe, and to a lesser extent in the Avondale nappe, some rocks contain remanent granulite-facies mineral assemblages of Grenville age. These mineral assemblages were modified to varying degrees by metamorphism during the Paleozoic. In the least deformed rocks, garnet coronas formed around all mafic minerals except clinopyroxene (Wagner and Crawford, 1975). The minerals of the coronas (gar-cpx-qtz-sodic plag) formed chiefly at the expense of orthopyroxene and calcic plagioclase, and indicate metamorphism at higher pressures and lower temperatures than the Grenville granulite-facies metamorphism. Where water was available, the mafic minerals inside the coronas were replaced by symplectites of hornblende, biotite, clinopyroxene, and quartz, assemblages typical of amphibolite facies. In addition, the garnet coronas are larger and more euhedral than in rocks that retain granulite-facies minerals such as hypersthene. Many recrystallized rocks contain large anhedral garnets from Grenville metamorphism, in addition to the small euhedral

Taconic age garnets. Compositions of minerals in the coronas indicate pressures of 1.0 ± 0.1 GPa and temperatures between 650° and 700°C for the Taconic metamorphism (Wagner and Srogi, 1987). Garnet coronas are generally not present in strongly deformed gneiss. This is particularly the case at Stop 5 near the CVFZ, where the most deformed rocks contain abundant epidote and little or no garnet.

In the Avondale nappe, reconnaissance mapping shows that, although there are scattered occurrences of Grenville granulite-facies gneiss with garnet coronas, in general the rocks are more deformed and more recrystallized than in the West Chester nappe.

Fine-grained diabase dikes with ophitic texture and chilled borders cut the Grenville gneiss of the West Chester nappe. They have coronas similar to those in the gneiss, replacing igneous minerals rather than metamorphic ones. In the Avondale nappe, some of the dikes have coronas and traces of ophitic texture, but more commonly they are completely recrystallized to a very garnetiferous fine-grained rock with an annealed equigranular texture. Where shearing was more intense, they are recrystallized to a hornblende-plagioclase rock. The best evidence they were originally dikes is that their chemical composition is nearly identical to undeformed dikes in the West Chester prong (Wagner, unpublished data).

The presence of coronas in both the gneiss and the shallowly emplaced dikes suggests that this part of the ancient North American continent was carried from upper crustal levels to depths of ≈ 35 km during the early Paleozoic. During thrusting and nappe formation, deformation destroyed the coronas, and the Grenville granulite-facies assemblages were largely replaced by amphibolite-facies minerals.

Structural History

The West Chester nappe is the leading edge of a wedge that was thrust continentward during emplacement of the Wilmington Complex. The CVFZ is the contact separating the imbricated North American crust from pelitic and psammitic metasediments of continental slope/rise affinity. Since the CVFZ juxtaposes rocks of different crustal levels and probably formed early in the Taconic orogeny, it contains a record of the structural development of this segment of the Piedmont.

<u>General Description.</u> The CVFZ shows significant variation in character along its length. In the northeast (Stops 5 and 6), it is a steep zone of mylonitized gneiss and schist dominated by strike-parallel shear fabrics. To the southwest (Stops 7, 8, and 10), the fault zone dips shallowly to the southeast and compressional features are most abundant. While strike slip fabrics are present to the southwest, they formed late in the deformation sequence and are not regionally penetrative.

Howard (1988), in a study conducted along the northeastern portion of the zone near Conshohocken, Pennsylvania, proposed that the CVFZ was initiated as a regional basement shear with mylonitization at epidote-amphibolite facies grade. Subsequently, a discrete fault formed, with vertical uplift and dextral layer-parallel slip.

Wiswall (1990a; 1990b; unpublished data) recognizes three deformational events (D1, D2, D3) based on studies of exposures along the CVFZ between the Poorhouse and Woodville nappes. Isoclinal folding occurred during D1 and was accompanied by prograde metamorphism to staurolite/kyanite grade south of the fault and to garnet grade north of the fault. Northwest-directed shearing and formation of nappes occurred during D2. Temperatures during D2 were at lower to middle amphibolite-facies grade in the hanging wall south of the CVFZ and at upper greenschist-facies grade in the footwall. D3 was a transpressional event which occurred at upper greenschist-facies grade on both sides of the fault zone. Refolding and oblique slip along the CVFZ during D3 produced the present map pattern. Fabric relationships and mineral assemblages suggest that D1 and D2 were a progressive deformation associated with the Taconic orogeny and that D3 was probably post-Taconic.

Structural Sequence Along Southwest CVFZ. Since the CVFZ juxtaposes different crustal levels, rocks on opposite sides of the fault may have different structural histories; thus, fabric labels used in this section are appended with "f" (footwall) or "h" (hanging wall). For example, S1h is the earliest tectonic foliation in the hanging wall, and F2f refers to second-generation folds in the footwall.

D1 structures. D1 in the hanging wall and footwall of the CVFZ was accompanied by prograde metamorphism which produced schistosity (S1f, garnet grade; S1h, staurolite/kyanite grade) in the metasediments. Radiometric dates obtained by Lapham and Basset (1964) suggest that this metamorphic event was associated with the Taconic orogeny. Variably developed gneissic layering in the basement massifs of the hanging wall probably also developed at this time. Evidence for this is the amphibolite-facies mineral assemblages which define the layering. Rare isoclinal folds in compositional layering, which are axial planar to S1f, have been recognized only in psammitic lithologies in the footwall.

D2 structures. Following the metamorphic peak, northwest-directed shearing occurred (D2). At depth, movement of the hanging wall was initiated in the Grenville gneiss on relatively narrow, ductile shear zones (S2h) (Figure 5) located near the base of the massif. These are presently the dominant fabric in the gneiss and may be correlative with the mylonitic foliation to the northeast described by Howard (1988). In the overlying Wissahickon schist, D2 shearing resulted in slip along S1h (Figure 5). If the Wissahickon-Grenville gneiss contact is tectonic in origin, it formed during D2 shearing but before recumbent folding.

S2h is deformed at all scales by northwest-vergent overturned to recumbent folds (F2h). Based on the consistent northwest vergence of F2h, the presence of overturned sections beneath the gneiss of the Woodville and Poorhouse nappes, and the geometry of mesoscopic F2h, we interpret this folding event to be nappe formation and emplacement. Mutually cross-cutting relationships between S2h and F2h suggest that folding replaced shearing during D2. The change in style of deformation from shearing to nappe formation presumably occurred as the rocks were transported to shallower crustal levels where cooling resulted in an increase in competence. However, the lack of retrogressed minerals of the S1 assemblage, new growth of epidote and sphene in S2h zones in the gneiss, and dynamic recrystallization of feldspar in S2h zones all suggest that the hanging wall remained at amphibolite-facies conditions during D2. These features suggest that D1 and D2 were progressive deformations associated with Taconic mountain building.

D2 shear fabrics and folds also occur in the footwall. However, since footwall rocks remained below the amphibolite facies throughout their history, early deformation in these rocks cannot be correlative in space with D2 structures in the hanging wall. The response of footwall rocks to D2 was controlled by lithology. In the pelitic rocks, a strong type II S-C mylonite fabric formed. Psammitic rocks were deformed by northwest vergent folds (F2f) similar to those in the hanging wall (Figure 5). D2 deformation in the footwall probably occurred during nappe formation at relatively shallow crustal levels.

In summary, D2 marks the formation of the CVFZ during the Taconic orogeny. Comparison of the structural sequences recognized in the footwall and hanging wall of the CVFZ suggests a model involving nappe formation and emplacement on shallow southeast dipping thrusts. This model is compatible with deformation during the Taconic orogeny elsewhere.

D3 structures. D3 produced a variety of structures. F3 folds formed within the hanging wall and footwall blocks. Shear fabrics are localized along the CVFZ trace. In the hanging wall, F3h (Figure 5) are upright to slightly northwest vergent and formed near the brittle-ductile transition with axes similar in orientation to F2 folds (20° S60°W). At the macroscopic scale, the West Chester nappe was refolded during this event, resulting in the present map pattern. F3f are near recumbent, sharp-crested, drag-type folds. The lower limbs of these folds are attenuated by shallow, south-dipping shear zones (F3h, Figure 5). Oblique slip crenulations (S3f, Figure 5) developed in pelitic lithologies along the CVFZ during combined normal and dextral movement.

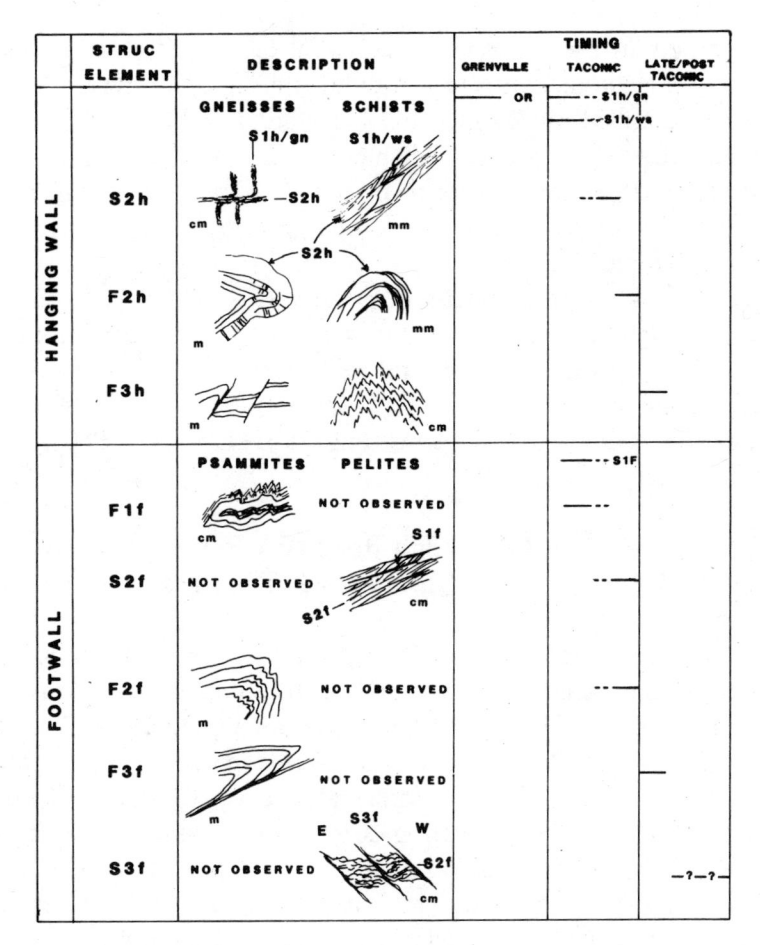

Figure 5. Structural elements and their sequence of development in rocks along the Cream Valley fault zone. Sketches are tracings from outcrop or thin section photographs. All views are to the southwest, except where indicated. Approximate scale indicates order of magnitude represented by view: mm = millimeter; cm = centimeter; m = meter.

STOP DESCRIPTIONS (4-10)

Stops 4 through 6 in the West Chester nappe are a southwest to northeast traverse from relatively undeformed rocks to mylonites adjacent to the CVFZ. Stops 4 and 6 are new exposures and have not been studied in detail. Stop 5 was the subject of part of an M.S. thesis at the University of Pennsylvania by Colin Howard (1988). These three exposures are about 5 miles east of the eastern margin of the area in the West Chester nappe described by Wagner and Crawford (1975).

Stops 7 through 10 show the structural relationships between the imbricated nappes and the slope/rise metasediments. Stop 7 is on the northern margin of the West Chester nappe. Stops 8 through 10 are outcrops in the footwall close to the trace of the CVFZ.

STOP 4 The "Blue Route" (I-476) south of U.S. Rte. 30 (Figure 3B)

The Grenville rocks exposed in the roadcuts at this stop are typical of gneiss in the West Chester nappe that was least deformed by Taconic metamorphism. It will form the basis for comparison with the more deformed and mylonitized rocks at today's remaining stops.

The rock is dominantly massive felsic gneiss that contains garnet coronas surrounding intergrowths of randomly oriented biotite laths, fibrous amphibole, quartz, apatite, plagioclase, and possibly scapolite. The garnets in the coronas are euhedral to subhedral and contain small felsic inclusions that are typical of the corona garnets throughout the West Chester nappe. By analogy with similar gneiss farther west, described by Wagner and Crawford (1975), the garnet must have once surrounded large mafic minerals (up to 4 mm in diameter).

An orange-beige colored granite appears to be intrusive into the gneiss at one of the cuts. It contains euhedral garnets that are similar to the corona garnets, but no other mafic minerals. Its age may, therefore, be pre-Taconic.

Diabase dikes cut across the felsic gneiss in at least two of the exposures. Both dikes are undeformed, with traces of ophitic texture, in thin section, in the form of randomly oriented plagioclase laths surrounded by abundant garnet and metamorphic amphibole.

A few subvertical shear zones less than 1 cm wide are also present. Pseudotachylite occurs associated with some shear zones in the gneiss (Armstrong, 1941; Howard, 1988), although none has yet been observed at this locality.

STOP 5 The "Blue Route" (I-476) at Old Gulph Road (PA Route 23) (Figure 3B)

This locality is about 1.8 mi. (2.7 km) NE of Stop 4. The trace of the Cream Valley fault is about 400–500 m farther north.

The rocks here (Howard, 1988) have a strong foliation and compositional banding, very different in appearance from the massive rocks at the last stop. Folding, almost non-existent in the more massive rocks, is present here. Howard (1988) describes the rocks as S-tectonites in which shear-sense indicators are rare. He divides the rocks into two broad categories: 1) mylonites, in which shearing is intense but folding is rare; and 2) protomylonites (matrix 10–15% of the rock), and crushed gneiss (matrix < 10%), in which shearing is less intense and folding is common. In the mylonites, dynamic recrystallization of micas has formed a shear foliation. Although biotite is the dominant mica, muscovite occurs locally. Late folds with vertical axes fold the mylonitic foliation.

Howard (1988), after studying both undeformed gneiss to the south and the deformed gneiss at this stop, defined the following sequence of events for the West Chester nappe:

(1) M1 granulite-grade metamorphism
(2) Intrusion of diabase dikes
(3) Onset of M2 upper-amphibolite grade metamorphism
(4) Felsic injection; formation of "hybrid-rock" of Armstrong (1941) (concentrated around present-day CVFZ)
(5) Initiation of CVFZ as a regional basement shear-zone
(6) Mylonitization at epidote-amphibolite grade
(7) Formation of a discrete fault in center of shear zone
(8) Pseudotachylite formation
(9) Vertical uplift along the fault plane
(10) Dextral layer-parallel slip

STOP 6 Four Falls Corporate Center. East of Rte. 23 just south of the I-76 (Schuylkill Expressway) overpass (Figure 3B).

There are two fault strands of the CVFZ between the large outcrop next to the corporate center and the Schuylkill River on the north. One, located under the small bridge just to the north of the corporate center, separates Grenville gneiss from a sliver of high-grade rocks of the Wissahickon Formation. The other fault, closer to the river, separates Wissahickon schist from Conestoga limestone.

The mylonitic foliation here is nearly vertical. There are tight intrafolial folds in which at least some of the axes are subhorizontal, in contrast to the near-vertical axes at Stop 5. Armstrong

(1941) described mylonitic foliation wrapping around unmylonitized lenses on all scales; similar features can be observed here.

STOP 7 Deborah's Rock Farm, Unionville 7.5 min. quadrangle

This outcrop, one of the largest natural exposures of the West Chester nappe, is on property owned by Deborah's Rock Farm (permission to visit must be obtained from Mr. Samuel Wagner). The site is reached via the farm's driveway, which intersects the south side of PA Rte. 162 (Figure 3C) just west of the bridge crossing the East Branch of Brandywine Creek. The CVFZ, lying approximately 1 km to the northwest, is separated from this exposure by a band of kyanite-bearing Wissahickon schist and the Poorhouse nappe (Figure 1).

The base of the West Chester nappe is exposed at a structural level similar to that at Stop 5; thus, comparisons in the change of deformational style along the fault can be made. The most significant differences here include the gentler and more variable dip of foliation and the greater variety of mesoscopic structures.

The oldest structure in the gneiss is a non-penetrative metamorphic foliation, which is best developed in the mafic lithologies. The foliation is defined by amphibolite grade minerals, suggesting a Paleozoic age. Following the metamorphic peak, ductile shearing produced thin (to 2 cm thick) ductile deformation zones (S2h, Figure 5) which dominate the rock fabric; the S2h surfaces were subsequently folded at least twice. Exposed in the outcrop is the hinge of a large, nearly recumbent F2h antiformal nappe with a shallowly south-dipping upper limb and steep lower limb.

Perhaps the most striking feature of this exposure is the complex and often perplexing contact between mafic and felsic lithologies. In general, these are the result of either deformation associated with nappe formation and emplacement (D2) or brittle to semi-brittle deformation associated with D3 tectonism. One striking example occurs at the northern end of the exposure about 20 meters above the river, where a fold, tentatively correlated with D2, folds S2h shear surfaces. The metamorphic foliation is nearly perpendicular to the folded surfaces (see F2h in gneiss column on Figure 5). An example of D3 brittle deformation occurs at the top of the exposure where folds and reverse faults with small displacements show transport to the northwest.

STOP 8 Waltz Road, Unionville 7.5 min. quadrangle

This outcrop is located on Waltz Rd. approximately 100 meters southeast of its junction with Sugarsbridge Rd. and US 322 (Figure 3C). It lies in the footwall approximately 0.4 km north of the CVFZ trace.

The CVFZ as mapped in this area is the northern contact of the Poorhouse nappe (Figure 1). Domainal analysis of dominant foliation orientation indicates that the CVFZ also separates foliation domains. To the north these are defined by point maxima, whereas on the south they exhibit a girdle distribution (Figure 6). Previous workers made no distinction between the rock here and the schist in the hanging wall; both are mapped as Wissahickon schist. However, the two lithologies are quite different in metamorphic grade and fabric. The schist between the West Chester and Poorhouse nappes (high grade Wissahickon schist) contains amphibolite-facies minerals, including centimeter-sized garnet, kyanite, staurolite, and muscovite along with abundant pegmatitic segregations. At this location the rock is a fine-grained schist with garnet, chlorite, muscovite, plagioclase, and quartz.

The rocks at this outcrop are button schists (Lister and Snoke, 1984) and are typical of rocks in the footwall adjacent to the CVFZ. Three foliations are present. The dominant one, dipping gently to the southeast, is S2f. On the east face (closest to the road), the metamorphic foliation (S1f) is deflected by S2f; asymmetry indicates top to the northwest sense of movement. The third foliation, S3f, is best observed on the north face of the outcrop. The trace of S3f shear surfaces on this face plunge toward the southwest. Deflection of S2f by these surfaces indicates oblique slip with top down-dip to the southwest. Note that mesoscopic folds are not present here.

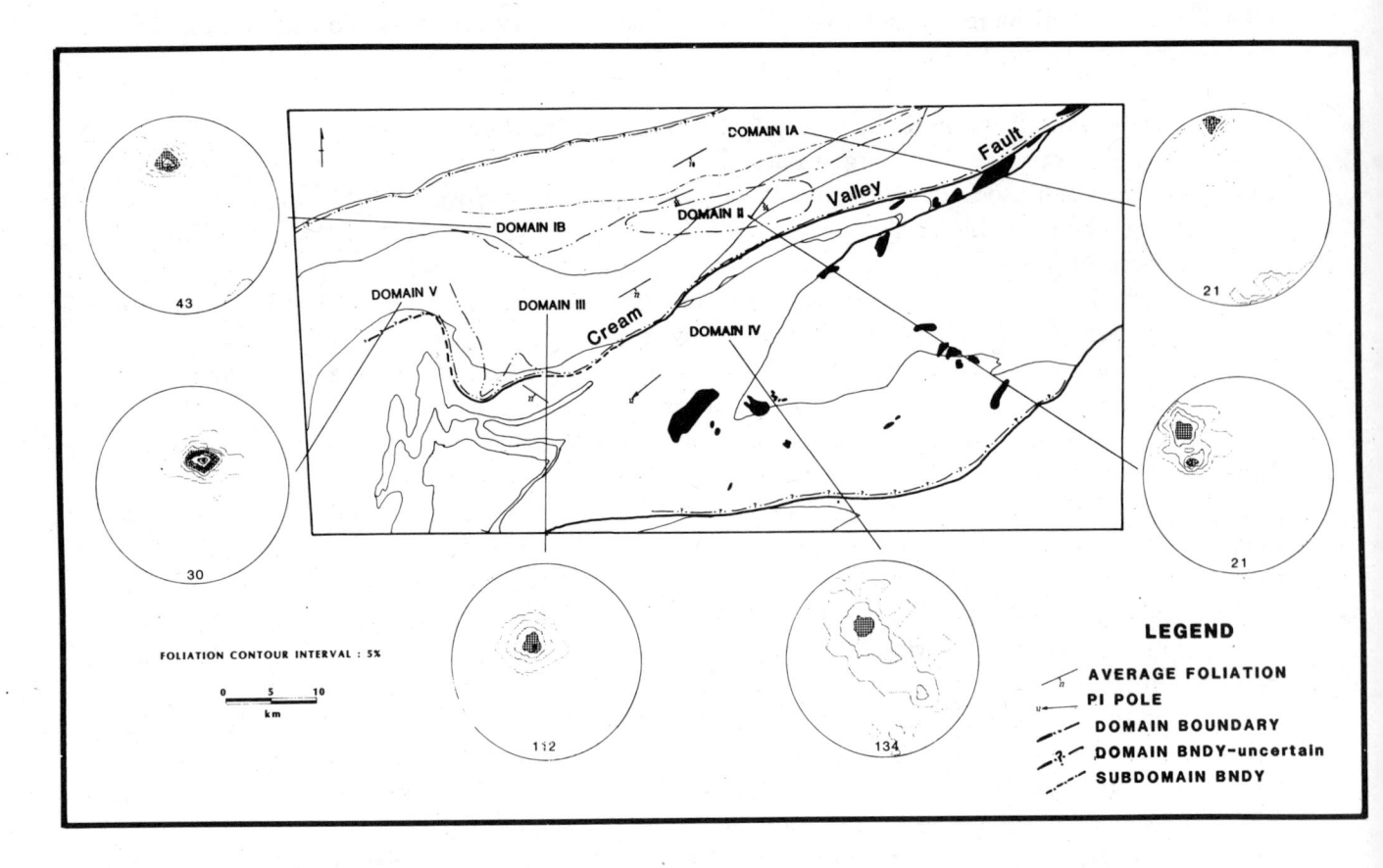

Figure 6. Map showing dominant foliation domains of the CVFZ.

STOP 9 Laurel Road, Coatesville 7.5 min. quadrangle

This stop is located on the northeastern bank of the West Branch of Brandywine Creek about 2.9 km south of Mortonville, PA (Figure 3C).

The rocks in this exposure, 0.6 km north of the CVFZ, have all the lithologic variations of the Peters Creek schist typical of this portion of the Piedmont. Most of the outcrop consists of interlayered pelites and psammites containing garnet, chlorite, biotite, muscovite, and quartz. Magnetite is ubiquitous. A greenstone, consisting primarily of epidote, chlorite, and plagioclase, occurs near road level just north of the crest of the hill.

This outcrop also shows the variations in structures that are present in the footwall near the fault. Pelitic rocks throughout the outcrop have a wavy schistosity reminiscent of type II S-C mylonites but clear distinction between S and C surfaces is difficult. In thin section, the dominant schistosity is defined by oriented muscovite, chlorite, and a strong grain-shape fabric in quartzofeldspathic layers. The older foliation (S1f) is preserved in porphyroclasts at a high angle to the dominant schistosity or is deflected into C-surfaces resulting in mesoscopic waviness. Thus, these rocks record high shear strain that, in some areas, obliterates S1f. At the northwestern end of the exposure, mesoscopic muscovite porphyroclasts ("mica fish") indicate a top to the northwest sense of motion. Thus, the dominant schistosity here is S2f. Note that the S3f fabric observed at Stop 8 is not present here.

In psammitic rocks, the compositional layering shows three generations of folds. The most common are F2f which refold F1f isoclines and are themselves refolded by F3f drag structures. F2f and F3f have similar geometries and development sequences as folds in the hanging wall (Figure 5).

Comparison of the style, vergence, and microstructures of F2f with F3f suggest northwest transport. This began at temperatures close to, but below, the metamorphic peak and continued through cooling and increasing competence of the rock.

Although the mineralogy of the pelitic schist corresponds to the garnet zone of the greenschist facies, several features suggest a polymetamorphic history. Two types of garnets occur: millimeter-scale grains with euhedral rims around anhedral cores that overgrow S2f; and larger, anhedral to subhedral grains that deflect S2f and often show some retrograde reaction to chlorite. Chlorite grains may be found that both parallel S2f and overgrow S2f. In quartzofeldspathic layers, euhedral chlorite grains with random orientation have pinned grain boundaries during annealing. These features suggest either that two metamorphic events occurred, both at upper greenschist conditions, or that the rocks remained at greenschist-facies conditions throughout deformation.

STOP 10 Buck and Doe Run Farm, Coatesville 7.5 min. quadrangle

The outcrops for this stop are reached by walking north along the dirt road from the intersection of PA 82 with Apple Grove Rd (Figure 3C).

This location lies along the southwestward continuation of the CVFZ. The Pennsylvania geologic map (Berg et. al., 1980) shows the CVFZ ending abruptly at the southwestern end of the Poorhouse nappe approximately 8 km along strike to the northeast. However, the steep metamorphic gradient and structural discontinuity here, which is similar to those at Stops 7 and 8, suggests that the fault continues.

A small outcrop in the roadbed and east road cut approximately 460 m north of the entrance gate at Apple Grove Road is typical of high grade Wissahickon schist except that staurolite is strongly retrograded to chlorite and sericite. D2 shearing in the Wissahickon schist resulted in slip along the metamorphic schistosity (S1h) at amphibolite grade. In these rocks, the S2h shear surfaces form a small acute angle with S1h, making it difficult to separate them mesoscopically. Metamorphic retrogression of the amphibolite-facies mineral assemblage occurred only along the fault, and the degree of retrogression decreases sharply to the south away from the fault. The association of retrograde reactions with the fault zone indicates post-D2h deformation at lower temperatures.

Another outcrop occurs approximately 150 m farther north; this rock is low-grade Wissahickon schist very similar to that observed at Stop 8. It contains upper-greenschist facies minerals and has three cross-cutting foliations. Comparison of this rock with the high-grade Wissahickon rocks to the south illustrates the steep metamorphic gradient and structural discontinuity which characterize the CVFZ in this area.

EVIDENCE FOR THRUSTING AT THE BASE OF THE WISSAHICKON FORMATION

Metamorphic and structural discontinuities occur at the base of the Wissahickon Formation. These discontinuities are the result of thrust emplacement of rocks of the Wissahickon Formation onto Grenville basement and its cover (Setters Formation and Cockeysville Marble) during the Taconic orogeny. We agree that the Wissahickon Formation south of the Avondale nappe is in thrust contact with the rock beneath it. However, we differ in our interpretation of apparent structural discontinuities around the Woodville nappe.

Metamorphic Discontinuity at the Base of the Wissahickon Formation

South of the Avondale nappe. The Wissahickon schist–Cockeysville Marble contact in this area is a metamorphic discontinuity. The higher-grade schist and gneiss of the Wissahickon Formation is above lower-grade marble. Uppermost amphibolite-facies gneiss of the Wissahickon Formation contains the assemblage Bi-Si-Gt-Kf-Pl-Q-Ilm ± secondary Mu, indicating that they were metamorphosed at temperatures above the second sillimanite isograd (Plank, 1988; Alcock, 1989a). Garnets in migmatitic gneiss appear to have grown into zones of Bi + Si, suggestive of the reaction:

$$Si + Bi + Q + Na\text{-}Pl = Gt + Ca\text{-}Pl + melt$$

Experimental investigations of the second sillimanite isograd (Chatterjee and Johannes, 1974) and melting in pelites (for example, Kerrick, 1972; LeBreton and Thompson, 1988) indicate that peak temperature was between 650°C and 750°C (Figure 7). Results of geothermo-barometry are consistent with the mineral assemblages present (Figure 7).

Two low-variance assemblages belonging to the $MgO\text{-}CaO\text{-}SiO_2\text{-}KAlO2\text{-}H_2O\text{-}CO_2$ system, Do-Q-Ph-Cc-Tr and Cc-Q-metamorphic Kf-Ph-Tr, are found in the Cockeysville Marble in a single outcrop south of Landenberg within 15 m of the high-grade Wissahickon gneiss (Alcock, 1989a). These assemblages define an isobaric invariant point at the intersection of the reactions

$$Do + Q + H_2O = Tr + Cc + CO_2 \qquad (A)$$
$$Ph + Cc + Q = Tr + Kf + Fluid \qquad (B)$$

Experiments place the isobaric invariant point at T ≈ 575°C (Eggert and Kerrick, 1981; Hoschek, 1973), assuming P = 4.0 kb, which is consistent with the pressure in the overlying Wissahickon Formation. Minor impurities in the phases (e.g., <1 wt% Fe and Al in tremolite) are assumed not to have affected reaction temperatures significantly. Calculations using an ideal solution model support the assumption.

The differences in metamorphic grade of the Wissahickon Formation and Cockeysville Marble indicate a metamorphic inversion across the contact. We conclude that peak metamorphism of the Wissahickon Formation occurred before the contact formed and interpret the discontinuity to be evidence that the Wissahickon Formation in this area was emplaced above the Cockeysville Marble as a thrust.

The area of the Woodville nappe. The metamorphic discontinuity across the Wissahickon-Cockeysville contact is not observed around the Woodville nappe. In contrast to the sillimanite-orthoclase assemblages south of the Avondale nappe, the Wissahickon schist here has staurolite and kyanite. Gt-Bi thermometry using the compositions of garnet rims and matrix biotite yields estimates of ≈ 560°C. Garnets often show a marked increase in Ca content near the rims, possibly reflecting an increase in pressure from ≈ 4.0 kb to ≈ 8.0 kb (GASP barometer, Koziol and Newton, 1988) with a slight increase in temperature (Alcock, 1989a, b). The assemblage Do-Cc-Q-Kf-Ph in the Cockeysville Marble would be stable under these conditions.

The absence of an observed metamorphic discontinuity does not preclude a thrust contact. Both formations may have experienced amphibolite-facies metamorphism independently before emplacement. Alternatively the late high-pressure metamorphism of the Wissahickon Formation in the Woodville area may have resulted after it was emplaced, possibly during the stacking of the Avondale nappe above the Woodville nappe along the Street Road fault. If so, the Wissahickon Formation in this area could have reached peak metamorphic conditions after its emplacement.

Structural Discontinuity at the Base of the Wissahickon Formation

In the northern part of the Woodville nappe, Grenville gneiss and rocks of the Setters For-

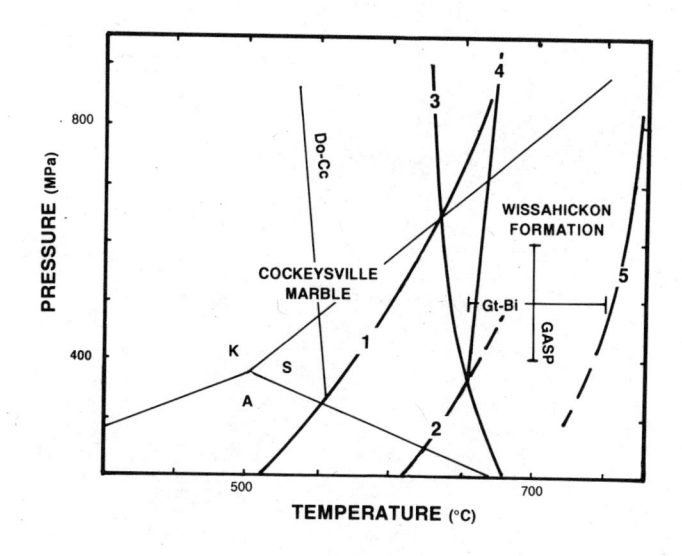

Figure 7. Pressure-temperature conditions affecting the Wissahickon Formation and Cockeysville Marble at Landenberg, PA. Assemblages in the Cockeysville Marble include those that would lie to the left of Curve 1, the estimated position of the intersection of reactions (A) (Eggert and Kerrick, 1981) and (B) (Hoschek, 1973). See text for discussion of reactions (A) and (B). Curve Do-Cc is temperature estimated by dolomite-calcite thermometry from thin sections without obvious signs of exsolution (Bickle and Powell, 1977). Assemblages in the Wissahickon Formation lie to the right of Curves 2 and 4 and relatively near Curve 5. Intersection of lines Gt-Bi and GASP give pressure and temperature estimated by thermo-barometry (Perchuk and Laverent'eva, 1983; Koziol and Newton, 1988 respectively). Curves are 2: Mu + Q = Si + Kf + H2O (Chatterjee and Johannes, 1974); 3: Minimum melting of granite; 4: Mu + Q + Na-Pl = Si + Ca-Pl + Liq (Kerrick, 1972); and 5: Si + Bi + Q + Na-Pl = Gt + Ca-Pl + Liq (LeBreton and Thompson, 1988).

mation and Cockeysville Marble are exposed in a N-S trending structure, bordered on the east, north, and west by the Wissahickon Formation and on the south by the Street Road fault (Figure 8). The structural interpretation of this area has remained controversial since it was mapped by Bliss and Jonas (1916), primarily because it is difficult to define the relationship of the Wissahickon Formation to the other units in the area (Bailey and Mackin, 1937; Mackin, 1962; McKinstry, 1961).

Wagner and Srogi (1987) have argued in favor of the Bailey-Mackin (1937) and Mackin (1962) model, which is based on a down-plunge projection. They believe there are two or more thrusts that emplaced different packages of pelitic rocks presently called "Wissahickon." One of these packages, with serpentinites near its base, lies above the West Chester nappe and plunges beneath the eastern limb (folded thrust surface) of the Woodville nappe (Figure 2B). The other was thrust over the upper limb of the Avondale nappe, where Alcock (1989a) has documented a metamorphic discontinuity.

The aeromagnetic map of the area (Fisher et al., 1979) has been used to support the Bailey-Mackin model. The West Chester and Poorhouse nappes are sharply delineated on the aeromagnetic map because of their low and flat magnetic character in contrast to the surrounding highly magnetic Wissahickon Formation. This is not the case for the Woodville dome. Highly magnetic patterns associated with the Wissahickon Formation exposed between the West Chester and Woodville nappes appear to extend some distance west into the Woodville nappe, where Grenville gneiss has been mapped at the surface, suggesting that the gneiss is thin there and underlain by rocks of the

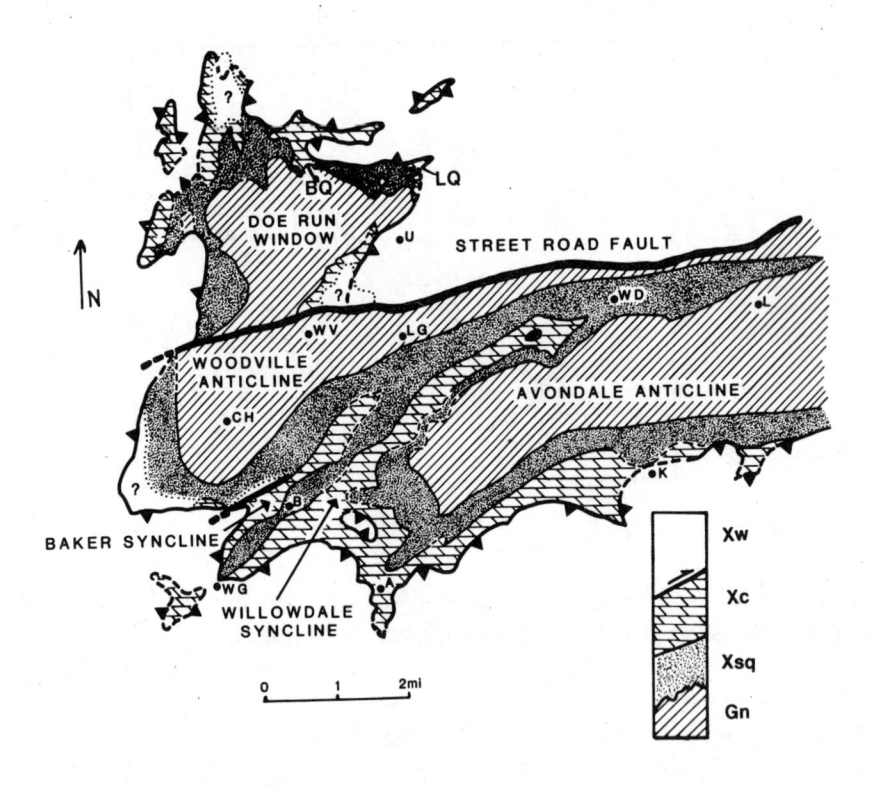

Figure 8. Geologic map of the Avondale and Woodville nappes after Alcock (1989a). The Doe Run thrust is shown as a heavy line with solid triangles. BQ (Buck and Doe Run Farm quarry) and LQ (Logan's quarry) are the sites of Stops 11 and 12. U: Upland.

Wissahickon Formation.

Alcock has proposed an alternative model based on thrust emplacement of the Wissahickon Formation (the Doe Run thrust of Bliss and Jonas, 1916) after F_R folding of the Grenville gneiss and Lower Glenarm Series rocks of the Avondale and Woodville nappes. (See Table 1 for an explanation of the fold terminology used in this section, and its relation to the deformation events described in the last section.) In this model, the Grenville gneiss, the Setters Formation, and Cockeysville Marble are exposed in the complex Doe Run window (Figure 8), bordered by the Doe Run thrust and Street Road fault. Alcock believes that the gneiss is thin and underlain by Wissahickon schist only along the northern margin of the Avondale nappe, south of the Street Road fault, thus explaining the aeromagnetic pattern. The present pattern of the exposed Doe Run thrust reflects post-thrust deformation and should not be used to model a pre-thrust structure (Figure 2A).

Alcock's (1989a) revisions to the geologic map (Figures 8–10) are based on his finding that the position of the Wissahickon Formation is not consistent with the Bailey-Mackin model. For example, at Upland (Figure 8) the Wissahickon Formation lies above the Cockeysville Marble where schist should be beneath the Woodville nappe. Along the western margin of the nappe, the Bailey-Mackin model predicts an upright sequence on the upper limb of the fold. Instead, the Wissahickon Formation *lies above and crosscuts* an inverted stratigraphic sequence of Cockeysville Marble below the Setters Formation which is below Grenville gneiss. Thus, Alcock (1989a) interprets the structural discontinuity to indicate that the Doe Run thrust crosscuts the recumbent folds and is therefore a later structure.

Table 1. Fold terminology as used in this section.

F_R	Recumbent folds, axes trend N30°E, plunge 30° SW.
$F_{3.1h}$	Open folds with nearly vertical axial planes, axes trend N70°E, plunge 10-20°SW. Visible in the outcrop pattern.
$F_{3.2h}$	Open to tight folds with near vertical axial planes. Axes trend approximately N-S, plunge is <10° to N or S.

N.B. F_R may or may not correspond to F2h of the previous section. F3h has been divided to include two distinct fold orientations.

STOP 11: Buck and Doe Run Farm Quarry, south of Rte. 82, ≈ 0.3 mi E of Apple Grove Road (Figure 3C)

Rte. 82 lies on a ridge of Wissahickon schist; blocks of garnetiferous schist are common in the road bank. Follow an old farm road south from Rte. 82 across a creek and into the woods past three quarries that expose Cockeysville marble and the Setters Formation. After examining float of mylonitic Grenville gneiss, common in the woods, we will return to the southern quarry. In the east and south walls, quartzite of the Setters Formation is exposed above the marble. Near the contact, recumbent F_R (in the marble) and open $F_{3.1h}$ (in the marble and Setters Formation) are present. Looking north, one can see a north-south-trending $F_{3.2h}$ antiform.

The quarry has exposed the lower limb of a recumbent fold where Cockeysville Marble is beneath the Setters Formation which is below Grenville gneiss. The marble is brought to the surface in an $F_{3.2h}$ antiform that trends N10°W with a nearly horizontal axis. Bedding in the marble and in the Setters Formation in the east wall of the quarry is oriented N30°W, ≈ 25°NE. The Wissahickon Formation, exposed in the ridge north and east of the quarry, has a N90°E, 30°S foliation that is discordant to structures in the quarry.

Two interpretations are possible: 1) The Wissahickon Formation is under the marble; or 2) the Cockeysville Marble and the Setters Formation pass under the Wissahickon Formation. While neither interpretation can be excluded on the basis of field evidence, the latter is preferred by Alcock (Figure 9) for the following reasons: 1) Cockeysville Marble lies beneath the Wissahickon Formation on the north slope of the ridge; 2) The south and southeast dipping foliation of the Wissahickon Formation at this location is ubiquitous in the area, as it does not conform to F_r structure, it must either be entirely reoriented by later events or unaffected by it; 3) The contact between the Setters Formation and Grenville gneiss both east and west of the quarry (Figure 9) is suggestive of synforms, for the Setters Formation and Cockeysville Marble to be over the Wissahickon Formation the structures would be antiforms; and 4) It is consistent with the observed discordant relations along the western margin of the structure. In this interpretation both $F_{3.1h}$ and $F_{3.2h}$ deform the thrust, creating the ridge of Wissahickon ($F_{3.1h}$ synform) and the "marble" valley that runs north from the quarry across Rte. 82 ($F_{3.2h}$ antiform).

STOP 12 Logan's Quarry, just N of Rte. 82, ≈ 1.75 mi E of Apple Grove Rd.

Logan's Quarry exposes several moderate-sized F_R folds in the Cockeysville Marble. These are best seen in the NE extension of the quarry where they are drag folds with top to the northwest. In the northwest part of the quarry, the marble dips at a low angle to the northwest and is folded gently along north-south trending axes ($F_{3.2h}$). The Wissahickon Formation occupies the higher elevations north, south, and east of the quarry. Recent road work just west of the quarry on Rte. 82 (Figure 10) has exposed a pelitic schist above Grenville gneiss, which is above the Setters Formation. This schist is tentatively identified as the Wissahickon Formation on the basis of its pelitic nature, anastomosing

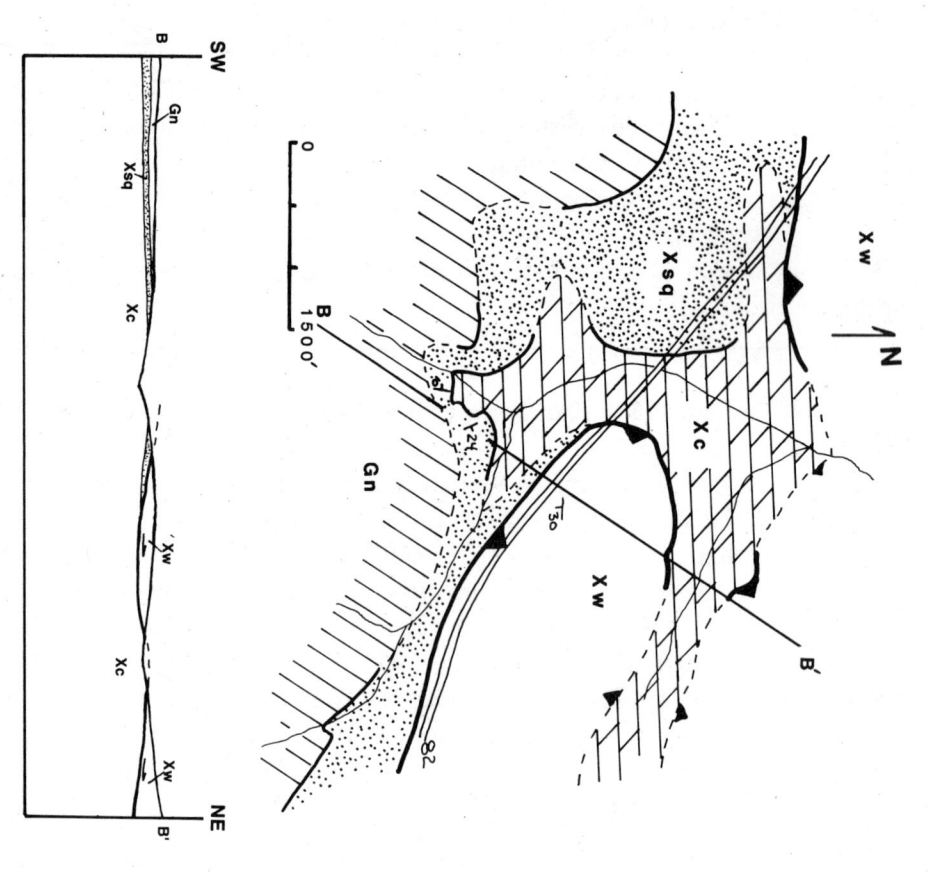

Figure 9. Detail of geologic map and interpretive cross section illustrating structural relations at the Buck and Doe Run quarry. Scale of the cross section is approximate. Abbreviations: Gn, Grenville Gneiss; Xsq, Setters Formation; Xc, Cockeysville Marble; Xw, Wissahickon Formation and related rocks. (No attempt is made to model the structure beneath the Cockeysville Marble, and relations of the Wissahickon to the Peters Creek schist and Cream Valley fault are not shown.)

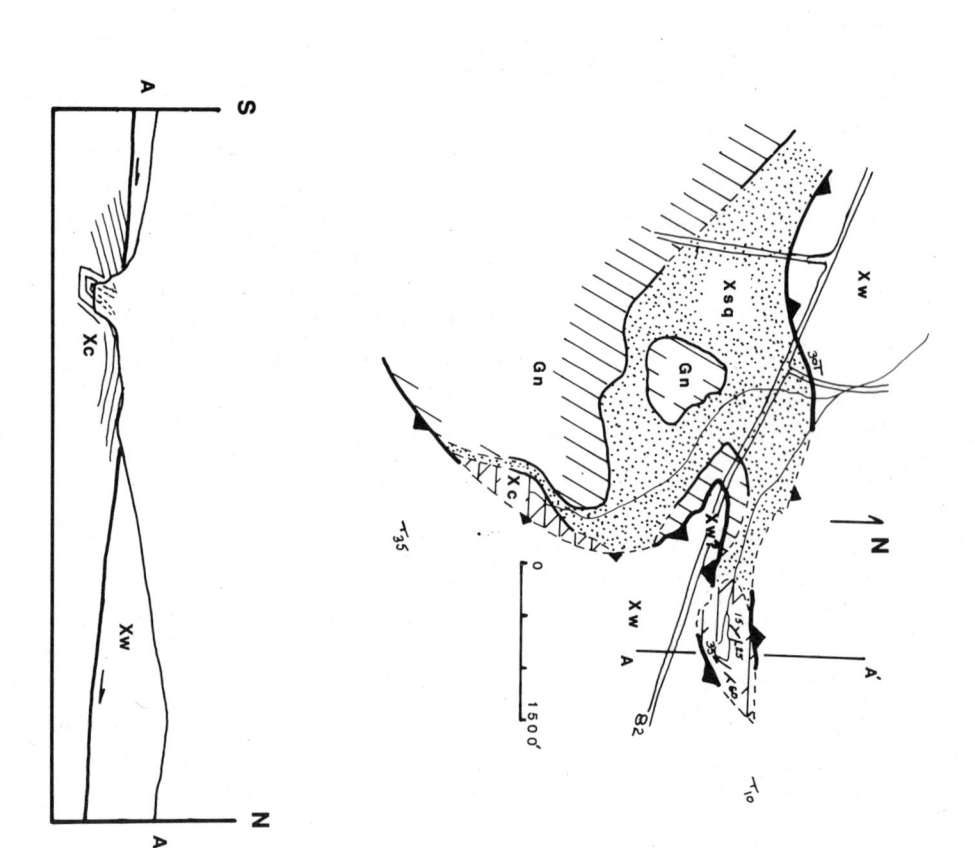

Figure 10. Detail of geologic map and interpretive cross section illustrating Alcock's interpretation of structural relations at Logan's quarry. Detail reflects the recent exposure of pelitic schist in the area west of quarry on Rte. 82, tentatively identified as Wissahickon Formation and indicated by Xw? and, therefore, differs from the area map. Abbreviations: Gn, Grenville Gneiss; Xsq, Setters Formation; Xc, Cockeysville Marble; Xw, Wissahickon Formation and related rocks.

114

cleavage, and Wissahickon-style garnets, but further study is required for a definite identification.

The controversial aspect of Alcock's interpretation (Figure 10) is placement of the Wissahickon Formation on the ridge to the north of Logan's Quarry above the Cockeysville Marble, even though the down-plunge projection places the Wissahickon Formation beneath the marble (Bailey and Mackin, 1937). However, a structural discontinuity across the Wissahickon-Lower Glenarm Series contact would mean that a down-plunge projection of the F_R folds in Logan's Quarry can be applied only to units beneath the Doe Run thrust and provides no information about the position of the Wissahickon Formation relative to those units. If the schist in contact with the Grenville Gneiss west of the quarry is Wissahickon, then the model presented here would be correct.

ACKNOWLEDGMENTS

We wish to thank M. L. Crawford and W. A. Crawford for many suggestions that helped to improve the manuscript and K. D. Woodruff for the location of geologic units in the Mill Creek nappe, based on recent mapping by the Delaware Geological Survey.

REFERENCES

Alcock, J., 1989a, Tectonic units in the Pennsylvania-Delaware Piedmont: Evidence from regional metamorphism and structure: unpublished Ph.D. thesis, University of Pennsylvania, 259 p.

——, 1989b, Changing tectonics of the Pennsylvania-Delaware Piedmont: Pressure-temperature-time histories of the Wissahickon formation: Geological Society of America Abstracts with Programs, v. 21, p. 1.

Armstrong, E., 1941, Mylonization of hybrid rocks near Philadelphia, Pennsylvania: Geological Society of America Bulletin, v. 52, p. 667–694.

Bailey, E. B., and Mackin, J. H., 1937, The recumbent folding in the Pennsylvania Piedmont; preliminary statement: American Journal of Science, v. 33, p. 187–190.

Barker, F., Arth, J. G., and Stern, T. W., 1986, Evolution of the Coast batholith along the Skagway Traverse, Alaska and British Columbia: American Mineralogist, v. 71, p. 632–643.

Bascom, F., and Miller, B. L., 1920, Elkton-Wilmington Folio. U.S. Geological Survey Atlas, Folio 211, 22 p.

——, and Stose, G. W., 1932, Description of the Coatesville and West Chester Quadrangles: U.S. Geological Survey Atlas, Folio 223, 15 p.

Berg, T. M., Edmunds, W. E., Geyer, A. R., et al., 1980, Geologic map of Pennsylvania: Pennsylvania Geological Survey, 4th ser., Map 1 1:250,000.

Bickle, M. J., and Powell, R., 1977, Calcite-dolomite geothermometry for iron-bearing carbonates: Contributions to Mineralogy and Petrology, v. 59, p. 281–292.

Bliss, E. F., and Jonas, A. I., 1916, Relation of the Wissahickon mica gneiss to the Shenandoah limestone and the Octararo schist of the Doe Run and Avondale region, Chester County, Pennsylvania: U.S. Geological Survey Professional Paper 98-B, 34 p.

Brick, E., 1980, Field and petrographic study of the Bringhurst pluton, northern Delaware: unpublished Senior thesis, University of Pennsylvania, 24 p.

Burns, L. E., 1985, The Border Ranges ultramafic and mafic complex, south-central Alaska: Cumulate fractionates of island-arc volcanics: Canadian Journal of Earth Science, v. 22, p. 1020–1038.

Calem, J. A., 1987, A petrologic study of migmatites of Brandywine Creek State Park, northern Delaware, and implications for the mechanism of migmatization: unpublished M.S. thesis, University of Pennsylvania, 111 p.

Chatterjee, N. D., and Johannes, W., 1974, Thermal stability and standard thermodynamic properties of synthetic 2M1-muscovite, $KAl_2AlSi_3O_{10}(OH)_2$: Contributions to Mineralogy and Petrology, v. 48, p. 89–114.

Crawford, M. L., and Crawford, W. A., 1980, Metamorphic and tectonic history of the Pennsylvania Piedmont: Journal of the Geological Society of London, v. 137, p. 311–320.

——, and Mark, L., 1982, Evidence from metamorphic rocks for overthrusting, Pennsylvania Piedmont: Canadian Mineralogist, v. 20, p. 333–347.

Drake, A. A., Jr., 1984, The Reading Prong of New Jersey and eastern Pennsylvania: An appraisal of rock relations and chemistry of a major Proterozoic terrane in the Appalachians: Geological Society of America Special Paper 194, p. 75–110.

Eggert, R. G., and Kerrick, D. M., 1981, Metamorphic equilibria in the siliceous dolomite system: 6 Kbar experimental data and geologic implications: Geochimica et Cosmochimica Acta, v. 45, p. 1039–1049.

Eggins, S., and Hensen, B. J., 1987, Evolution of mantle-derived, augite-hypersthene granodiorites by crystal-liquid fractionation: Barrington Tops Batholith, eastern Australia: Lithos, v. 20, p. 295–310.

Fisher, G. W., Higgins, M. W., and Zietz, I., 1979, Geological interpretations of aeromagnetic maps of the crystalline rocks in the Appalachians, Northern Virginia to New Jersey: Report of Investigations No. 32, Maryland Geological Survey, 43 p.

Foland, K. A., and Muessig, K. W., 1978, A Paleozoic age for some charnockitic-anorthositic rocks: Geology, v. 6, p. 143–146.

Grauert, B., Crawford, M. L., and Wagner, M. E., 1973, U-Pb isotopic analysis of zircons from granulite and amphibolite facies rocks of the West Chester prong and Avondale anticline, southeastern Pennsylvania: Yearbook of the Carnegie Institution of Washington, v. 72, p. 290–293.

——, and Wagner, M. E., 1975, Age of the granulite facies metamorphism of the Wilmington Complex, Delaware-Pennsylvania Piedmont: American Journal of Science, v. 275, p. 683–691.

——, ——, and Crawford, M. L., 1974, Age and origin of amphibolite-facies rocks of the Avondale anticline (southeastern Pennsylvania) as derived from U-Pb isotopic studies on zircon: Yearbook of the Carnegie Institution of Washington, v. 73, p. 1000–1003.

Hager, G. M., and Thompson, A. M., 1975, Fault origin of major lithologic boundary in Delaware Piedmont: Geological Society of America Abstracts with Programs, v. 7, p. 68–69.

Hamre, J., 1983, The contact between the Wilmington Complex and the Wissahickon near Brandywine Creek State Park: unpublished Senior thesis, University of Pennsylvania, 10 p.

Hoschek, G., 1973, Die reaktion phlogopit + calcit + quarz = tremolit + kalifeldspar + H_2O + CO_2: Contributions to Mineralogy and Petrology, v. 39, p. 231–237.

Howard, C. T., 1988, The structural and metamorphic history of the Cream Valley fault zone near Conshohocken, PA: unpublished M.S. thesis, University of Pennsylvania, 281 p.

Kerrick, D. M., 1972, Experimental determination of muscovite + quartz stability with $P_{H2O} < P_{total}$: American Journal of Science, v. 272, p. 946–958.

Koziol, A. M., and Newton, R. C., 1988, Redetermination of the anorthite breakdown reaction and improvement of the plagioclase-garnet-Al_2SiO_5-quartz geobarometer: American Mineralogist, v. 73, p. 216–223.

Lapham, D. M., and Basset, W. A., 1964, K-Ar dating of rocks and tectonic events in the Piedmont of southeastern Pennsylvania: Geological Society of America Bulletin, v. 75, p. 661–668.

LeBreton, N., and Thompson, A. B., 1988, Fluid absent (dehydration) melting of biotite in metapelites in the early stages of crustal anatexis: Contributions to Mineralogy and Petrology, v. 99, p. 226–237.

Lister, G. S., and Snoke, A. W., 1984, S-C mylonites: Journal of Structural Geology, v. 6, p. 617–638.

Loney, R. A., and Himmelberg, G. R., 1983, Structure and petrology of the La Perouse Gabbro intrusion, Fairweather Range, southeastern Alaska: Journal of Petrology, v. 24, p. 377–423.

Luborsky, P., 1983, Petrology of Route 141 outcrop between Routes 52 and 48, Wilmington, Del.: unpublished report, University of Pennsylvania, 45 p.

Mackin, J. H., 1962, Structure of the Glenarm Series in Chester County, Pennsylvania: Geological Society of America Bulletin, v. 73, p. 403–409.

Mark, L. E., 1977, Petrology and metamorphism in the Marcus Hook Quadrangle, southeastern Pennsylvania: unpublished M.A. thesis, Bryn Mawr College, 57 p.

McKinstry, H., 1961, Structure of the Glenarm Series in Chester County, Pennsylvania: Geological Society of America Bulletin, v. 72, p. 557–578.

Morse, S. A., 1982, A partisan review of Proterozoic anorthosites: American Mineralogist, v. 67, p. 1087–1100.

Pavlides, L., 1981, The central Virginia volcanic-plutonic belt: An island arc of Cambrian(?) age: U. S. Geological Survey Professional Paper 1231-A, 31 p.

Perchuk, L. L., and Laverent'eva, I. V., 1983, Experimental investigation of exchange equilibria in the system cordierite-garnet-biotite, in S. K. Saxena, ed., Kinetics and Equilibrium in Mineral Reactions, Springer-Verlag, New York, p. 199–239.

Plank, M. O., 1988, Metamorphism in the Wissahickon Formation of Delaware and adjacent areas of Maryland and Pennsylvania: unpublished M.S. thesis, University of Delaware, 126 p.

Rdesinski, A., 1983, The contact between the Wilmington Complex gneisses and the Arden pluton: unpublished Senior thesis, University of Pennsylvania, 18 p.

Sinha, A. K., Sans, J. R., Hanan, B. B., and Hall, S. T., 1980, Igneous rocks of the Maryland Piedmont, in Wones, David R., ed., The Caledonides in the USA: Virginia Polytechnic Institute and State University, Memoir 2, p. 131–136.

Srogi, L., 1982, A new interpretation of contact relationships and early Paleozoic history of the Delaware Piedmont: Geological Society of America Abstracts with Programs, v. 14, p. 85.

——, 1988, The petrogeneisis of the igneous and metamorphic rocks in the Wilmington Complex, Pennsylvania-Delaware Piedmont: unpublished Ph.D. thesis, University of Pennsylvania, 613 p.

——, and Lutz, T. M., 1987a, Chemical variation in a plutonic rock suite as a result of equilibrium crystallization and crystal-liquid mixing [abs.]: EOS (American Geophysical Union Transactions), v. 68, p. 442.

——, and ——, 1987b, Chemical variation in a plutonic rock suite as a result of equilibrium crystallization and late-stage melt migration [abs.] EOS (American Geophysical Union Transactions), v. 68, p. 1522.

——, Wagner, M. E., and Lutz, T. M., 1985, Dehydration partial melting in the granulite facies: An example from the Wilmington complex, Delaware-Pennsylvania Piedmont [abs.]: EOS (American Geophysical Union Transactions), v. 66, p. 398.

Sultan, M., Batiza, R., and Sturchio, N. C., 1986, The origin of small-scale geochemical and mineralogic variations in a granitic intrusion, a crystallization and mixing model: Contributions to Mineralogy and Petrology, v. 93, p. 513–523.

Thompson, A. M., 1975, Anorthosite in the Piedmont of northern Delaware: Geological Society of America Abstracts with Programs, v. 7, p. 124–125.

Valentino, D. W., 1986, High-temperature deformation of gneiss in Delaware County, southeastern Pennsylvania: Geological Society of America Abstracts with Programs, v. 18, p. 73.

——, 1988, Rosemont shear zone, central Appalachian Piedmont: Dextral displacement of the Wilmington Complex: Geological Society of America Abstracts with Programs, v. 20, p. 77.

Wagner, M. E., 1982, Taconic metamorphism at two crustal levels and a tectonic model for the Pennsylvania-Delaware Piedmont: Geological Society of America Abstracts with Programs, v. 14, p. 640.

——, and Crawford, M. L., 1975, Polymetamorphism of the Precambrian Baltimore Gneiss in southeastern Pennsylvania: American Journal of Science, v. 275, p. 653–682.

——, and Srogi, L. A., 1987, Early Paleozoic metamorphism at two crustal levels and a tectonic model for the Pennsylvania-Delaware Piedmont: Geological Society of America Bulletin, v. 99, p. 113–126.

——, ——, and Brick, E., 1987, Bringhurst gabbro and banded gneiss of the Wilmington Complex, Bringhurst Woods Park, northern Delaware: Geological Society of America Centennial Field Guide—Northeastern Section, p. 25–28.

Ward, R. F., 1959, Petrology and metamorphism of the Wilmington Complex, Delaware, Pennsylvania, and Maryland: Geological Society of America Bulletin, v. 70, p. 1425–1458.

Wiener, R. W., McLelland, J. M., Isachsen, Y., and Hall, L. M., 1984, Stratigraphy and structural geology of the Adirondack Mountains, New York: Review and synthesis: Geological Society of America Special Paper, v. 194, p. 1–56.

Wiswall, C. G., 1990a, Tectonic history of a terrane boundary based on structural analysis in the Pennsylvania Piedmont: Northeast Geology, v. 12, p. 73–81.

——, 1990b, Westward extension of the Cream Valley fault, southeastern Pennsylvania Piedmont: Geological Society of America Abstracts with Programs, v. 22, p. 79.

Woodruff, K. D., and Thompson, A. M., 1975, Geology of the Wilmington area, Delaware: Delaware Geological Survey, Geologic Map Series, No. 4, 1:24,000.

5
STRATIGRAPHY OF UPPER PROTEROZOIC AND LOWER CAMBRIAN SILICICLASTIC ROCKS, SOUTHWESTERN VIRGINIA AND NORTHEASTERN TENNESSEE

Dan Walker
Kentucky Geological Survey
228 Mining and Mineral Resources Building
University of Kentucky, Lexington, KY 40506-0107

Edward L. Simpson
Department of Physical Sciences
Kutztown University, Kutztown, PA 19530

INTRODUCTION

Rocks of the Chilhowee Group (uppermost Proterozoic to Lower Cambrian; Simpson and Sundberg, 1987; Walker and Driese, in press) record the transition from a continental rift/incipient ocean system (Ocoee Supergroup and Grandfather Mountain Formation) to a passive-margin setting (Ashe and Alligator Back Formations), associated with the opening of the Iapetus (Proto-Atlantic) Ocean (Hatcher, 1972; 1978; Rankin, 1975; 1976). Thus, understanding the Chilhowee Group is important to the reconstruction of the early Iapetus Ocean.

In eastern Tennessee and southwestern Virginia, recent studies by Cudzil (1985), Skelly (1987), Simpson (1987), and Walker (1990) have extended the understanding of depositional settings, biostratigraphic relationships, and tectonic history of the Chilhowee Group. This field trip concentrates on the Chilhowee Group and examines: 1) depositional environments; 2) biostratigraphic zonation using body and trace fossils; 3) correlation within the Chilhowee Group and between other stratigraphic units, employing lithostratigraphic and sequence stratigraphic principles; and 4) the tectonic implications of the distribution of facies and petrology.

DEPOSITIONAL ENVIRONMENTS AND LITHOSTRATIGRAPHY

The Chilhowee Group possesses a complex stratigraphic nomenclature (Figure 1) and is exposed in discontinuous strike-belts from Alabama to Newfoundland. In the southern Appalachian region, the Chilhowee Group is confined primarily to the westernmost Blue Ridge. East of this tectono-stratigraphic boundary, the Chilhowee Group and possible equivalent units are typically exposed in structural windows within overlying eastern Blue Ridge or Inner Piedmont strata (e.g., Tablerock Thrust Sheet of Grandfather Mountain; Bryant and Reed, 1960; 1970) and are associated with internal basement massifs (e.g., Sauratown Mountains of North Carolina and Pine Mountain Belt of Georgia and Alabama; Figure 2; Hatcher, 1987). In the southern Appalachian region, the Chilhowee Group consists of a 600–1200 m thick sequence with a three-fold stratigraphy throughout most of the study area. The three stratigraphic units consist of: 1) feldspathic conglomerate, feldspathic and quartzose sandstone, and subordinate volcanics of the lower interval (Cochran/

AGE	North Georgia and Alabama	Southeastern Tennessee	Hot Springs window, North Carolina	Northeastern Tennessee	Southwestern Virginia	Northwestern Virginia
EARLY CAMBRIAN	Shady Dolomite	Shady Dolomite	Shady Dolomite	Shady Dolomite	Shady Dolomite	Tomstown Dolomite
	CHILHOWEE GROUP — Weisner Formation	Helenmode Formation	Erwin Formation — Helenmode Member	Erwin Formation — Helenmode Member	Erwin Formation	Antietam Quartzite
		Hesse Quartzite	Hesse Quartzite Member	Hesse Quartzite Member		
	Wilson Ridge Formation	Murray Shale	Murray Shale Member	Murray Shale Member		
		Nebo Quartzite	Nebo Quartzite Member	Nebo Quartzite Member		
	Nichols Shale	Nichols Shale	Hampton Shale	Hampton Shale	Hampton Shale	Harpers Formation
	Cochran Formation	Cochran Formation	Unicoi Formation	Unicoi Formation	Unicoi Formation	Weaverton Quartzite
						Loudon Formation
-?- PROTEROZOIC	base of section always faulted out	Ocoee Supergroup / Sandsuck Formation	Ocoee Supergroup / Sandsuck Formation	Mount Rogers Volcanic Group or Grenville basement	Mount Rogers Volcanic Group	Catoctin Greenstone or Swift Run Fm. or injection complexes

Figure 1. Stratigraphic nomenclature for the Chilhowee Group. Modified from Schwab (1972) and Mack (1980).

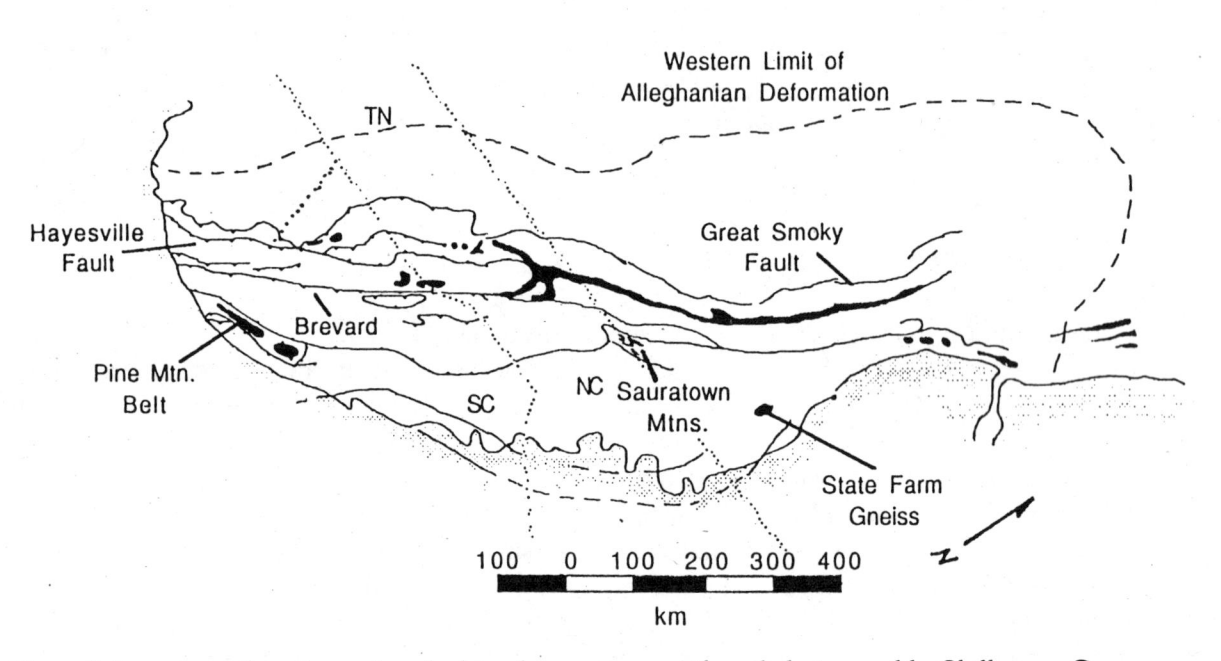

Figure 2. Location of southern Appalachian basement massifs and their possible Chilhowee Group cover sequences (from Hatcher, 1984).

Unicoi); 2) micaceous siltstone and shale with subordinate quartzose sandstone of the Nichols/ Hampton Formations; and 3) quartz sandstone with lesser amounts of siltstone and shale of the Erwin Formation (Figure 3; Schwab, 1970; 1971; 1972; Whisonant, 1974; Mack, 1980; Cudzil and Driese, 1987; Skelly, 1987; Simpson and Eriksson, 1989; 1990; Walker, 1990). The basal Chilhowee Group overlies the Mount Rogers Volcanic Group (southwestern Virginia), Grenvillian crystalline basement (southwestern and central Virginia and northeastern Tennessee), and the Ocoee Supergroup (southern and central East Tennessee) and is comprised of the Cochran (southern belts) and Unicoi Formations (northern belts; Figure 4).

Previous stratigraphic and petrologic studies (Figure 3) of the Chilhowee in the western Blue Ridge (Schwab, 1970; 1971; 1972; Whisonant, 1974), as well as sedimentologic (facies analysis) studies conducted in Virginia (Simpson and Eriksson, 1989; 1990; Simpson, in press) and eastern Tennessee (Cudzil, 1985; Skelly, 1987; Cudzil and Driese, 1987; Walker, 1990), have led to some tentative interpretations of provenance and depositional environments. In most of these interpretations, the lowermost Cochran/Unicoi interval is interpreted to be a fluvial to coastal-alluvial environment overlain by strand-plain deposits of the uppermost Cochran/Unicoi formations. Rocks of the Nichols/Hampton and Erwin Formations are interpreted to record shallow-marine deposition on a storm-dominated shelf with stratial geometries produced by fluctuating relative sea-level (Mack, 1980; Cudzil and Driese, 1987; Simpson and Eriksson, 1990; Walker, 1990). More specific details of depositional environments are discussed in the stop descriptions.

Pre-Chilhowee Group Strata of the Western Blue Ridge

Pre-Chilhowee Group strata of the western Blue Ridge include a series of regionally discontinuous units which possess complex and poorly understood stratigraphic relationships (Figures 1 and 4). The Chilhowee Group overlies three major Proterozoic units including the Ocoee Supergroup (southeastern and east-central Tennessee and west central North Carolina), Grenvillian basement (northeastern Tennessee and southern Virginia), and the Mount Rogers Formation (southern Virginia).

Ocoee Supergroup

Since its original definition as the Ocoee Slate and Conglomerate by Safford (1856; 1869), the Ocoee has been variously ranked as a formation, group (Keith, 1895; 1896), series (Stose and Stose, 1944; 1949) and finally a supergroup by the U.S.G.S. The Ocoee Supergroup is a large body of dominantly terrigenous clastic sedimentary rocks, estimated to exceed 12 km in thickness at some localities (Figure 5A; Rast and Kohles, 1986). Although many workers (e.g., Rast and Kohles, 1986) generally regard the Ocoee as being devoid of both fossils and volcanics, recent studies have demonstrated restricted occurrences of microfossils (acritarchs) and metamorphosed mafic igneous bodies which provide limited information on the history of this basin-fill sequence (Knoll and Keller, 1979; Misra and Lawson, in press). Recent paleontological discoveries indicate that the upper part of the Walden Creek Group is possibly Silurian or younger (Unrug and Unrug, 1990). Body fossils have been obtained by disaggregating shale and argillite surrounding limestone bodies interpreted as olistoliths. Fossils reported include: microcrinoids, fenestrate bryozoans, ostracodes, trilobites, and agglutinated foraminifera. The Miller Cove fault separates the Chilhowee Group and Sandsuck Formation from these localities. Unrug and Unrug (1990) suggest that the shale overlying the Wilhite Formation in the area has been incorrectly mapped as the Sandsuck Formation, and that the shale, as well as the underlying Walden Creek Group strata, is Silurian or younger in age.

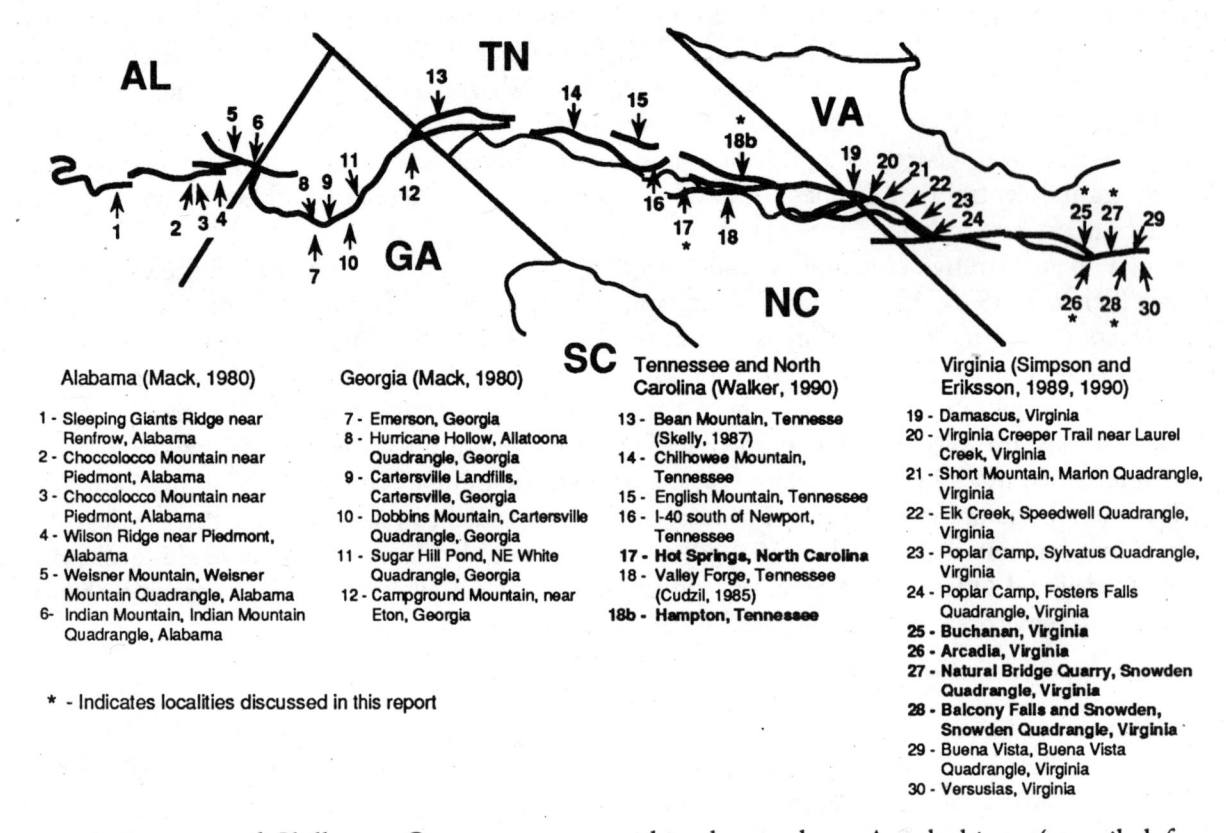

Figure 3. Location of Chilhowee Group exposures within the southern Appalachians (compiled from Mack, 1980; Cudzil, 1985; Cudzil and Driese, 1987; Skelly, 1987; Simpson and Eriksson, 1989, 1990; Walker, 1990).

Alabama (Mack, 1980)
1 - Sleeping Giants Ridge near Renfrow, Alabama
2 - Choccolocco Mountain near Piedmont, Alabama
3 - Choccolocco Mountain near Piedmont, Alabama
4 - Wilson Ridge near Piedmont, Alabama
5 - Weisner Mountain, Weisner Mountain Quadrangle, Alabama
6 - Indian Mountain, Indian Mountain Quadrangle, Alabama

Georgia (Mack, 1980)
7 - Emerson, Georgia
8 - Hurricane Hollow, Allatoona Quadrangle, Georgia
9 - Cartersville Landfills, Cartersville, Georgia
10 - Dobbins Mountain, Cartersville Quadrangle, Georgia
11 - Sugar Hill Pond, NE White Quadrangle, Georgia
12 - Campground Mountain, near Eton, Georgia

Tennessee and North Carolina (Walker, 1990)
13 - Bean Mountain, Tennesse (Skelly, 1987)
14 - Chilhowee Mountain, Tennessee
15 - English Mountain, Tennessee
16 - I-40 south of Newport, Tennessee
17 - **Hot Springs, North Carolina**
18 - Valley Forge, Tennessee (Cudzil, 1985)
18b - **Hampton, Tennessee**

Virginia (Simpson and Eriksson, 1989, 1990)
19 - Damascus, Virginia
20 - Virginia Creeper Trail near Laurel Creek, Virginia
21 - Short Mountain, Marion Quadrangle, Virginia
22 - Elk Creek, Speedwell Quadrangle, Virginia
23 - Poplar Camp, Sylvatus Quadrangle, Virginia
24 - Poplar Camp, Fosters Falls Quadrangle, Virginia
25 - **Buchanan, Virginia**
26 - **Arcadia, Virginia**
27 - **Natural Bridge Quarry, Snowden Quadrangle, Virginia**
28 - **Balcony Falls and Snowden, Snowden Quadrangle, Virginia**
29 - Buena Vista, Buena Vista Quadrangle, Virginia
30 - Versusias, Virginia

* - Indicates localities discussed in this report

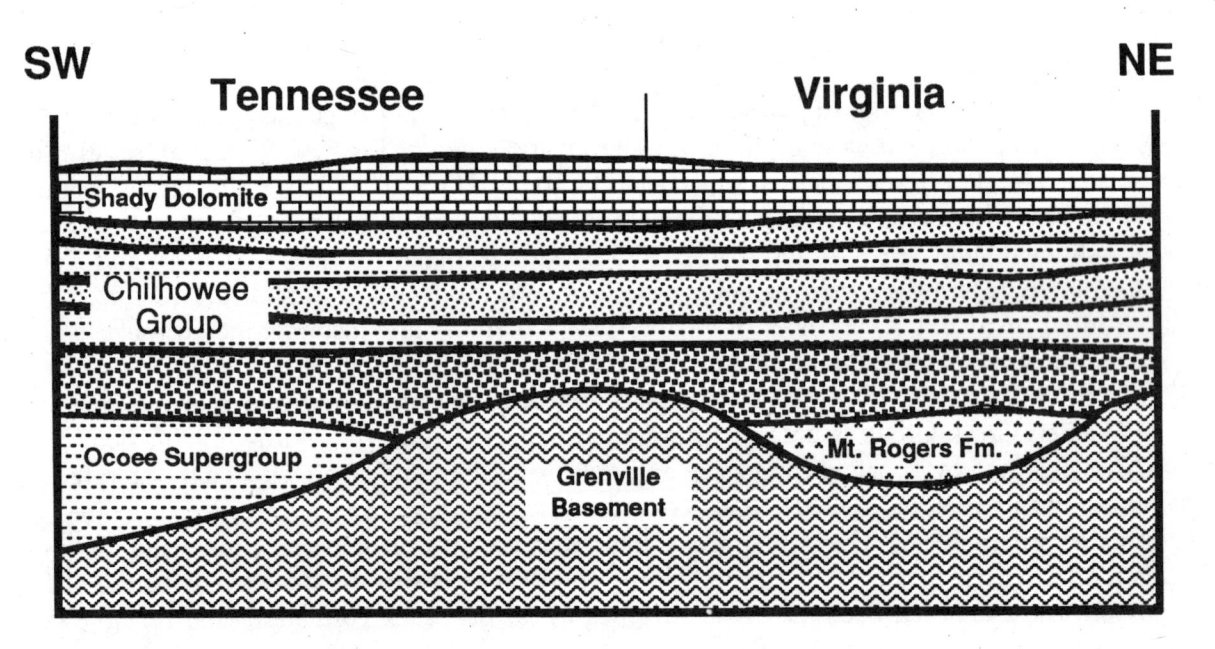

Figure 4. Generalized stratigraphic cross section showing regional variations in the nature of pre-Chilhowee Group strata (modified from Schwab, 1972).

Mapping and paleontologic work conducted to the northwest of the Unrug and Unrug study area, between the Pigeon and French Broad rivers (Keller, 1980) and in the Del Rio district of Cocke County (Ferguson and Jewell, 1951), indicates that the Walden Creek Group and Chilhowee Group form a continuous stratigraphic sequence and that the Walden Creek Group is Vendian in age (Knoll and Keller, 1979). Although these localities reside in a separate structural block than the Unrug and Unrug localities, the significance of the recent fossil discoveries with respect to the age of the Chilhowee Group is unclear. In absence of independent confirmation of these discoveries in areas where stratigraphic relationships between these various units is clearer, we maintain that the Walden Creek Group is in stratigraphic contact with the Chilhowee Group and is Vendian in age. Corroboration of Unrug and Unrug's finding would necessitate a reappraisal of this interpretation, but would not directly affect the age assignments of the Chilhowee Group.

A large percentage of the siliciclastic sediment preserved within the Ocoee Supergroup (especially the Great Smoky Group) was derived apparently from the Late Proterozoic Laurentian craton to the northeast (Thomas, 1977). The younger Snowbird Group sediment was derived from continental blocks exposed to the east and southeast (Hadley and Goldsmith, 1963; Thomas, 1977; Rast and Kohles, 1986). Thomas (1977) proposed that this source terrane was Grenvillian granitic-gneissic basement of the eastern Great Smoky Mountains.

Amphibolites interpreted as metamorphosed mafic volcanic flows have been sampled within the lower portion of the Ocoee Supergroup near Ducktown, Tennessee. Geochemical analyses indicate relict trace element signatures similar to MORB basalts (Misra and Lawson, in press). The Walden Creek Group, containing greater than 3 km of dominantly siliciclastic strata, is the most heterolithic unit of the Ocoee Supergroup (Hadley and Goldsmith, 1963). Minor limestone and dolostone (Yellow Breeches Member of the Wilhite Formation) may indicate relatively widespread but intermittent carbonate deposition (Hadley and Goldsmith, 1963). Directly above the Wilhite Formation are fine-grained siliciclastic deposits of the Sandsuck Formation (Hadley and Goldsmith, 1963).

Mount Rogers Formation

Unlike the Ocoee Supergroup to the southwest, the Mount Rogers Formation contains large quantities of volcanic and volcano-sedimentary rock. From base to top the Mount Rogers Formation is subdivided into three members (Figure 5B): 1) interbedded basalt, rhyolite, and siliciclastic rocks; 2) massive rhyolite flows; and 3) a predominantly sedimentary sequence of arkose, rhythmite, laminated pebbly mudstone, and diamictite (Rankin, 1970; 1975; Blondeau and Lowe, 1972; Schwab, 1976; Wehr and Glover, 1985). As discussed by Rankin (1968; 1970; 1975; 1976) and Rankin et al. (1969), stratigraphic, mineralogic, chemical, and isotopic-geochronologic evidence indicates that the volcanic part of the Mount Rogers Formation is a small, distinct portion of the much more regionally extensive bimodal and anorogenic magmatic suite. This suite includes volcanic rocks of the Grandfather Mountain Formation, which together with the Crossnore, Beech, Striped Rock, and other plutons comprise the 680–710 Ma Crossnore Plutonic Series (Odom and Fullagar, 1984).

The occurrence of cross-stratified feldspathic sandstone and pebbly sandstone led Schwab (1976; 1986) to propose an alluvial origin for some deposits within the Mount Rogers Formation. The terrestrial origin of some portions of the Mount Rogers is further substantiated by the occurrence of widespread, thick rhyolite ash flows of subaerial origin (Rankin, 1970). Conversely, fine-grained, laminated facies with dropstones, diamictites, and turbidites occur in the upper member, indicating glacially influenced, shallow marine or lacustrine deposition (Schwab, 1976; 1986). At the base, an erosional surface with considerable relief strongly suggests that the Mount Rogers Formation was dominated by deposition in fault-bounded basins (Wehr and Glover, 1985).

BIOSTRATIGRAPHY AND AGE CONSTRAINTS ON THE CHILHOWEE GROUP

The Chilhowee Group is probably late Proterozoic (Vendian) to Early Cambrian (Placentian-equivalent or younger) age (Figure 6). There is strong facies control on the distribution of trace and body fossils in the Chilhowee Group. In fact, the problem of precisely dating the Chilhowee Group may ultimately prove insurmountable (Walker and Driese, in press).

Dating the Cochran-Unicoi interval is especially critical because it is underlain by the Ocoee Supergroup containing Vendian acritarchs (Knoll and Keller, 1979), and is overlain by younger Chilhowee formations such as the Murray Shale, which contain reliable indicators of an Atdabanian-equivalent or younger age (e.g., Laurence and Palmer, 1963; Wood and Clendening, 1982). Hence, a part of the lower Placentian-equivalent (pre-Tommotian-equivalent) stage may be represented by a thick sequence of fluvial or alluvial deposits.

The Chilhowee Group is completely devoid of any carbonate (limestone or dolostone) deposits. Therefore, the possibility of extracting shelly microfossils characteristic of the Placentian-equivalent stage appears remote. Only trace fossils offer much hope of allowing a more refined biostratigraphic zonation of the Precambrian-Cambrian boundary in the southern Appalachians, but with limited resolution (Figure 7; Crimes, 1987). Trace fossil assemblages are facies dependent; therefore, care must be taken to compare strata representing similar depositional settings when attempting to draw conclusions regarding trace fossil diversity.

The timing of continental breakup may also prove critical. The occurrence of synrift strata of the Unicoi Formation in northeastern Tennessee and southern Virginia suggests that the southern Appalachian margin may have been too youthful and terrigenous clastic-dominated at approximately 590 to 570 Ma to have accumulated a stratigraphic record with associated fossils that can be dated with precision. Age constraints are defined by: 1) the occurrence of Vendian acritarchs in the subjacent Sandsuck, Wilhite, and Shields formations of the Ocoee Supergroup (Knoll and Keller, 1979); 2) the first occurrence of Palaeophycus traces in the basal Cochran and Unicoi Formations (Simpson and Sundberg, 1987; Walker and Driese, in press); 3) the first occurrences of *Skolithos* and Planolites traces in the overlying Nichols and Hampton Formations (Simpson and Sundberg, 1987; Walker and Driese, in press); 4) the abundance of well-developed arthropod (Rusophycus, Cruziana) as well as other diagnostic traces (Diplocraterion) in the uppermost Nebo and the overlying Murray Formation (Simpson and Sundberg, 1987; Walker and Driese, in press); 5) the recalculated age of 539 + 30 Ma for the Murray Shale (based Rb-Sr ratios as determined from glauconite; Hurley et al., 1960; Cormier, pers. comm., 1989); and 6) reported occurrences of upper Placentian-equivalent or younger (Atdabanian-equivalent or younger) body fossils recovered from the Murray Shale, which include trilobites, ostracodes, inarticulate brachiopods, hyolithids, and acritarchs (Laurence and Palmer, 1963). Based on the recent suggestions of Crimes (1987), trace fossils can be used to assist in correlating the Precambrian-Cambrian boundary interval in stratigraphic sequences where diagnostic body fossils are lacking. Consequently, a late Vendian? to early Placentian-equivalent (sub-Tommotian-equivalent) age is assigned to the Cochran and Unicoi Formations. An early late Placentian-equivalent (early to late Tommotian-equivalent) age is assigned to the Nichols and Hampton Formations and the lower and middle Nebo Formation. Finally, a late Placentian-equivalent

A

Latest Proterozoic to Early Cambrian	Chilhowee Group		
Late Proterozoic	OCOEE SUPERGROUP	Walden Creek Group	Sandsuck Formation / Wilhite Formation / Shields Formation / Licklog Formation
		Unclassified Formation	Sandstones of Webb Mountain and Big Ridge
		Cades Sandstone	Rich Butt Sandstone
		Snowbird Group	Metcalf Phyllite / Pigeon Siltstone / Roaring Fork Sandstone / Longarm Quartzite / Wading Branch Formation
Middle Proterozoic	Grenville Basement		

B

Latest Proterozoic to Early Cambrian	Chilhowee Group	
Late Proterozoic	MOUNT ROGERS FM.	upper member - glaciogenic strata
		middle member - interbedded latites and rhyolites
		lower member - cobble conglomerates with minor basalt and rhyolite
Middle Proterozoic	Grenville Basement	

Figure 5. Stratigraphy of A, the Ocoee Supergroup west of the Greenbrier fault (modified from Hadley and Goldsmith, 1963) and B, the Mount Rogers Formation (modified from Rankin, 1967).

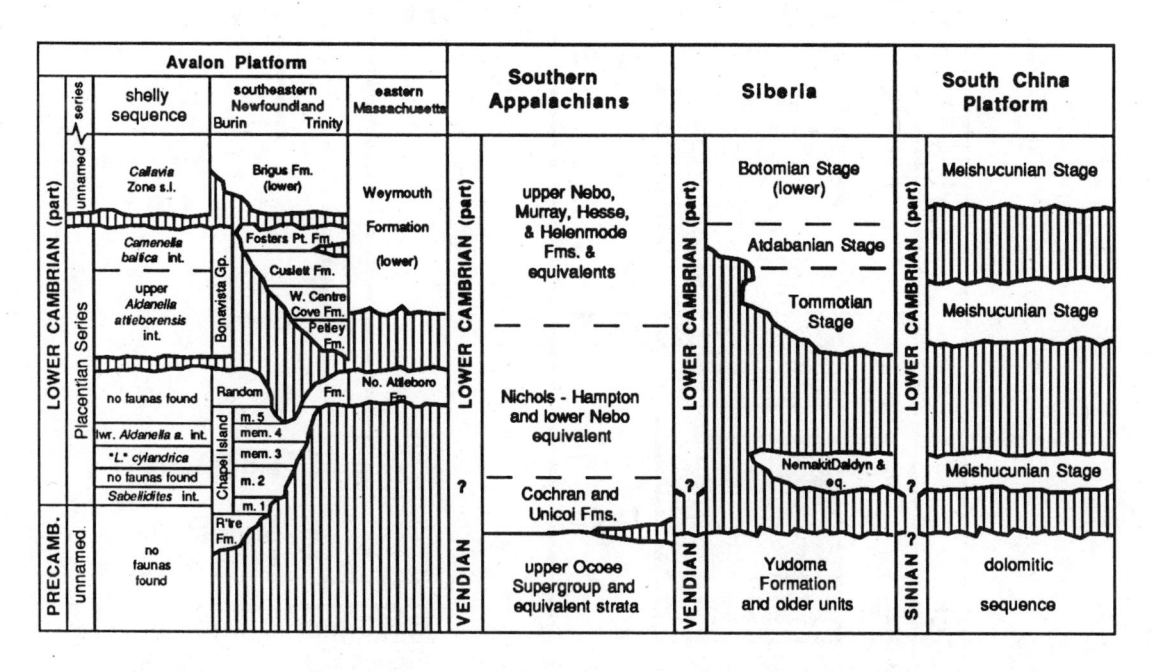

Figure 6. Comparison of age assignments for the Chilhowee Group with Placentian-equivalent strata of the Avalon Platform, Siberia, and South China Platform. Modified from Landing (1988) and Walker and Driese (in press).

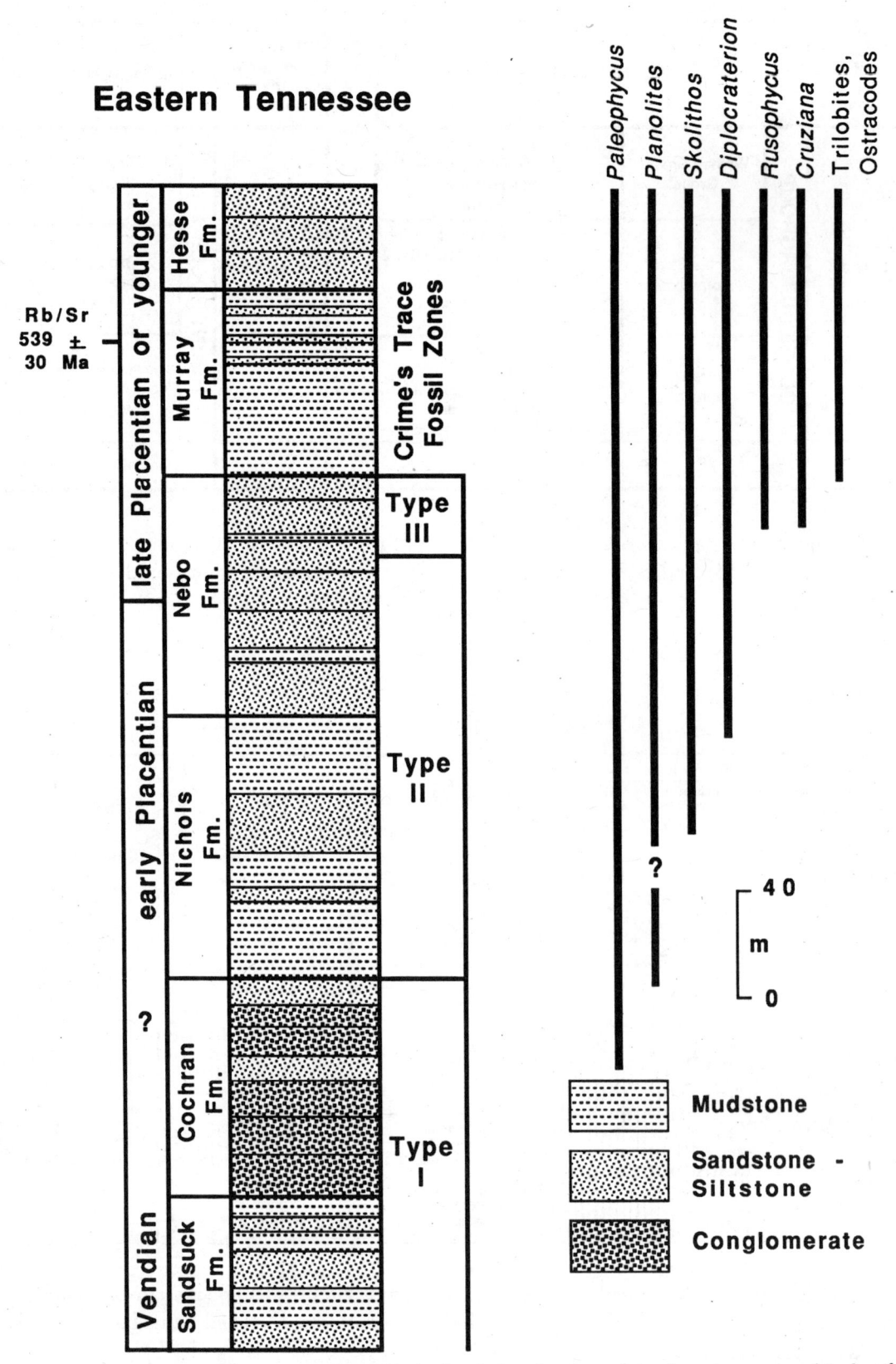

Figure 7. Composite section showing trace and body fossil distribution plotted against a graphic log for the Chilhowee Group in eastern Tennessee. The Precambrian (Vendian) and Cambrian (Placentian-equivalent) age assignments are based on an integrated approach utilizing trace and body fossils (modified from Walker and Driese, in press).

128

or younger (Atdabanian-equivalent or younger) age is assigned to the upper Nebo, Murray, Hesse, and Helenmode Formations. The Precambrian-Cambrian boundary is probably located somewhere within the uppermost portion of Cochran-Unicoi interval. Unfortunately, this sequence is dominated by coarse-grained braided fluvial facies, and so it may ultimately prove impossible to locate the boundary more precisely because of a lack of marine facies in this critical time interval.

If the Unicoi Formation of southwestern Virginia precisely correlates (in-age) with the Cochran-Unicoi interval in Tennessee, then the occurrence of Rusophycus in the middle Unicoi Formation of southwestern Virginia (Simpson and Sundberg, 1987) suggests that the upper Cochran-Unicoi interval in Tennessee is late Placentian-equivalent (Tommotian-equivalent) age. Similarly, if the Hampton Formation of southwestern Virginia precisely correlates (in age) with the Nichols-Hampton interval in Tennessee, then the occurrence of a hyolith? in the lower Hampton Formation (Simpson and Sundberg, 1987) would indicate that the Nichols-Hampton, as well as all of the overlying Nebo Formation, is of latest Placentian-equivalent (late Tommotian-equivalent to early Atdabanian-equivalent) age. More conservative stage assignments for eastern Tennessee have been proposed by Walker and Driese (in press) because of the probability of at least some diachronism (on a regional basis) between Chilhowee formational units (Figure 7). Any stage assignments are therefore subject to possible revision, if and when more body or trace fossil data become available.

SEQUENCE STRATIGRAPHIC PRINCIPLES

The refinement of sequence stratigraphic principles by Exxon workers permits high resolution intrabasinal and extrabasinal correlation (see Van Wagoner et al., 1990 and references therein). Harris and Eriksson (1990) and Jackson et al. (1990) have successfully applied these principles to nonfossiliferous Precambrian strata. Application of sequence stratigraphic principles to nonfossiliferous strata such as the Lower Cambrian (Simpson and Eriksson, 1990) may permit finer-scale subdivision than attributable to conventional paleontological means.

A sequence is defined as a relatively conformable, genetically related secession of strata bounded by unconformities or their correlative conformable surfaces (Mitchum, 1977). Parasequences are a conformable secession of genetically related beds bounded by marine flooding surfaces or their correlative conformable surfaces (Van Wagoner et al., 1990). Sequences can potentially be subdivided into systems tracts based on parasequence stacking patterns, position in the sequence, types of bounding surfaces, geometry, and facies associations (Figure 8A; Van Wagoner et al., 1988; Posamentier et al., 1988). A type 1 sequence (basal unconformity associated with stream rejuvenation) has a lowstand systems tract that consists of a basin-floor slope fan and a lowstand wedge (Van Wagoner et al., 1990); the lowstand wedge develops oceanward of the shelf break (Figure 8A). Type 2 sequences (basal unconformity produced by subaerial exposure not associated with stream rejuvenation) have shelf-margin systems tracts developed instead of the lowstand systems tract. Both the lowstand and the shelf margin systems tract form during a relative sea-level fall. The transgressive systems tract forms during a relative sea-level rise and is characterized by retrogradational parasequences and deepening upward of successively younger parasequences (Figure 8A). The base of the transgressive systems tract is the transgressive surface; the top is delimited by the marine flooding surface. A highstand systems tract is composed of aggradational parasequences early and progradational parasequences latter. The highstand systems tract develops during or after maximum flooding and relative sea-level fall. The marine flooding surface marks the base of the highstand systems tract, whereas the top is the next overlying system. Condensed sections develop during the transgressive systems tract into the early highstand systems tract because the shelf is terrigenous sediment starved (Van Wagoner et al., 1990).

In a limited geographical area, Simpson and Eriksson (1990) have applied sequence stratigraphic principles to the Chilhowee Group (Figure 8B). The lowermost sequence of the passive-margin starts in the upper Cochran/Unicoi Formations and terminates in a sequence boundary in the upper Erwin Formation (Figure 8B; Stop 2-1 and 2-3; Simpson and Eriksson, 1990). The upper Cochran/Unicoi Formations record an initial transgressive systems tract and typically are overlain by a condensed section at the base of the Hampton Formation. The highstand systems tract commences after the condensed section and includes most of the Erwin Formation up to the sequence boundary (Figure 8B). The uppermost Erwin Formation is grouped into a younger transgressive systems tract that continues into the overlying Shady Dolomite (Figure 8B). Superimposed on the basal sequence are a maximum of five, smaller-scale parasequences. The complex interplay of different relative sea-level changes has also been reported from the Shady Dolomite (Barnaby and Read, 1990).

CHILHOWEE GROUP AND ITS POSSIBLE EQUIVALENTS

Several structural windows occur within the Blue Ridge and Piedmont which expose crystalline basement and their autochthonous lower Paleozoic cover sequences. Chief among these are the Grandfather Mountain (North Carolina), Sauratown Mountain (North Carolina), and Pine Mountain (Alabama and Georgia) windows (Bryant and Reed, 1970; Brown, 1970; Heyn, 1984; Hooper, 1986; Hatcher, 1987; 1988).

Grandfather Mountain Window

The southwestern portion of the Grandfather Mountain window exposes a second thrust sheet termed the Tablerock thrust (Bryant and Reed, 1970). This thrust sheet forms a prominent klippe which caps Tablerock Mountain, and contains an extensive interval of quartzite, feldspathic quartzite, and phyllite (Figure 9A), which has been assigned to the Chilhowee Group based on lithologic and stratigraphic similarity (Bryant and Reed, 1970). Because of its isolated tectonic position, rocks of the Tablerock thrust sheet have not been correlated with specific formations within the Chilhowee Group of the western Blue Ridge. Bryant and Reed (1970), therefore, subdivided the sequence into two quartzites, separated by a persistent blue phyllite. The lower quartzite varies from 250 m to 700 m in thickness and contains medium- to fine-grained quartzite and feldspathic quartzite with numerous interbeds of green sericitic phyllite. Interbedded within the quartzite and feldspathic quartzite are intervals of quartz-pebble and feldspar-pebble conglomerate similar to the Cochran/ Unicoi Formations of the western Blue Ridge. Overlying the lower quartzite occur rocks of the phyllite unit which are described as finely laminated, dark blue-gray to blue, sericite with thin lenses of granoblastic quartz. The thickness of this unit is extremely variable, ranging from a few meters to as much as 130 m, with an average thickness of less than 50 m. The upper quartzite is comprised of 400–800 m of thin- to thick-bedded, fine- to medium-grained quartzite and feldspathic quartzite, containing occasional intervals of deformed *Skolithos* tubes, and is conformably overlain by rocks of the Shady Dolomite. Cross-bedding, defined by heavy mineral laminae, occurs throughout the Chilhowee Group and may lend itself to further correlation, via facies analysis and heavy mineral association with these units.

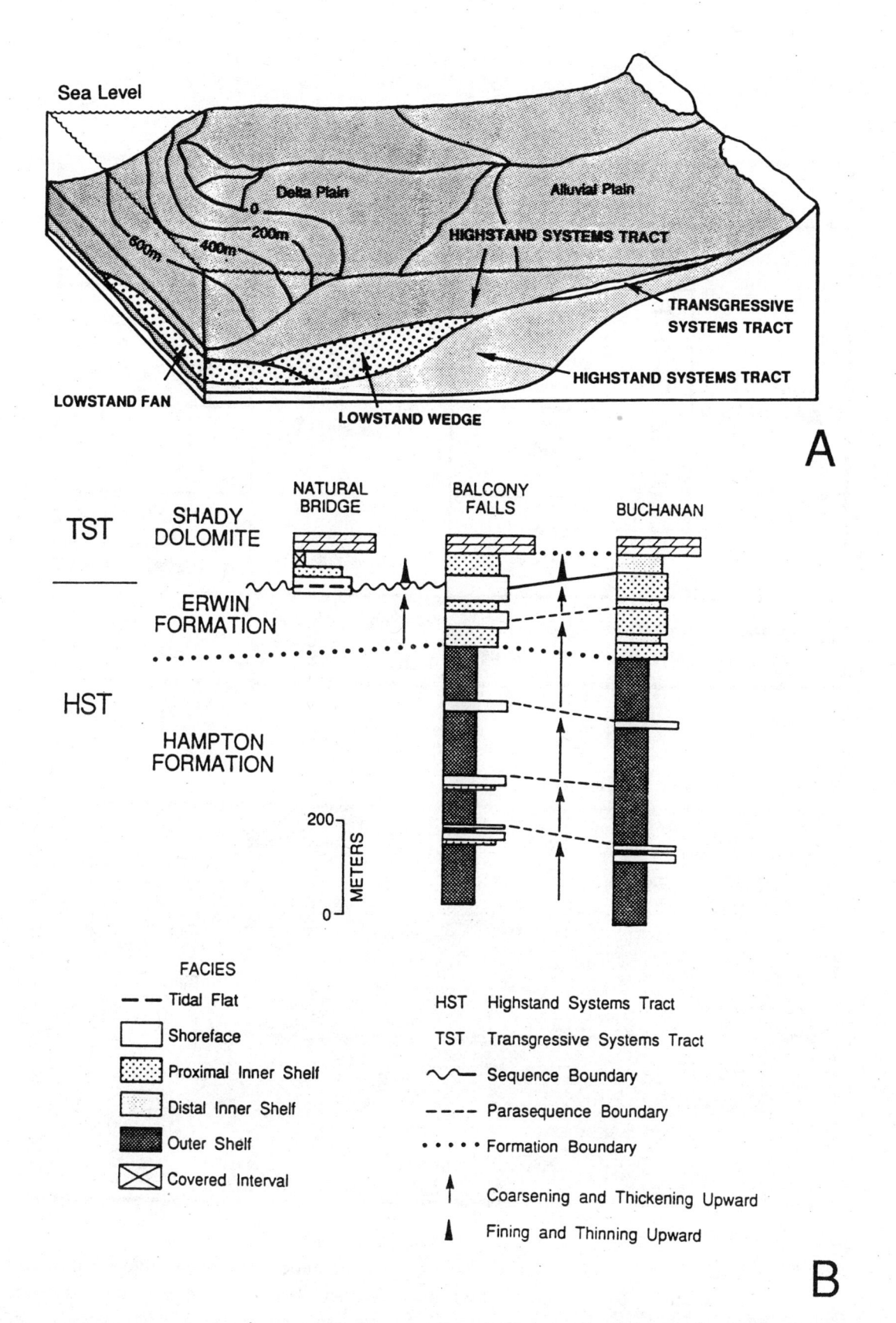

Figure 8. Sequence stratigraphic diagrams. A, Cartoon of different systems tracts developed on a shelf (modified from Posamentier et al., 1988). B, Parasequences and sequence boundaries within the Chilhowee Group (modified from Simpson and Eriksson, 1990).

AGE	Grandfather Mountain, NC		Pine Mtn. Belt Ala / Ga		Sauratown Mtns., NC
EARLY CAMBRIAN	CHILHOWEE GROUP	Shady Dolomite	Manchester Formation	Chewalaca Marble	absent
		upper quartzite		interbedded quartzite and schist	quartzite of Pilot Mtn. and Hanging Rock
		phyllite			pelitic schist and phyllite
		lower quartzite		Hollis Quartzite	quartzite and feldspathic quartzite / arkose
MIDDLE OR LATE PROTEROZOIC		truncated by Tablerock thrust	Woodland Gneiss (Alabama)	Sparks Schist (Georgia)	Grenvillian Basement

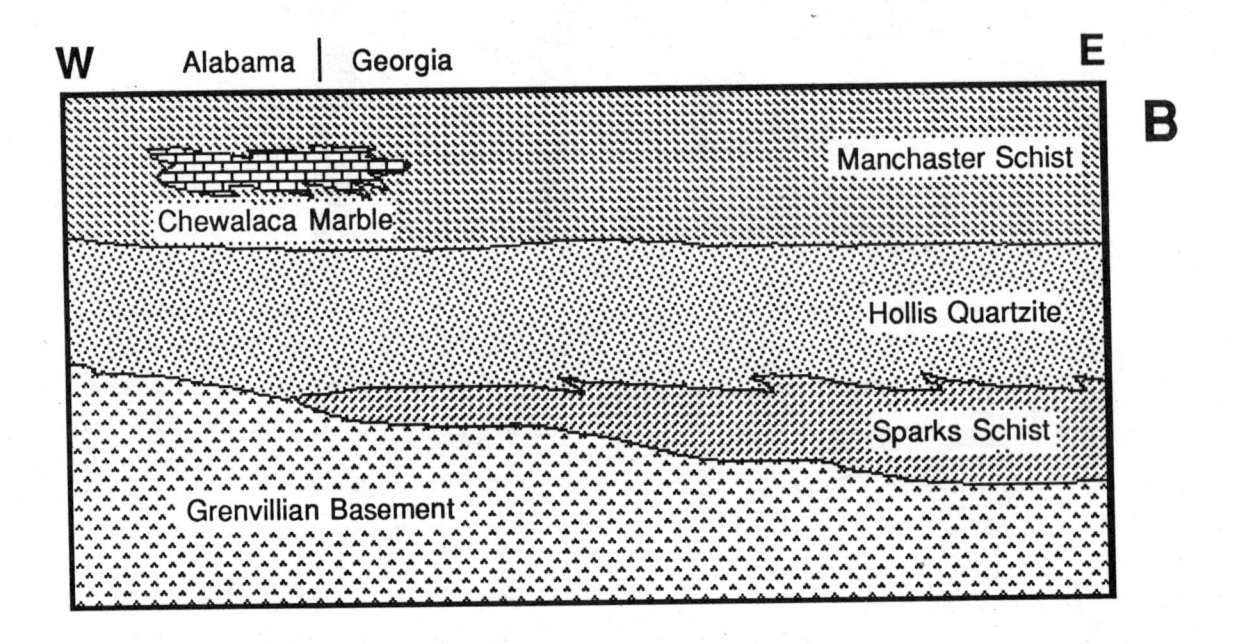

Figure 9. Possible Chilhowee equivalents: A, Stratigraphic nomenclature of possible Chilhowee Group equivalents of the eastern Blue Ridge and Inner Piedmont (compiled from Bryant and Reed, 1970; Schwab, 1977; Hooper, 1986; Heyn, 1984). Stratigraphic correlation shown here is speculative and has not yet been confirmed. See Walker (1990) for discussion. B, Generalized stratigraphic cross section for the Pine Mountain Belt, Alabama and Georgia (compiled from Clarke, 1952; Bentley and Neathery, 1970; Hooper, 1986).

Pine Mountain Window

The Pine Mountain window is structurally complex, framed in part by the pre-metamorphic Box Ankle fault, the post-metamorphic Towaliga fault, and the syn- to post-metamorphic Goat Rock–Bartletts Ferry fault system (Schamel and Bauer, 1980; Sears and Cook, 1984; Hooper, 1986; Hatcher, 1987; 1988). Within this window, Grenvillian age basement is overlain by a thin cover sequence composed of the Hollis Quartzite, Sparks and Manchester Schists, and Chewacla Marble (Figure 9B; Clarke, 1952; Bentley and Neathery, 1970) which comprise the Pine Mountain Series (Hooper, 1986); the overall gross lithology has resulted in many workers hypothesizing a stratigraphic correlation with the Chilhowee-Shady-Rome interval of the western Blue Ridge (Rankin, 1975; 1976; Thomas, 1977; 1983; Hatcher, 1987). The stratigraphic distribution of schists and marble within the area seem compatible with a westward sediment source, further substantiating the hypothesized Chilhowee Group affinity for the Pine Mountain Series (Figures 6–9). While the deformation within the area is pervasive, some primary sedimentary structures are preserved within the Manchester Schist (Clarke, 1952) and may be of some utility in determining the paleogeographic and paleoenvironmental history of the area.

Sauratown Mountain Window

Framed by the Forbush fault, a pre-metamorphic fault which may be a southeastern equivalent of the Hayesville-Fries thrust, the Sauratown Mountain Window contains a second inner window (termed the Hanging Rock Inner window by Hatcher et al., 1988), which exposes Grenvillian crystalline basement and its associated cover sequence (Bryant and Reed, 1960; Hatcher, 1987). This cover sequence consists of a basal arkosic unit, overlain by phyllite and muscovite schist, which in turn is overlain by a quartzite which forms the prominent cliffs of Pilot Mountain and Hanging Rock Bluff (Fig 9A; Bryant and Reed, 1961; Walker et al., 1989; Walker, 1990). While the gross stratigraphic succession resembles that of the Chilhowee Group of the western Blue Ridge, the deformation and metamorphism in the area have hindered attempts to correlate these two successions. Comparison of depositional processes and stratigraphic thicknesses of similar lithologies observed at Pilot Mountain and near Valley Forge, Tennessee, do not appear to be consistent with the interpretation that the quartzite of Pilot Mountain represents some eastern equivalent of the Chilhowee Group of east Tennessee. The observed lithostratigraphic similarity between these two sequences may be a manifestation of similarities of source rock and depositional setting. Whereas chronostratigraphic equivalence may not be applicable, this type of similarity would be consistent with interpretation of the quartzite at Pilot Mountain as representing deposition along an offshore, rifted microcontinent or similar terrane, as proposed by Thomas (1977).

Because the entire Pilot Mountain sedimentary sequence rests on Grenville basement (Hatcher, 1984; 1987; Hatcher et al., 1988; McConnell et al., 1986), a North American affinity appears certain. Palinspastic reconstruction of the southern Appalachian orogen indicates that the sedimentary sequences exposed within the Sauratown Mountain window, the Grandfather Mountain window, and the Unaka belt occupy the same relative positions (with respect to the Laurentian continental margin) today as they did when they were first deposited. The quartzites exposed within the Sauratown Mountains window probably represent either Late Proterozoic or Early Cambrian deposition along a sea-floor high associated with an isolated basement terrane during an early marine incursion into the late rift or early drift phase of the Iapetus basin (Walker et al., 1989).

Talladega Belt

The Kahatchee Mountain Group (Tull, 1982), approximately the same age as the uppermost Ocoee Supergroup and Chilhowee Group, occurs within the Talladega belt of Alabama and Georgia (Guthrie, 1985; 1989). Five formations compose the Kahatchee Mountain Group (in ascending order): 1) the Waxahatchee Slate, a dark-gray to dark-green, micaceous, feldspathic, very-fined sandstone or siltstone with minor calcareous slate and conglomerate; 2) the Sawyer Limestone, a sequence of interbedded limestone and dolostone which has algal stromatilites and becomes more arenaceous near the top (Guthrie, pers. comm., 1990); 3) the Brewer Formation, a lithologically diverse sequence of maroon phyllite and micaceous, calcareous, arkosic sandstone and conglomerate; 4) the Stumps Creek Formation, a gray-green, micaceous, feldspathic, arenaceous siltstone and fine-grained sandstone; and 5) the Wash Creek Slate, interbedded, laminated siltstone and quartz sandstone with lenticular and flaser bedding, wave and current ripples, and other primary sedimentary features (Guthrie, 1985). The coarser sandstones of the upper portion of the Wash Creek Formation have been described as medium- to thick-bedded quartz and feldspathic sandstone (Guthrie, 1989). The Wash Creek Formation is overlain by carbonate of the Jumbo Dolomite (Tull, 1982). The gross lithologic similarities between these rocks and the Chilhowee Group prompted Guthrie (1985) to propose that the Brewer, Stumps Creek, Wash Creek Formations (of the Kahatchee Mountain Group) and overlying Jumbo Dolomite were correlative with the Unicoi, Hampton, Erwin Formations and Shady Dolomite. This correlation was corroborated by the discovery of Lower to Middle Cambrian archeocyathids in the Jumbo Dolomite of Alabama (Tull et al., 1988).

EVOLUTION OF THE NORTH AMERICAN–IAPETUS MARGIN

As with all major ocean basins, the initial Iapetus formation was marked by a major rifting event which occurred between 690 and 570 Ma (Odom and Fullagar, 1984). Consideration of the stratigraphic and geochronological elements yields the following tentative history for the evolution of the North American–Iapetus continental margin in the southern Appalachians.

1) Grenville Orogeny (1.1 Ga)

2) Initiation of thermal perturbation and injection of Crossnore Plutonic Series (690–710 Ma; Odom and Fullagar, 1984) and possible coeval development of a system of asymmetric, alternately facing half-grabens. During this event, major attenuation of the 1.1 Ga Grenvillian basement occurred, resulting in the formation of discontinuous rift basins and associated volcanic centers, isolated basement blocks, and an irregular continental margin (Hatcher, 1972; 1987; Rankin, 1975; 1976; Thomas, 1977; 1983; Walker et al., 1989). Continued extension culminated with the initiation of oceanic crust formation, marking the inception of the Iapetus Ocean and the formation of twin, opposing, passive continental margins and associated micro-continents (Thomas, 1977; 1983). The North American representatives of this basin system would then include the Ocoee Supergroup (Snowbird and Great Smoky Groups) and the middle and lower Mount Rogers, Grandfather Mountain, Swift Run, and Mechum River Formations.

3) Cessation of thermal activity resulted in a transition from active volcanism and crustal extension to a brief period of tectonic quiescence. This stage is recorded by the fine-grained siliciclastic and carbonate lithologies of the Walden Creek Group (Tennessee embayment) and the Sawyer Limestone (Talladega belt), and the glaciogenic sediments of the upper Mount Rogers Formation (Virginia promontory).

4) Renewed thermal activity results in renewed volcanic activity (recorded by basalts of the lower Unicoi and Catoctin Formations) and associated uplift recorded by alluvial-fan and fluvial

strata of the Cochran and Unicoi Formations.

5) Cessation of rifting in the Tennessee and Virginia area and sea-level rise result in a change from fluvial to marine sedimentation. As a result, upper Cochran/Unicoi sediments were probably deposited on attenuated continental crust along a thermally subsiding continental margin. Active extension then shifted west to the Mississippi Valley–Rough Creek–Rome and Birmingham rift system, possibly as a result of a spreading-center shift from the Blue Ridge–Pine Mountain rift to the Ouachita rift along the Alabama-Oklahoma transform fault (Thomas, in press). Subsidence, most likely attributable to thermal processes, and continuing Early Cambrian sea-level rise produced an overall transgressive sequence, represented by the upper two-thirds of the Chilhowee Group (Hampton-Erwin equivalents). Interplay between subsidence and eustacy results in periodic progradation of shoreline sediment and the burial of basal Chilhowee "paleohighs." Paleocurrent analyses indicate sediment derivation from westward, principally cratonic sources and a transition from a sedimentation pattern dominated by point sources associated with topographic irregularities possibly inherited from rifting (Cochran and Unicoi formations), to a line-source pattern (easterly flow) characteristic of more mature passive-margin sedimentation (Erwin and equivalent formations; Schwab, 1970; 1971; 1972; Whisonant, 1974; Skelly, 1987; Simpson and Eriksson, 1989; 1990; Walker, 1990).

6) Stabilization of the margin and consequent reduction in terrigenous influx results in the development of a widespread carbonate shelf system represented by the Shady Dolomite and equivalent strata.

SUMMARY

Basal Chilhowee formations were deposited on attenuated continental crust along a thermally subsiding continental margin (Cochran Formation of the Tennessee embayment) or in response to active crustal extension (Unicoi Formation of the Virginia promontory); in contrast, upper units (e.g., Erwin and equivalent formations) were deposited on a thermally subsiding continental margin (Fichter and Diecchio, 1986; Simpson and Eriksson, 1989; 1990; Walker, 1990).

Although the state of our understanding of the stratigraphic evolution of upper Proterozoic and Lower Cambrian strata of the Blue Ridge of the southern Appalachians has greatly expanded in recent years, many areas remain problematic. The salient points mentioned previously, when considered in the new light shed by recent studies of modern tectonic analogues, provide improved understanding of the evolution of the Iapetus Margin as recorded by the Chilhowee Group in the southern Appalachians. Dominate among these is the concept that factors inherited from the "rift" phase greatly influence "drift" depositional and stratigraphic patterns, including: 1) along strike changes in basement configuration; 2) point source vs. line source paleocurrent patterns; 3) variation in stratigraphic thicknesses (possibly due to variation in subsidence related to regional differences in the timing and degree of crustal attenuation); 4) regional distribution of depositional environments; and 5) regional variation in the framework grain mineralogy of basal Chilhowee Group sequences.

STOP DESCRIPTIONS

DAY 1

Day 1 stops will highlight the Unicoi Formation. In this area the Unicoi Formation: 1) disconformably overlies the Sandsuck Formation of a stratigraphically thinner Walden Creek Group

(Hot Springs window, North Carolina; Oriel, 1950); 2) nonconformably overlies Grenvillian basement (northeasternmost Tennessee and southwestern Virginia; King and Ferguson, 1970; Simpson and Eriksson; 1989); and 3) disconformably overlies the Upper Proterozoic Mount Rogers Formation in south-central Virginia (Rankin, 1967; Thomas, 1977; Simpson and Eriksson, 1989). Stops today are designed to illustrate different aspects of basal Chilhowee Group sedimentology and stratigraphy. Stops will emphasize: 1) facies recording the fluvial-to-marine transition; 2) along-strike variations in the nature of "basement" upon which basal Chilhowee Group strata were deposited; and 3) variations in regional thickness of Chilhowee across the Virginia promontory. These aspects are important to understand the development of the Laurentia-Iapetus margin. At the end of each field stop description are discussion questions that relate directly to the interpretations presented.

STOP 1-1 Unicoi Formation, Hot Springs, NC

Rocks of the Snowbird and Walden Creek Groups in the Ocoee Supergroup, of the Chilhowee Group, and of the Shady Dolomite are exposed within the Hot Springs window (Figure 10). Framed by two thrust faults, which place Middle Proterozoic crystalline rocks over Ocoee Supergroup (along southern margin) and Ocoee Supergroup on the Chilhowee Group and Shady Dolomite (northern margin), this window was first mapped in detail by Oriel (1950). This exposure is unique in several ways; the most significant is the occurrence of basal Unicoi Formation above the Sandsuck Formation. This locality represents the transition from the Chilhowee Group to the southwest, where Cochran Formation overlies the Ocoee Supergroup (Chilhowee Mountain), and to the northeast, where the Unicoi Formation rests on Grenville-age basement or Mount Rogers Group (Doe River Gorge and southern Virginia). As shown in Figure 1, the Chilhowee Group of the Tennessee embayment possesses a different internal stratigraphic nomenclature reflecting, in part, a more distal depositional setting. The occurrence of the Unicoi Formation above the thinner Walden Creek Group is inferred to represent a more proximal position with respect to the Virginia promontory. In the Hot Springs window the Unicoi Formation is a 430 m thick sequence, and is coarser and more lithic rich than its southern counterpart (Cochran Formation) but not as thick or coarse-grained as the Unicoi Formation to the northeast. Basalt flows within the Unicoi Formation across much of the Virginia promontory are absent here.

Pebble to small cobble conglomerate characterizes the lower Unicoi Formation; conglomerate fines upward into pebbly, feldspathic sandstone with minor interbedded pebble to granule conglomerate. In the coarser units, cross-stratification is weakly developed, but the gross bedding geometry forms medium-scale, trough cross-stratification. The formational contact may be in part structural, as evidenced by strong cleavage developed within the Sandsuck Formation and the apparent flattening of pebbles in the basal conglomerate of the Unicoi Formation. The basal conglomerate of the Unicoi contains clasts identical to the underlying Sandsuck Formation, supporting a stratigraphic contact with deformation localized at the boundary due to strength differences.

Approximately 220 m above the base, a marked transition to quartz sandstone takes place. Above this horizon, the Unicoi Formation is characterized by interbedded well-sorted, medium-grained, medium- to small-scale trough cross-stratified quartz sandstone and moderately sorted, coarse-grained to granule felspathic sandstone. The texturally and mineralogically immature sediments of the basal Unicoi are interpreted to represent deposition in distal portions of a coastal alluvial system. Trough-crossbedding resulted from the migration of bedforms, probably as transverse bars. The more mature quartz sandstone and feldspathic sandstone of the upper portion of the Unicoi Formation record nearshore deposition in a marine setting, with cross-stratification produced by migration of subtidal sand bars and tidal channels.

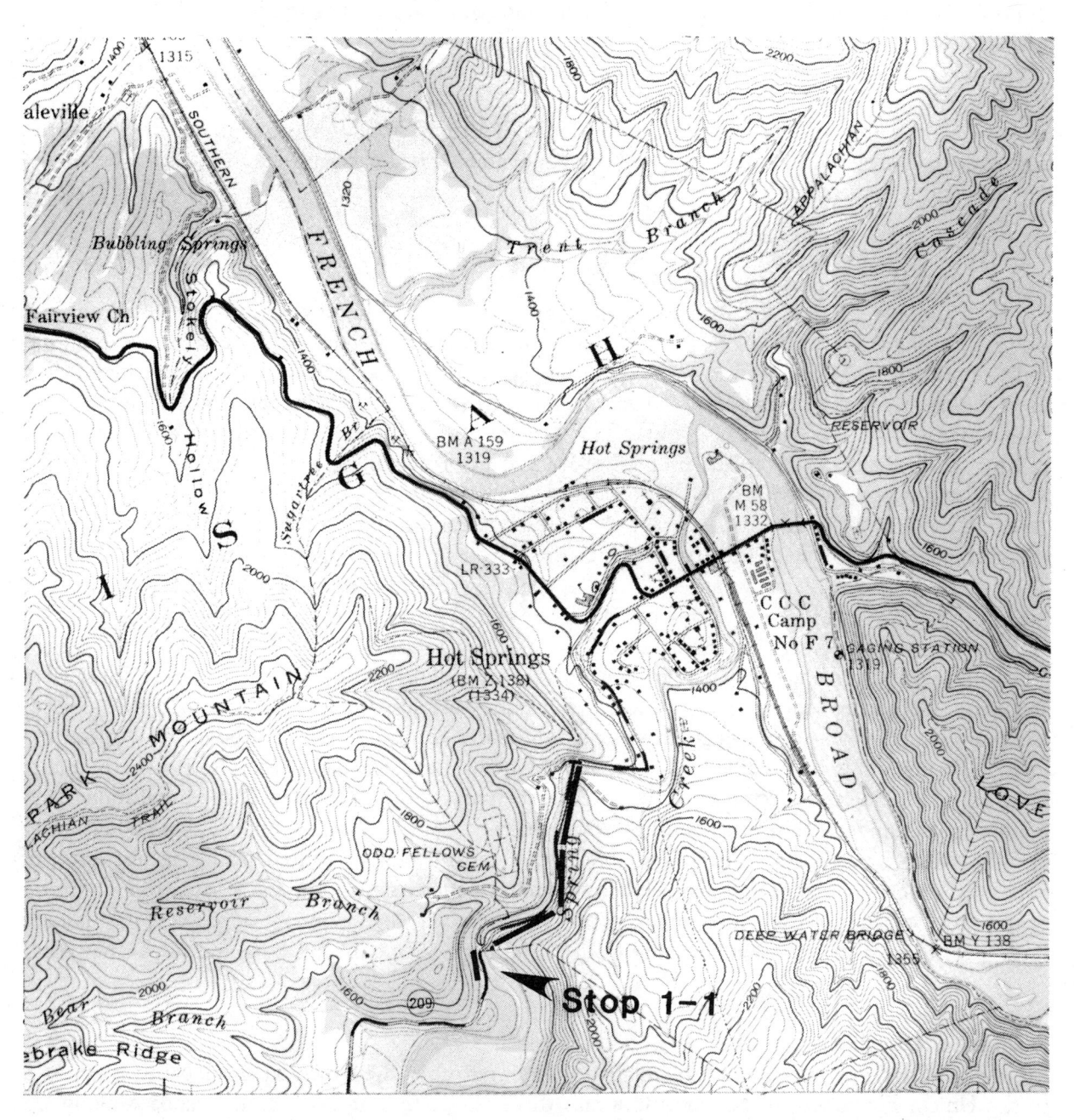

Figure 10. Location map of Stop 1-1. Portion of Hot Springs 7.5 minute quadrangle, NC. Scale is 1:24,000.

Discussion Questions

1) Is the contact between the Unicoi Formation and Sandsuck Formation structural or stratigraphic? If stratigraphic, does the occurrence of Sandsuck clasts in the basal conglomerate of the Unicoi indicate a major unconformity or does it represent the simple entrainment of semi-lithified Sandsuck "clasts" by high-velocity sediment laden unidirectional currents associated with a prograding fluvial system?

2) Does the marked contrast between the lower Unicoi and upper Unicoi represent a fluvial-to-marine transition?

3) Does this locality represent a position along the northeastern margin of the Tennessee Embayment as suggested by the stratigraphic relationships discussed between the Chilhowee Group and its various "basements"?

STOP 1-2 Unicoi Formation, Doe River Gorge, TN

The Unicoi Formation is exposed along an abandoned railroad grade within the Doe River Gorge, approximately 1.5 km southeast of Hampton, Tennessee (Figure 11). The Chilhowee Group and the underlying Middle Proterozoic crystalline basement are exposed with the Blue Ridge thrust sheet. Unlike many nearby exposures where the Unicoi Formation contains intercalated basalt flows, this sequence is devoid of volcanic rocks. This locality was originally measured and described by P. B. King, L. E. Smith, and C. A. Nelson in 1942, and published along with a regional geologic map by King and Ferguson (1970).

The Unicoi Formation measures 305 m and nonconformably overlies quartz monzonite and gneiss termed collectively as the "Basement rocks of Iron Mountain" by King and Ferguson (1970). The actual contact is deeply weathered but the schistosity within the crystalline rock terminates at the contact with the overlying pebbly conglomerate. Basal strata of the Unicoi Formation, as in the Hot Springs window, can be characterized as a poorly sorted, lithic-rich, medium-scale trough cross-bedded conglomerate which fines upward into medium- to coarse-grained, pebbly feldspathic and lithic sandstone. Approximately 220 m above the base a distinctive transition to well-sorted, medium-grained, small-scale trough cross-stratified quartz sandstone occurs. Above this horizon the Unicoi Formation is an interbedded sequence of quartz and feldspathic sandstones with quartz sandstone lithologies becoming more prevalent upsection. As before, the texturally and mineralogically immature sediments of the basal Unicoi are interpreted as representing deposition in distal portions of a coastal alluvial system. Trough cross-bedding resulted from the migration of bedforms, probably as transverse bars. The more mature quartz sandstone and feldspathic sandstone of the upper portion of the Unicoi are interpreted as representing nearshore deposition in a marine setting, with cross-stratification resulting from migration of subtidal sand bars and tidal channels.

Discussion Questions

1) What, if anything, does the relative thinner nature of the Unicoi at this locality as opposed to the Unicoi Formation of the Iron Mountain thrust sheet (305 m vs. 385 m), exposed along US 321 2.5 miles to north, indicate regarding the relative palinspastic position of these two localities? How does it relate to the overall geometry of the passive margin clastic "wedge" to which this sequence has been historically assigned?

2) Is there any significance to the regional, patchy distribution of the intercalated basalt flows common in this interval across much of northeasternmost Tennessee and southern Virginia?

3) Where is the top of the Unicoi at this locality?

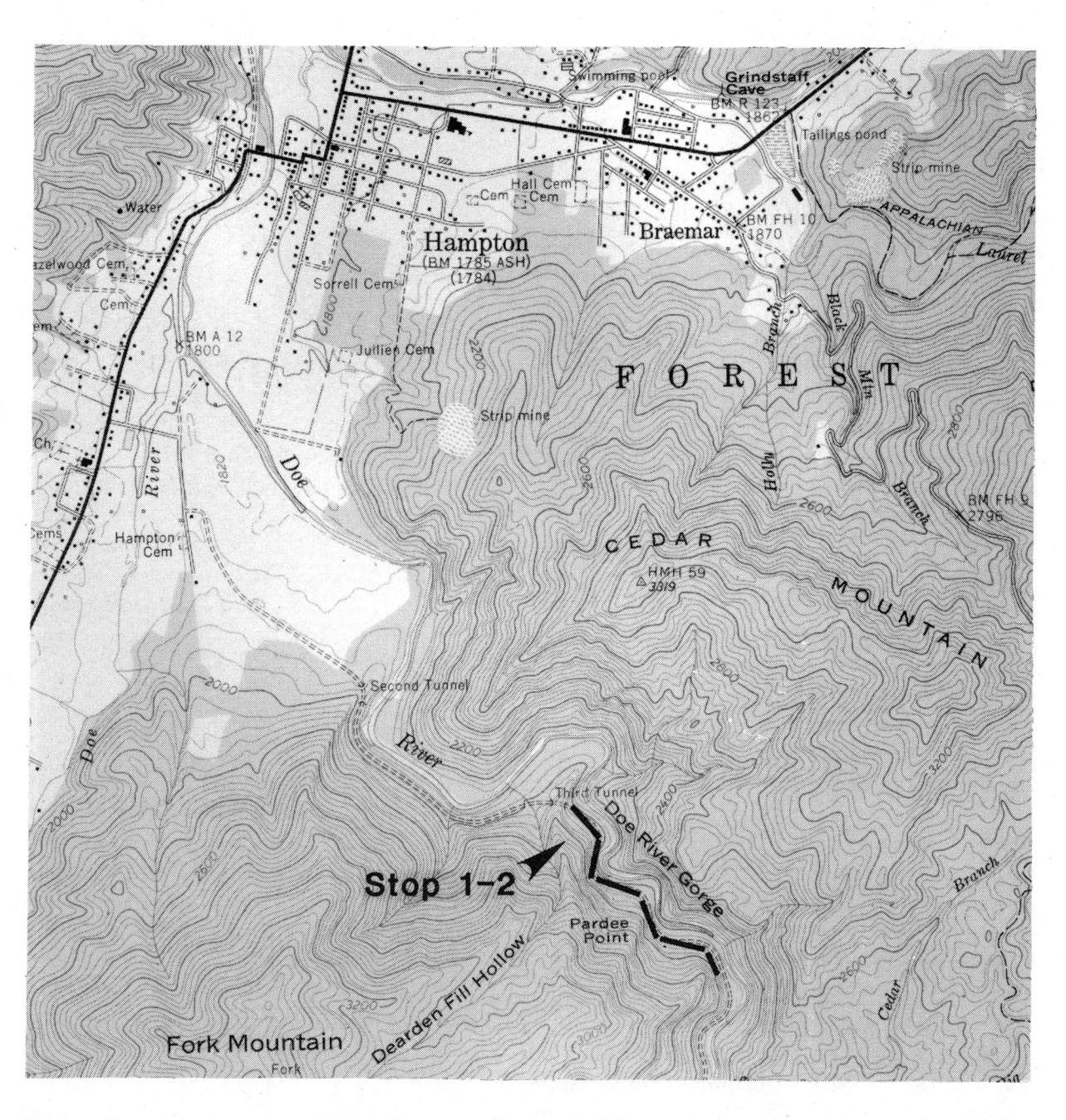

Figure 11. Location map of Stop 1-2. Portion of Elizabethton 7.5 minute quadrangle, TN. Scale is 1:24,000.

4) How does this section (as well as the section seen earlier at the Hot Springs window) compare to the Unicoi Formation exposed in Virginia?

DAY 2

These stops deal with the five different facies (representing outer shelf to tidal flat deposition) across the storm- and wave-dominated Chilhowee shelf. Sequence stratigraphic principles will be discussed at various stops. Stop 2 contrasts the Unicoi Formation in central Virginia with the Unicoi Formation in Tennessee.

Stop 2-1 Erwin Formation, Buchanan, VA

Two facies, distal inner and proximal inner shelf, are recognized at this locality (Figure 12). These facies are stacked into two parasequences. During this stop, we will concentrate on the uppermost parasequence.

Distal inner-shelf facies

This facies is typified by medium- to fine-grained sandstone, siltstone, and mudstone. Beds are sharp-based and vary in thickness from approximately 1 to 30 cm. Common sedimentary features include: 1) microhummocky cross stratification in fine-grained sandstone (MHCS); 2) horizontal stratification graded from medium to fine sandstone (HS); 3) symmetrical wave ripples (SR); 4) massive, medium-grained sandstone (MS); 5) form-discordant ripples (FR); 6) current-ripple cross-laminations (CR); 7) parallel-laminated, fine sandstone, siltstone, and mudstone (PL); and 8) mudstone (M). These structures are arranged into beds with the following vertical transitions: 1) MHCS - PL; 2) HS - SR - PL - M; 3) HS - PL; 4) MS - HS - PL; 5) FR - PL; 6) CR - FR - M; and 7) SR - PL. Bioturbation occurs in each bed type. Planolites beverliensis, Cruziana sp., and Rusophycus sp. have been identified from the outcrop. The distal inner-shelf facies in the upper part of the section lacks mudstone interbeds. Lack of mudstone is common throughout the Chilhowee Group at the transition into the overlying Shady Dolomite or equivalents.

Microhummocky cross-stratification in this facies is comparable to storm beds described by Dott and Bourgeois (1982). Horizontally stratified beds with wave-rippled tops are similar to storm-produced beds described by Kreisa (1981), Aigner (1985), and Soegaard and Eriksson (1985). Horizontal stratification was produced by upper-flow regime, unidirectional currents. Wave-rippled tops are the product of reworking of the horizontally stratified sand. As wave base rises, current velocities are reduced and suspension settling of silts and muds takes place. Wave-generated structures within this facies are interpreted to be produced by storms; therefore, these distal inner-shelf sediments accumulated above storm-wave base and below fairweather-wave base (see Simpson and Eriksson, 1990, for additional details).

Proximal inner-shelf facies

The proximal inner-shelf facies consists of medium- to thick-bedded, fine- to coarse-grained sandstone with subordinate siltstone and mudstone. Sandstone beds range in thickness from 0.5 to 9.5 m and are massive (structureless). Beds display crude grading and rare horizontal stratification; beds are sharp based and capped by either hummocks or coarse-grained, straight-crested ripples (CGR) (Figure 13A, B). Hummocks are gradational with the underlying bed, whereas CGR are sharp based and underlain by fine sand. CGR wavelengths range from 42 to 87 cm and heights vary from 3.5 to 9.0 cm. Within this section, CGR occur as four sets separated by discontinuous mudstone

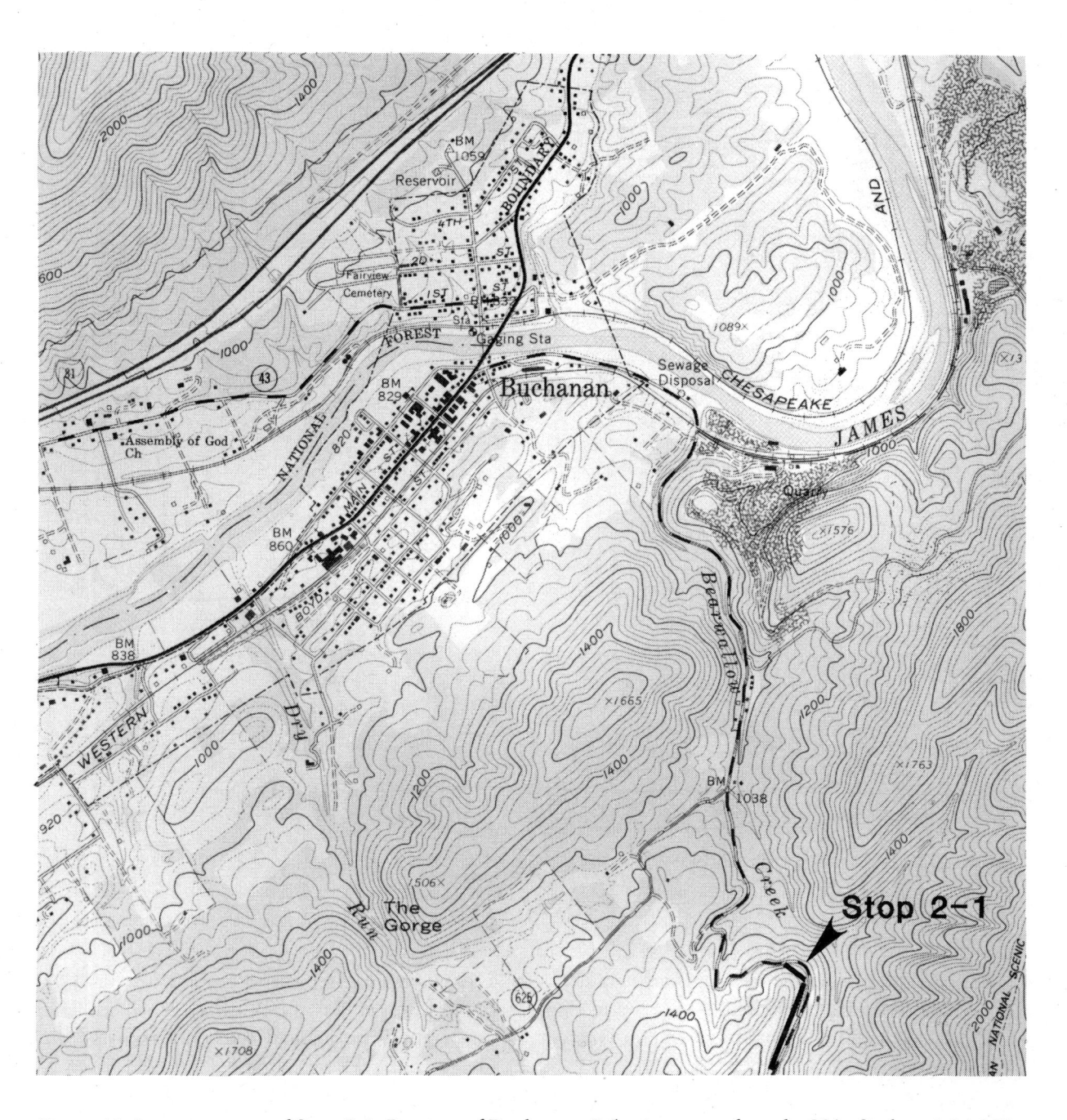

Figure 12. Location map of Stop 2-1. Portion of Buchanan 7.5 minute quadrangle, VA. Scale is 1:24,000.

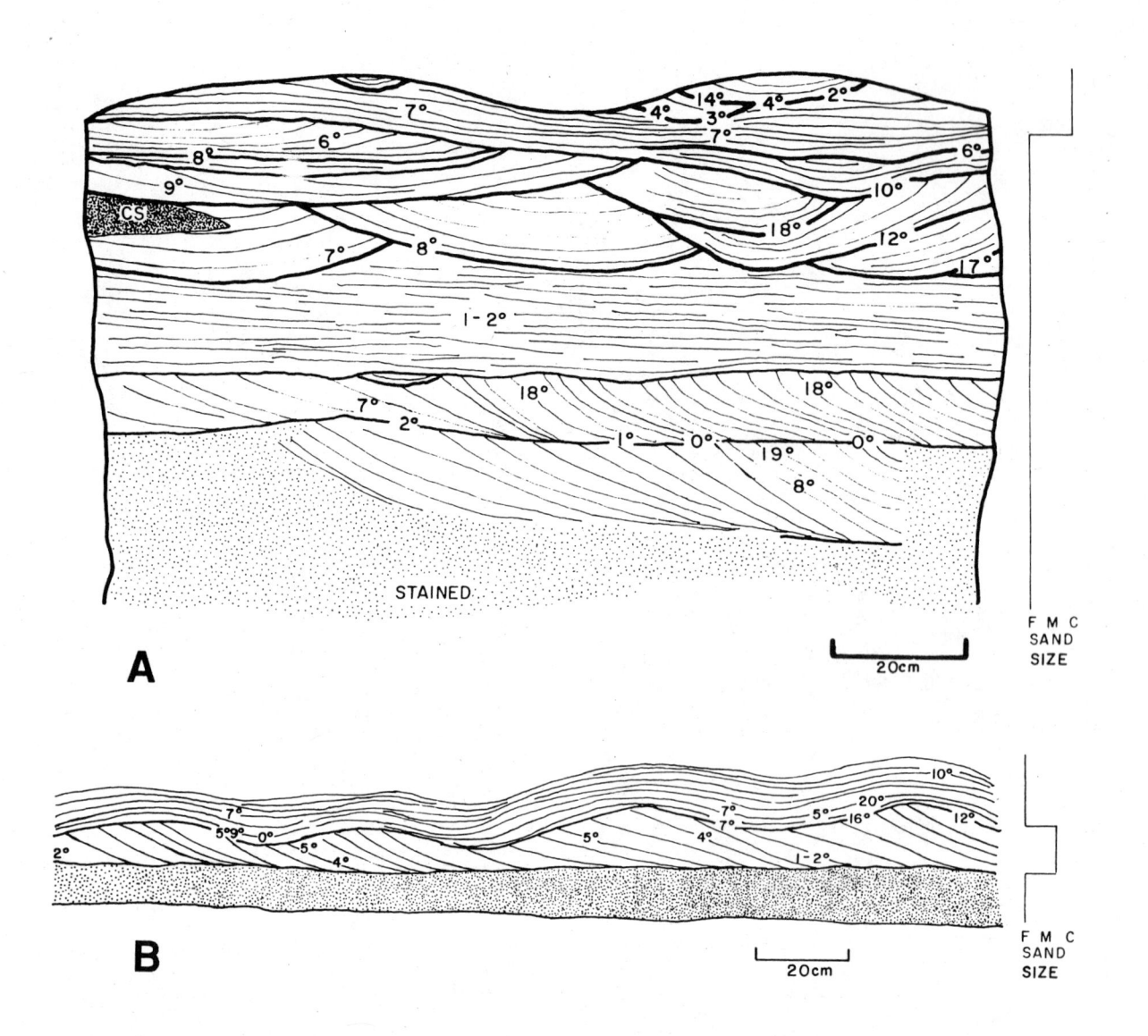

Figure 13. Internal structure of coarse-grained ripple deposits exposed at Stop 2-1. A, Trough cross-bedding reworked into a wave ripple cap. B, Internal structures of a combined flow, coarse-grained symmetrical ripple. Numbers on cross stratification and set boundaries are dip angles.

drapes. Paleocurrents from CGR indicate northwest-southeast oscillatory flow directions. Thinner beds within this facies contain hummocky cross stratification capped by a hummocky top.

Thick massive beds in this section are interpreted to be the result of rapid sedimentation from suspension with destruction of primary structures (Simpson, 1987). Massive to graded-horizontally stratified beds similar to beds in this section have been described from Miocene to Holocene deposits by Kumar and Sanders (1976); this bed type is attributed to storm events and produced by rapid suspension fallout coupled with tractional processes. CGR capping some massive beds are vortex ripples and can be used to reconstruct wave parameters (see Simpson and Eriksson, 1985; 1990). Calculated wave periods range from 2.4 to 13.0 seconds. Maximum calculated water depths are 50 to 66 m. Hunter et al. (1988) and Leckie (1988) have described similar wave ripples from modern shelves and the rock record. Leckie (1988) reports CGR are typically associated with transgressive surfaces. Hummocky beds in this facies are comparable to hummocky sequences described by Dott and Bourgeois (1982) and other workers and are caused by storms. The origin of hummocky cross stratification is problematic and is believed to be either generated by oscillatory flow or combined-flow currents. This facies records both fair-weather and storm processes and is therefore assigned to the proximal inner shelf setting.

Discussion Questions

1) Can the Chilhowee Group and associated strata be correlated more precisely via sequence stratigraphic methods than by conventional paleontological means? Can the Chilhowee Group be correlated into the Murphy and Talladaga belts and the eastern Blue Ridge?

2) Where are the sequence boundaries located in the stratigraphic section? What is the significance of the massive coarse-grained quartz sandstone?

3) Can trace fossils be used to provide a regional framework for the Chilhowee Group when body fossils are absent?

4) What is the significance of preserved bedforms on bedding plane tops?

STOP 2-2: Arcadia, VA

This stop will examine outer-shelf facies in the Hampton Formation and the problematic transition from the Unicoi into the Hampton Formation at Arcadia, Virginia (Figure 14). Shoreface facies is described in Stop 2-3.

Unicoi-Hampton Formation transition

Typically, the transition from the Unicoi Formation into the overlying Hampton Formation is marked by the presence of a shallow-marine unit. At this section a lower basalt flow is present within the basal part of the section. Overlying units are feldspathic conglomerate with interbedded mudstone. Mudstone of the Hampton Formation directly overlies the conglomeratic interval of the Unicoi Formation.

The feldspathic conglomerates are fluvial (?) and overlain by outer shelf deposits of the Hampton Formation. Two hypotheses can be invoked for this transition: 1) the transition is a structural contact and 2) the transition is "real," indicating fault activity into the Hampton time as suggested by Spencer for the area to the west of Stops 2-3 and 2-4 along the James River.

Outer-shelf facies

The outer-shelf facies consists of fine- to medium-grained sandstone, siltstone, and mudstone (Figure 15); subordinate coarse-grained sandstone is present. Beds vary in thickness from 1 to 25 cm;

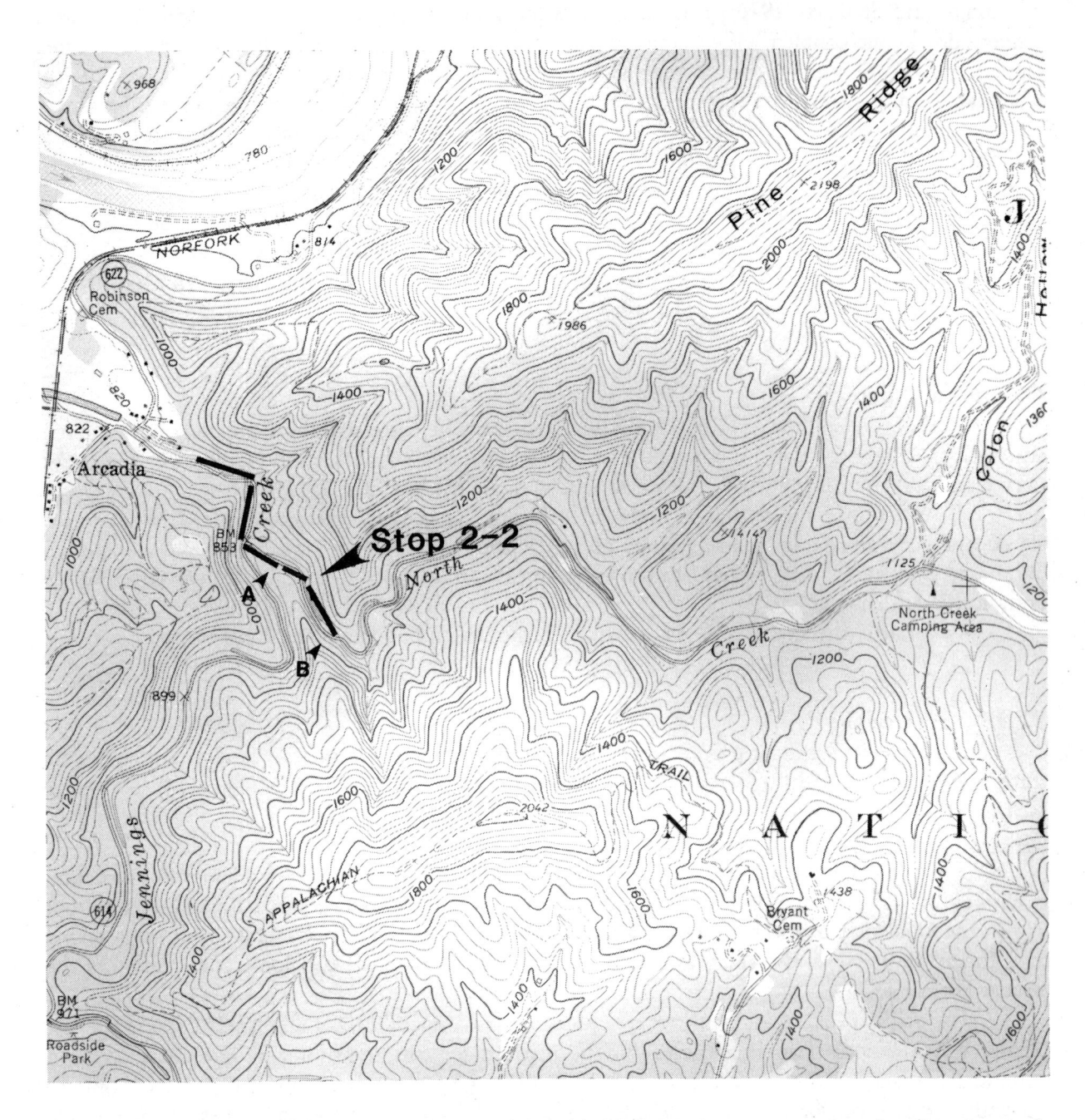

Figure 14. Location map of Stop 2-2. Portion of Arnolds Valley 7.5 minute quadrangle, VA. Arrow A points to the top of Hampton Formation outcrop. Arrow B identifies the Unicoi locality. Scale is 1:24,000.

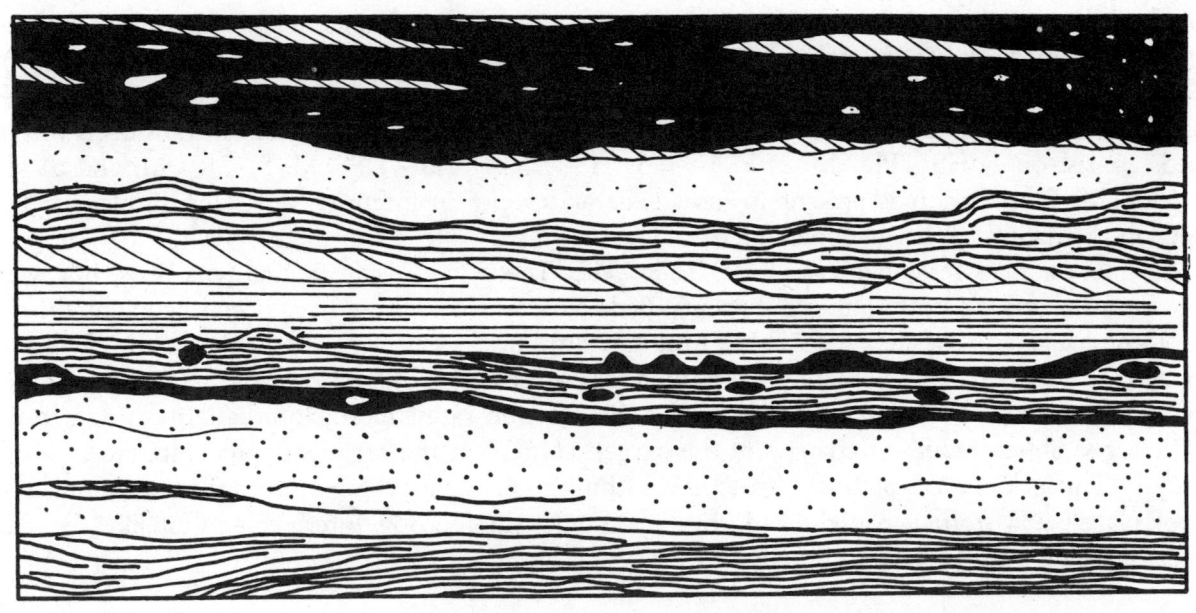

Figure 15. Field sketch of outer-shelf deposits exposed at Stop 2-2. A, Stratification styles within the outer-shelf sediments, circular structures are burrows. B, Stratification styles within outer-shelf deposits. Heavy diagonal lines indicate covered intervals.

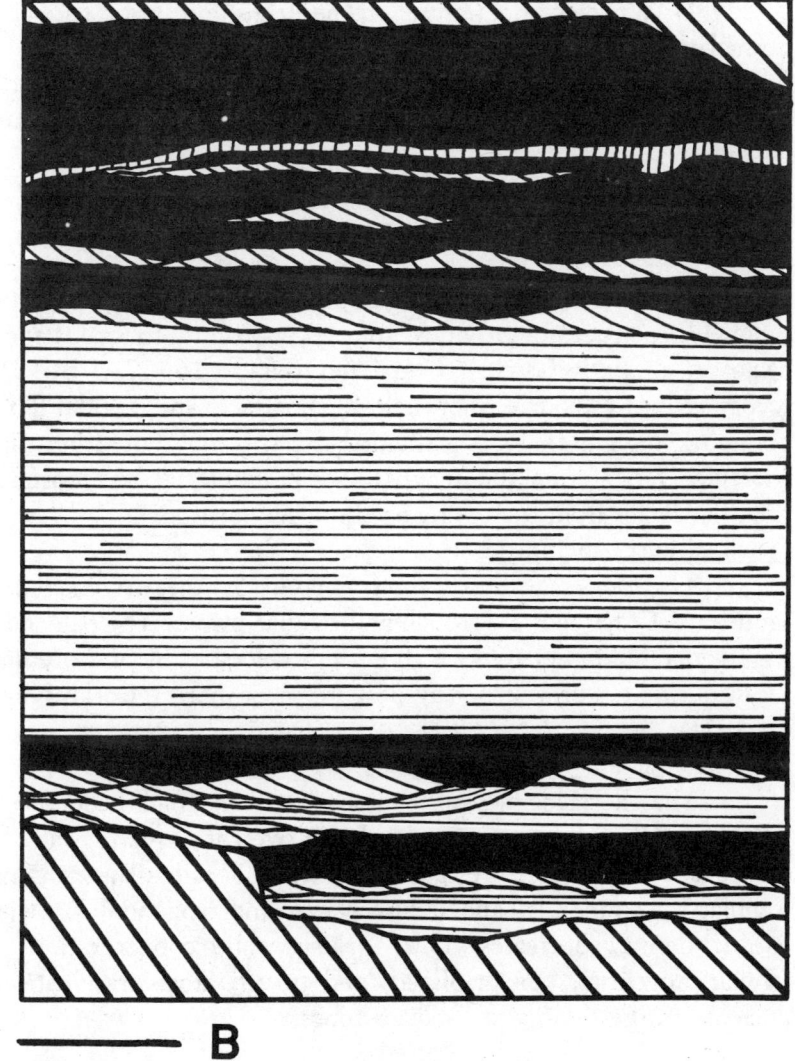

coarse-grained beds vary from 10 to 45 cm. Within coarse-grained beds, trough cross-bedding predominates, and beds are often capped by preserved bedforms. Sedimentary structures in fine- to medium-grained sandstones include: 1) horizontal stratification (HS); 2) massive-graded beds (MS); 3) current-ripple cross laminations (CR); 4) parallel lamination consisting of alternating fine-grained sandstone, siltstone, and mudstone (PL); and 5) massive mudstone (M). Vertical transitions within beds consist of: 1) HS - PL - M; 2) PL - M; 3) HS - M; 4) MS - HS - M; 5) CR - M; and 6) HS - CR - M. Bioturbation is present in most beds and varies in intensity vertically throughout the section. Planolites sp. is the dominant trace fossil.

These beds resemble Tae, Tabe, Tbde, Tbe, Tbcde, and Tcde Bouma-type beds, but were produced by waning storm-induced currents. Beds comparable to these have been documented from Holocene and ancient deposits. These beds are attributed to storm-generated geostrophic flows moving parallel to the shoreline (Simpson and Eriksson, 1990). Differences in the amount of bioturbation may be attributed to either low sedimentation rates (more bioturbation during periods of lesser sediment input) or changes in the oxygen content at the sediment water interface.

The lack of wave-produced structures within this facies indicates that sediment accumulation took place below storm-wave base and, thus, in an outer shelf setting. Interbedded shoreface deposits lack *Skolithos* tubes present in Stop 2-3.

Discussion Questions

1) What is the nature of the transition from the Unicoi Formation to the Hampton Formation? Is the contact stratigraphic or structural?

2) Are the thin-bedded sandstones, turbidites or storm deposits?

3) Why is the contact between the shoreface and outer shelf deposits of the Hampton abrupt?

STOP 2-3: Natural Bridge Quarry, VA

At Natural Bridge Quarry, superb bedding plane exposures of tidally generated bedforms and sedimentary structures provide a unique opportunity to reconstruct the process active during the tidal cycle on an Early Cambrian tidal flat (Figure 16, 17A, B). Within the quarry, three facies are recognized: 1) proximal inner shelf, 2) shoreface, and 3) tidal flat (Simpson, in press). The sequence bounding unconformity is exhumed on the main quarry surface and separates the underlying shoreface facies from the overlying tidal flat facies. A shoreface retreat unconformity separates tidal-flat facies from the overlying shoreface facies.

Proximal inner-shelf facies

The proximal inner shelf deposits consist of medium-bedded, medium-grained quartz sandstone. Cross beds are trough and tabular-planar. The paleocurrent mode is northerly. Cross beds were generated by relatively high current velocities by onshore-directed flow produced during initial setup during a storm or during post-storm onshore reworking by fair-weather waves (cf. Clifton et al., 1971).

Shoreface facies

The shoreface facies is composed of very thick-bedded, coarse-grained sandstone and forms the base of the quarry surface and the overlying quarry wall. *Skolithos linearis* is present in high densities. Primary structures are absent when *Skolithos linearis* density is high and grain size is coarse. Tabular set boundaries and diffuse planar and tangential foresets appear as grain size and *Skolithos linearis* density decreases. Low *Skolithos* tube concentration permits complete preservation of tabular-tangential, tabular-planar, and trough cross-stratification; the upper quarry wall shows a decrease in the density of *Skolithos* tubes.

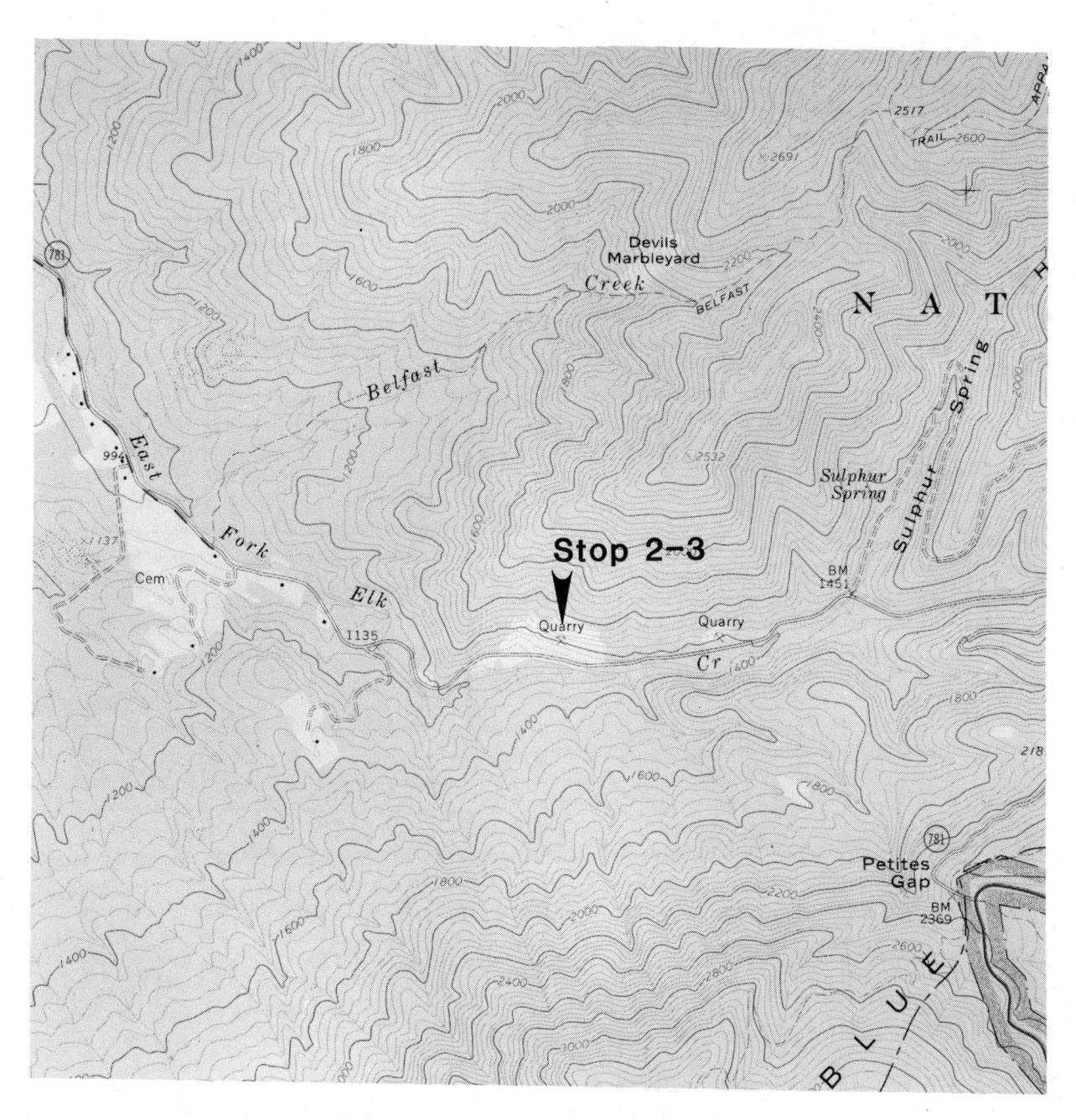

Figure 16. Location map of Stop 2-3. Portion of Snowden 7.5 minute quadrangle, VA. Scale is 1:24,000.

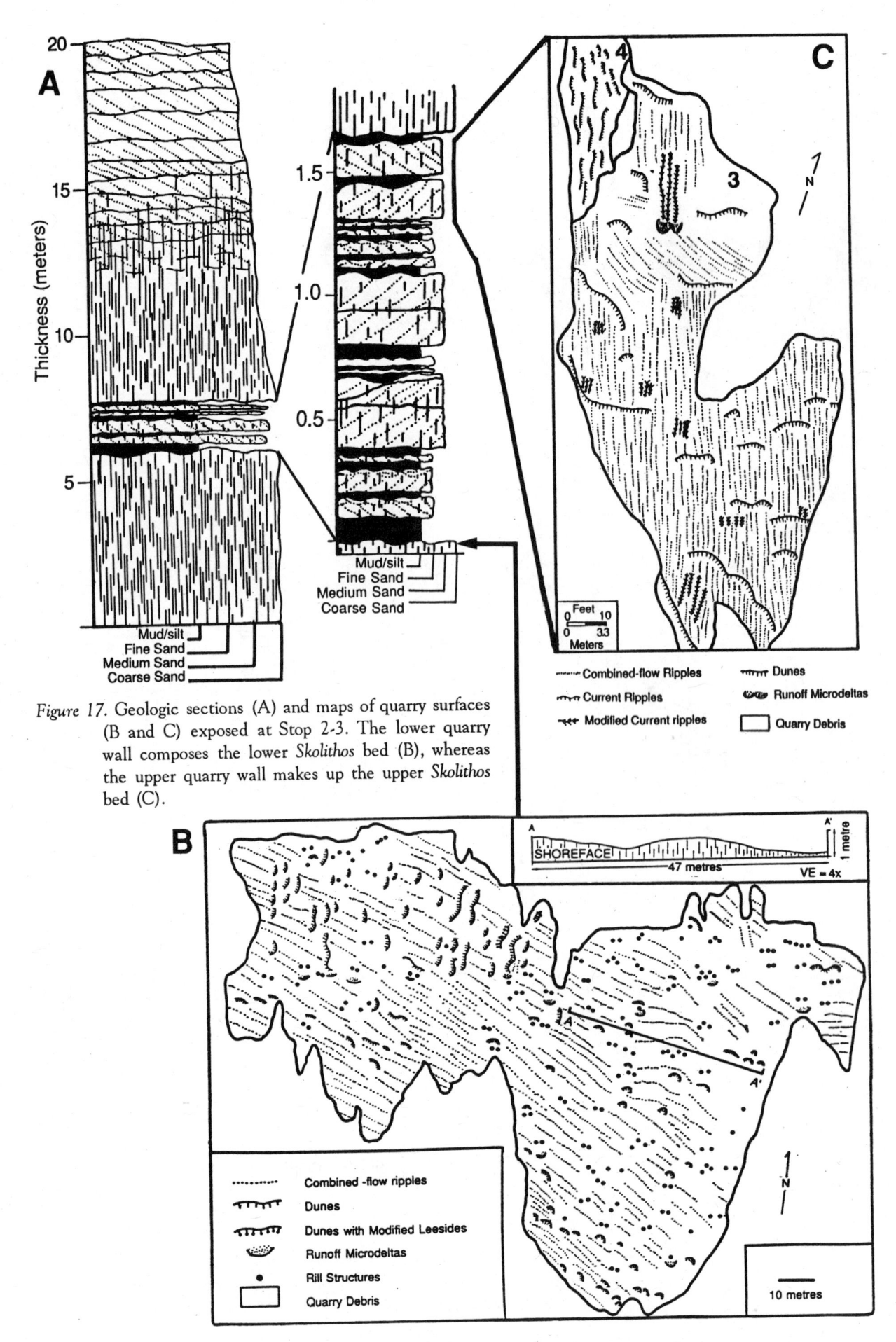

Figure 17. Geologic sections (A) and maps of quarry surfaces (B and C) exposed at Stop 2-3. The lower quarry wall composes the lower *Skolithos* bed (B), whereas the upper quarry wall makes up the upper *Skolithos* bed (C).

This facies was deposited in a high-energy shoreface setting influenced by shore-parallel currents of uncertain affinity, but probably longshore (see Simpson and Eriksson, 1990, for detailed discussion). *Skolithos linearis* is diagnostic of the *Skolithos* ichnofacies, which is best developed in a shallow-marine, subtidal to intertidal setting on a mobile substrate. Using the distribution of Diapatra cuprea from modern shoreline settings as an analog, Skoog et al. (in press) demonstrate that *Skolithos linearis* in this setting reflects the invasion of tubes from the overlying intertidal environment coupled with the lower limits of its biological distribution. The decrease in abundance of *Skolithos* tubes in the upper quarry wall records the development of a transgressive systems tract (Simpson and Eriksson, 1990).

Tidal-flat facies

The tidal-flat facies consists of interbedded thin- to medium-bedded quartz sandstone and mudstone. Quarrying operations exhumed three types of bedding planes: 1) thin-bedded sandstone covered with combined-flow, wave, and interference ripples; 2) medium-bedded sandstones topped by small to medium-scale dunes, ripples, modified ripples, and runoff microdeltas; and 3) the sequence-boundary surface capped by small- to medium-scale dunes, ripples, rills, runoff microdeltas, and preserved paleotopography. Bases of some thin- and some medium-bedded sandstones display Planolites burrows and dessication cracks. Preserved dunes with steepened leesides have abundant current ripples in their troughs. Ichnofauna has a low diversity and includes Monocraterion sp. and *Skolithos linearis*.

Reconstruction of the ebb through flood tidal phases is possible by integrating superimposed sedimentary structures with information from modern Holocence tidal flats (see Simpson, in press). The tidal cycle is subdivided in four phases:

1. Flood phase: Northwest-directed dunes on the main quarry surface where activated and record the maximum flood velocity; coeval with or after dune migration ceased, current ripples developed on the stoss sides of dunes.

2. Flood slack-water phase: Maximum flooding of the tidal flat occurred during this phase along with near-total reduction in tidally generated currents. Muds draped pause planes of some bedforms. In exposed areas, waves reworked sediment into oscillatory and combined-flow ripples. On the main quarry surface, waves that generated the combined-flow ripples were not of sufficient strength to rework dunes. Shoaling waves probably produced the observed northeast asymmetry of the combined-flow ripples.

3. Ebb phase: During the ebb phase, water level dropped and ebb channel became active. Ebb channel may have been separate from flood because of the scarcity of concurrent opposed sedimentary structures, with the exception of bed 4. Flood slack-water mud drapes were buried by sand transported during this phase, producing tidal bundles (Allen and Homewood, 1984).

4. Ebb slack-water phase: Shallow water runoff features and modification of existing bedforms occurred during this stage. Late ebb runoff was funneled along dune troughs modifying the leesides; sand transported along the troughs produced runoff microdeltas (cf. Dalrymple, 1984). Late-stage runoff produced rills that modified bedforms. Sandstone dikelets at the bases of beds were generated by either synaeresis or subaerial desiccation of mudstone beds. Incomplete cracking may be caused by limited exposure during the tidal cycle (cf. Allen, 1983).

Paleotidal range estimates are problematic (see Simpson, in press). The best estimate of tidal range can be gleaned from the preserved paleotopography which yields at minimum an upper microtidal to lower mesotidal range.

Discussion Questions

1) How can the tidal range be determined in this section? What method is best?

2) What is the significance of changing abundances in *Skolithos* tubes?

3) Why are the bedforms preserved, and do the exposed bedforms reflect more than one tidal cycle?

Stop 2-4 James River, VA (Optional)

This stop examines more accessible outcrops of the shoreface facies similar to those exposed in the Natural Bridge Quarry (Figure 18). Two possible outcrop localities can be examined: 1) at the confluence of the James and Maury Rivers; and 2) Route 501. Varying degrees of bioturbation by *Skolithos* can be observed at both localities. Cross-bed styles are generally trough to tabular tangential (Figure 19), when *Skolithos* tubes are low density.

ACKNOWLEDGMENTS

This field guide results from Ph.D. dissertations conducted at University of Tennessee (DW) and Virginia Tech (ES). We both thank Steve Driese and Ken Eriksson for help for various interpretations. The Appalachian Basin Industrial Associates funded both authors. DW thanks the Southeastern Section of the Geological Society of America and the Discretionary Fund of the Department of Geological Sciences, University of Tennessee. ES acknowledges the Geological Society of America, the Mobil Foundation, Sigma Xi, and the Kutztown Research Committee for funding. Jim Drahovzal, David Harris, Bruce Rowell, Terry Hamilton-Smith, and Wendy Simpson reviewed an early draft of the field guide.

REFERENCES

Aigner, T., 1985, Storm depositional systems: Dynamic stratigraphy in modern and ancient shallow-marine sequences: Lecture notes in Earth Science 3, New York, Springer Verlag, 173 p.

Allen, J. R. L., 1983, Sedimentary structures: their character and physical basis: Developments in Sedimentology no. 30, Elsevier, New York, 663 p.

Allen, P. A., and Homewood, P., 1984, Evolution and mechanics of a Miocene tidal sandwave: Sedimentology, v. 31, p. 63–81.

Barnaby, R. J., and Read, J. F., 1990, Evolution of the Lower to Middle Cambrian carbonate platform: Shady Dolomite, Virginia Appalachians: Geological Society of America, v. 102, p. 391–404.

Bentley, R. D., and Neathery, T. L., 1970, Geology of the Brevard fault zone and related rocks of the Inner Piedmont of Alabama: Alabama Geological Society Guidebook, no. 8, 119 p.

Blondeau, K. M., and Lowe, D. R., 1972, Upper Precambrian glacial deposits of the Mount Rogers Formation, central Appalachians, U.S.A.: 24th International Geological Congress, Montreal Proceedings, Series 1, p. 323–332.

Brown, W. R., 1970, Investigations of the sedimentary record in the Piedmont and Blue Ridge of Virginia, in Fisher, G. W., Pettijohn, F. J., Reed, J. C., Jr., and Weaver, K. N., eds., Studies in Appalachian Geology, Central and Southern: New York, John Wiley Interscience, p. 336–349.

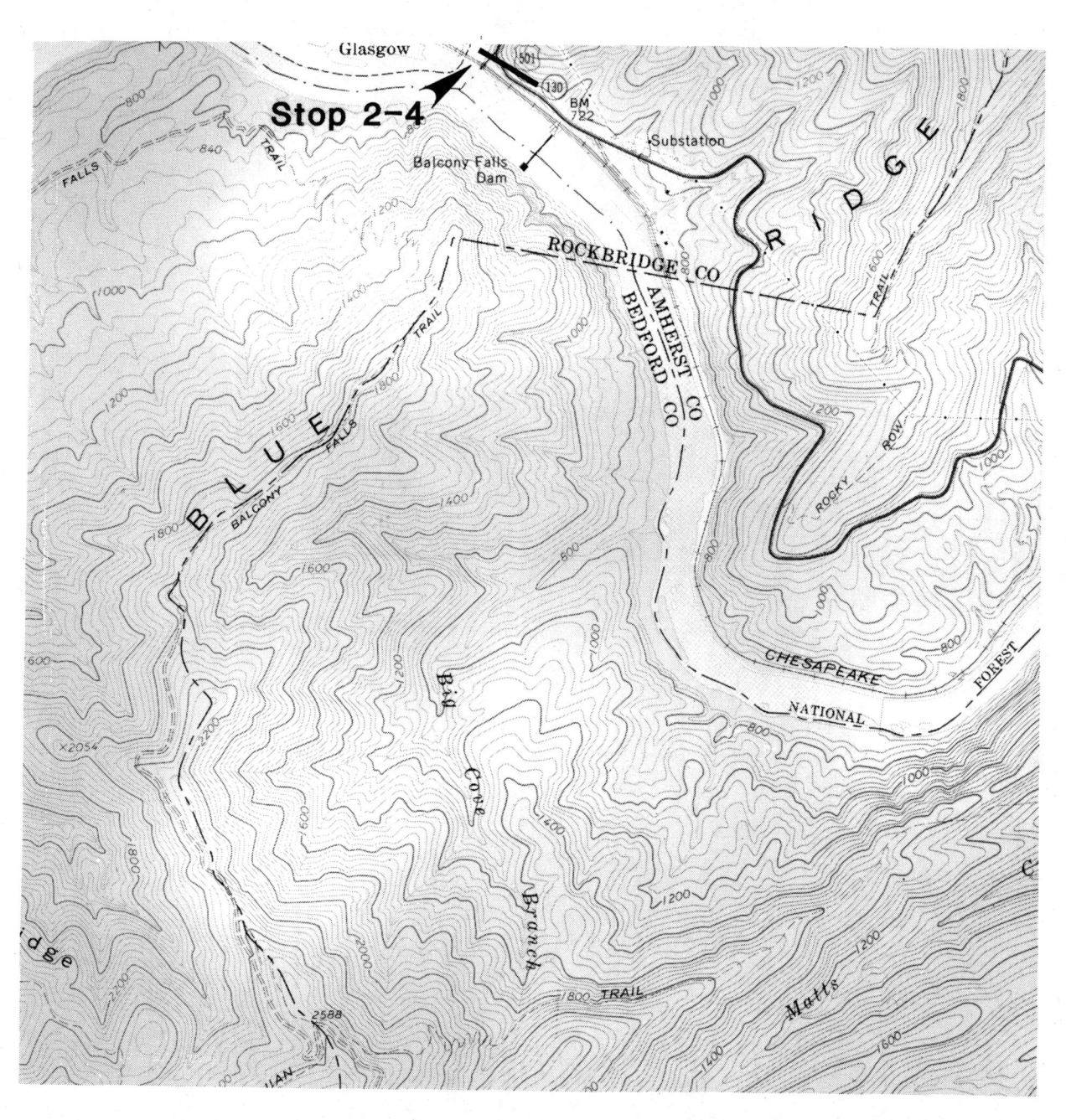

Figure 18. Location map of Stop 2-4. Portion of Snowden 7.5 minute quadrangle, VA. Scale is 1:24,000.

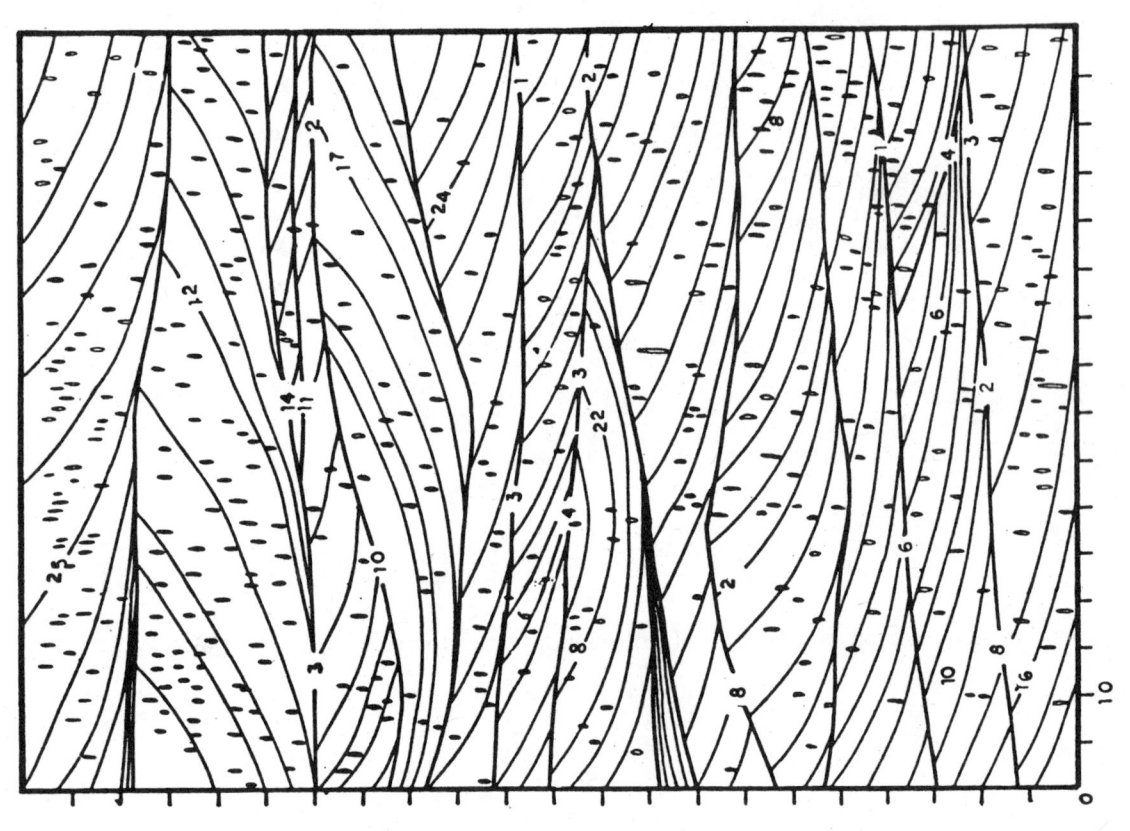

Figure 19. *Skolithos* tubes and tabular tangential cross-stratification exposed at Stop 2-4. Diagram is field sketch of stratification style. Numbers indicate dip of cross strata and set boundaries. Photograph shows *Skolithos* tubes and diffuse set boundaries.

Bryant, B., and Reed, J. C., Jr., 1960, Road log of the Grandfather Mountain area, North Carolina: Carolina Geological Society Field Trip Guidebook, 21 p.

——, and ——, 1961, The Stokes and Surry Counties quartzite area, North Carolina—A window?: U.S. Geological Survey Professional Paper 424-D, p. D61–D63.

Clarke, J. W., 1952, The geology and mineral resources of the Thomaston Quadrangle, Georgia: Georgia Geological Survey, Bulletin 59, 103 p.

Clifton, H. E., Hunter, R. E., and Phillips, R. L., 1971, Depositional structures and processes in the non-barred, high-energy nearshore: Journal Sedimentary Petrology, v. 41, p. 651–670.

Crimes, T.P, 1987, Trace fossils and correlation of late Precambrian and Early Cambrian strata: Geological Magazine, v. 124, p. 97–119.

Cudzil, M. R., 1985, Fluvial, tidal and storm sedimentation in the Chilhowee Group (Lower Cambrian), northeastern Tennessee: unpublished M.S. thesis, University of Tennessee, Knoxville, 164 p.

——, and Driese, S. G., 1987, Fluvial, tidal and storm sedimentation in the Chilhowee Group (Lower Cambrian), northeastern Tennessee: Sedimentology, v. 34, p. 861–883.

Dalrymple, R. W., 1984, Runoff microdeltas: A potential emergence indicator in cross-bedded sandstones: Journal of Sedimentary Petrology, v. 54, p. 825–830.

Dott, R. H., Jr., and Bourgeois, J., 1982, Hummocky stratification: Significance of its variable bedding sequence: Geological Society of America Bulletin, v. 93, p. 663–680.

Ferguson, H. W., and Jewell, W. B., 1951, Geology and barite deposits of the Del Rio district, Cocke County, Tennessee: Tennessee Division of Geology Bulletin 57, 235 p.

Fichter, L. S., and Diecchio, R. J., 1986, Stratigraphic model for timing the opening of the Proto-Atlantic Ocean in northern Virginia: Geology, v. 14, p. 307–309.

Guthrie, G. M., 1985, The Kahatchee Mountain Group and upper Precambrian–Lower Cambrian western margin evolution, in Tull, J. F., Bearce, D. N., and Guthrie, G. M., eds., Early evolution of the Appalachian miogeocline: Upper Precambrian–lower Paleozoic stratigraphy of the Talladega slate belt: Alabama Geological Society, 22nd Annual Field Trip Guidebook, p. 11–20.

——, 1989, Geology and marble resources of the Sylacauga Marble District: Geological Survey of Alabama Bulletin 131, 81 p.

Hadley, J. B., and Goldsmith, R., 1963, Geology of the eastern Great Smoky Mountains, North Carolina and Tennessee: U.S. Geological Survey Professional Paper 349-B, 118 p.

Harris, C. W., and Eriksson, K. A., 1990, Allogenic controls on the evolution of storm to tidal shelf sequences in the Early Proterozoic Uncompahgre Group, southwest Colorado, USA: Sedimentology, v. 37, p. 189–213.

Hatcher, R. D., Jr., 1972, Developmental model for the southern Appalachians: Geological Society of America Bulletin, v. 83, p. 2735–2760.

―――, 1978, Tectonics of the western Piedmont and Blue Ridge, southern Appalachians: Review and speculation: American Journal of Science, v. 278, p. 276–304.

―――, 1984, Southern and central Appalachian basement massifs: Geological Society of America Memoir 194, p. 149–153.

―――, 1987, Tectonics of the southern and central Appalachian Internides: Annual Reviews in Earth and Planetary Sciences, v. 15, p. 337–362.

―――, 1988, Sauratown Mountains window—Problems and perspective, in Hatcher, R. D., Jr., ed., Structure of the Sauratown Mountains Window, North Carolina: Carolina Geological Society 1988 Meeting, North Carolina Geological Survey, Raleigh, p. 1–19.

―――, Hooper, R. J., Heyn, T., McConnel, K. I., and Costello, J. O., 1988, Geometric and time relationships of thrusts in the crystalline southern Appalachians, in Mitra, G., and Wojtal, S., eds., Geometry and mechanisms of Appalachian thrusting: Geological Society of America Special Paper 222, p. 185–196.

Heyn, T., 1984, Stratigraphic and structural relationships along the southwestern flank of the Sauratown Mountains anticlinorium: unpublished M.S. thesis, University of South Carolina, Columbia, 192 p.

Hooper, R. J., 1986, Geologic studies at the east end of the Pine Mountain Window and adjacent Piedmont, central Georgia: unpublished Ph.D. dissertation, University of South Carolina, Columbia, 284 p.

Hunter, R. E., Dingler, J. R., Anima, R. J., and Richmound, B. M., 1988, Coarse-grained-sediment bands on the inner shelf of southern Monterey Bay, California: Marine Geology, v. 80, p. 81–98.

Hurley, P. M., Cormier, R. F., Hower, J., Fairborn, H. W., and Pinson, W. H., Jr., 1960, Reliability of glauconite for age measurement by K-Ar and Rb-Sr methods: American Association of Petroleum Geologists Bulletin, v. 44, p. 1793–1808.

Jackson, M. J., Simpson, E. L., and Eriksson, K. A., 1990, Facies and sequence stratigraphic analysis in an intracratonic thermal-relaxation basin: The early Proterozoic, lower Quilalar Formation and Ballara Quartzite, Mount Isa Inlier, Australia: Sedimentology, v. 73.

Keith, A., 1895, Description of the Knoxville Sheet: U.S. Geological Survey, Geological Atlas, Folio 16.

―――, 1986, Description of the Loudon sheet (Tennessee): U.S. Geological Survey, Geological Atlas, Folio 25, 6p.

Keller, F. B., 1980, Late Precambrian stratigraphy, depositional history and structural chronology of part of the Tennessee Blue Ridge: unpublished Ph.D. dissertation, Yale University, New Haven, 353 p.

King, P. B., 1964, Geology of the central Great Smoky Mountains, Tennessee: U.S. Geological Survey Professional Paper 349-C, 148 p.

———, and Ferguson, H. W., 1970, Geology of northeasternmost Tennessee: U. S. Geological Survey Professional Paper 311, 136 p.

Knoll, A. H., and Keller, F. B., 1979, Late Precambrian microfossils from the Walden Creek Group, Ocoee Supergroup, Tennessee: Geological Society of America Abstracts with Programs, v. 11, p. 185.

Kreisa, R. D., 1981, Storm-generated sedimentary structures in subtidal marine facies with examples from the middle Ordovician of southwestern Virginia: Journal of Sedimentary Petrology, v. 51, p. 823–848.

Kumar, N., and Sanders, J. E., 1976, Characteristics of shoreface deposits: Modern and ancient examples: Journal of Sedimentary Petrology, v. 46, p. 145–162.

Landing, E., 1988, Lower Cambrian of eastern Massachusetts: Stratigraphy and small shelly fossils: Journal of Paleontology, v. 62, p. 661–695.

Laurence, R. A., and Palmer, A. R., 1963, Age of the Murray Shale and Hesse Quartzite on Chilhowee Mountain, Blount County, Tennessee: U.S. Geological Survey Professional Paper 475–C, p. C53–C54.

Leckie, D., 1988, Wave-formed, coarse-grained ripples and their relationship to HCS: Journal of Sedimentary Petrology, v. 58, p. 607–622.

Mack, G. H., 1980, Stratigraphy and depositional environments of the Chilhowee Group (Cambrian) in Georgia and Alabama: American Journal of Science, v. 280, p. 497–517.

McConnel, K. L., Hatcher, R. D., Jr., and Sinha, A. K., 1986, Rb/Sr geochronology and areal extent of Grenville basement in the Sauratown Mountains massif, western North Carolina: Geological Society of America Abstracts with Programs, v. 18, p. 253.

Misra, K. C., and Lawson, J. S., in press, Paleotectonic setting of massive sulfide deposits of the Ducktown mining district, Tennessee, U.S.A.: Proceedings IAGOD Symposium, Lulea, Sweden, August 18–22, 1986.

Mitchum, R. M., 1977, Seismic stratigraphy and global changes of sea level, Part 1: Glossary of terms used in seismic stratigraphy, in Payton, C. E., ed., Seismic Stratigraphy—Applications to Hydrocarbon Exploration: American Association of Petroleum Geologists Memoir 26, p. 205–212.

Odom, A. L., and Fullagar, P. D., 1984, Rb-Sr whole-rock and inherited zircon ages of the plutonic suite of the Crossnore Complex, Southern Appalachians, and their implications regarding the time of opening of the Iapetus Ocean: Geological Society of America Special Paper 194, p. 255–261.

Oriel, S. S., 1950, Geologic map Hot Springs area, Madison County, North Carolina: North Carolina Department of Conservation and Development, 1:24,000.

Posamentier, H. W., Jervey, M. T., and Vail, P. R., 1988, Eustatic controls on clastic deposition I—Conceptual framework, in Wilgus et al., eds., Sea-level changes: An integrated approach: Society of Economic Paleontologists and Mineralogists Special Publication 42, p. 109–124.

Rankin, D. W., 1967, Guide to the geology of the Mount Rogers area, Virginia, North Carolina and Tennessee: Carolina Geological Society Field Trip Guidebook, 48 p.

——, 1968, Volcanism related to tectonism in the Piscataquis volcanic belt, an island of early Devonian age in north-central Maine, in Zen, E-An, White, W. S., Hadley, J. B., and Thompson, J. B., eds., Studies in Appalachian Geology, Northern and Maritime: John Wiley Interscience, New York, p. 355–369.

——, 1970, Stratigraphy and structure of Precambrian rocks in northwestern North Carolina, in Fischer, G. W., Pettijohn, F. J., Reed, J. C., Jr., and Weaver, K. N., eds., Studies in Appalachian Geology: Central and Southern: John Wiley Interscience, New York, p. 227–245.

——, 1975, The continental margin of eastern North America in the southern Appalachians: The opening and closing of the proto-Atlantic Ocean: American Journal of Science, v. 275–A, p. 298–336.

——, 1976, Appalachian salients and recesses: Late Precambrian continental breakup and the opening of the Iapetus Ocean: Journal of Geophysical Research, v. 81, p. 5605–5619.

——, Stern, T. W., Reed, J. C., Jr., and Newell, M. F., 1969, Zircon ages of felsic volcanic rocks in the upper Precambrian of the Blue Ridge, Appalachian Mountains: Science, v. 166, p. 741–744.

Rast, N., and Kohles, K. M., 1986, The origin of the Ocoee Supergroup: American Journal of Science, V. 286, p. 593-616.

Safford, J. M., 1856, A geological reconnaissance of the state of Tennessee: Tennessee Geological Survey, 1st Biennial Report, 164 p.

——, 1869, Geology of Tennessee: Nashville, TN, 550 p.

Schamel, S., and Baker, D. T., 1980, Remobilized Grenville basement in the Pine Mountain window, in Wones, D. R., ed., The Caledonides in the U.S.A.: Virginia Polytechnical Institute and State University, Department of Geological Sciences, p. 131–316.

Schwab, F. L., 1970, Origin of the Antietam Formation (Late Precambrian–Lower Cambrian), central Virginia: Journal of Sedimentary Petrology, v. 40, p. 354–366.

——, 1971, Harpers Formation, central Virginia: A sedimentary model: Journal of Sedimentary Petrology, v. 41, p. 139–149.

——, 1972, The Chilhowee Group and the Late Precambrian–Early Paleozoic sedimentary framework in the central and southern Appalachians, in Lessing, P., Hayhurst, R. I., Barlow, J. A., and Woodfork, L. D., eds., Appalachian Structures: Origin, Evolution, and Possible Potential for New Exploration Frontiers, a Seminar: West Virginia University and West Virginia Geological and Economic Survey, p. 59–101.

——, 1976, Depositional environments, provenance, and tectonic framework: Upper part of the late Precambrian Mount Rogers Formation, Blue Ridge province, southwestern Virginia: Journal of Sedimentary Petrology, v. 46, p. 3–13.

——, 1977, Grandfather Mountain Formation: Depositional environment, provenance, and tectonic setting of late Precambrian alluvium in the Blue Ridge of North Carolina: Journal of Sedimentary Petrology, v. 47, p. 800–810.

——, 1986, Upper Precambrian–Lower Paleozoic clastic sequences, Blue Ridge and adjacent areas, Virginia and North Carolina: Initial rifting and continental margin development, Appalachian orogen, in Textoris, E. A., ed., SEPM guidebooks—southeastern United States third annual meeting: Society of Economic Paleontologists and Mineralogists, p. 1–42.

Sears, J. W., and Cook, R. B., Jr., 1984, An overview of the Grenville basement complex of the Pine Mountain window, Alabama and Georgia, in Bartholomew, M. J., ed., The Grenville event in the Appalachians and related topics: Geological Society of America Special Paper 194, p. 281–287.

Simpson, E. L., 1987, Sedimentology and tectonic implications of the late Proterozoic to Early Cambrian Chilhowee Group in southern and central Virginia: unpublished Ph.D. dissertation, Virginia Polytechnic Institute and State University, Blacksburg, 298 p.

——, in press, An exhumed, lower Cambrian tidal flat: The Antietam Formation, central Virginia, U.S.A., in Smith, D., ed., Clastic Tidal Sedimentation: Canadian Society of Petroleum Geologists Memoir No. 16.

——, and Eriksson, K. A., 1985, Paleoenvironmental constraints provided by paleowave reconstructions: Early Cambrian Chilhowee Group drift-stage sedimentation in the central Appalachians: Geological Society of America Abstracts with Programs, v. 17, p. 718.

——, and ——, 1989, Sedimentology of the Unicoi Formation in southern and central Virginia: Evidence for Late Proterozoic to Early Cambrian rift-to-passive margin transition: Geological Society of America Bulletin, v. 101, p. 42–54.

——, and ——, 1990, Early Cambrian progradational and transgressive sedimentation patterns in Virginia: An example of the early history of a passive margin: Journal of Sedimentary Petrology, v. 60, p. 84–100.

——, and Sundberg, F. A., 1987, Early Cambrian age for synrift deposits of the Chilhowee Group of southwestern Virginia: Geology, v. 15, p. 123–126.

Skelly, R., 1987, Depositional environments of the Chilhowee group, Bean Mountain, Tennessee: unpublished M.S. thesis, University of Tennessee, Knoxville, 208 p.

Skoog, S., Venn, C., and Simpson, E. L., in review, Three-dimensional distribution of Diopatera cuprea from a modern tidal flat: Implications for the distribution of *Skolithos* in Lower Cambrian rocks of Virginia: Geological Society of America Abstracts with Programs, v. 23.

Soegaard, K., and Eriksson, K. A., 1985, Evidence of tide, storm and wave interaction on a Precambrian siliciclastic shelf: The 1,700 M.Y. Ortega Group, New Mexico: Journal of Sedimentary Petrology, v. 85, p. 672–684.

Stose, G. W., and Stose, A. J., 1944, The Chilhowee group and Ocoee series of the southern Appalachians: American Journal of Science, v. 242, p. 367–390, 401–416.

——, and ——, 1949, Ocoee Series of the southern Appalachians: Geological Society of America Bulletin, v. 60, p. 267–320.

Thomas, W. A., 1977, Evolution of the Appalachian-Ouachita salients and reentrants from reentrants and promontories in the continental margin: American Journal of Science, v. 277, p. 1233–1278.

——, 1983, Continental margins, orogenic belts, and intracratonic structures: Geology, v. 11, p. 270–272.

Tull, J. F., 1982, Stratigraphic framework of the Talladega slate belt, Alabama Appalachians, in Bearce, D. N., Black, W. W., Kish, S. A., and Tull, J. F., eds., Tectonic studies in the Talladega and Carolina slate belts, southern Appalachian orogen: Geological Society of America Special Paper 191, p. 3–18.

——, Harris, A. G., Repetski, J. E., McKinney, F. K., Garrett, C. B., and Bearce, D. N., 1988, New paleontologic evidence constraining the age and paleotectonic setting of the Talladega slate belt: Geological Society of America Bulletin, v. 100, p. 1291–1299.

Unrug, R., and Unrug, S., 1990, Paleontological evidence of Paleozoic age for the Walden Creek Group, Ocoee Supergroup, Tennessee: Geology, v. 18, p. 1041–1045.

Van Wagoner, J. C., Posamentier, H. W., Mitchum, R. M., Vail, P. R., Sarg, J. F., Loutit, T. S., and Hardenbol, J., 1988, An overview of sequence stratigraphy and key definitions, in Wiligus, C. W., et al., eds., Sea level changes: an integrated approach: Society of Economic Paleontologists and Mineralogists Special Publication 42, p. 39–45.

——, Mitchum, R. M., Campion, K. M., and Rahmanian, V. D., 1990, Siliciclastic sequence stratigraphy in well logs, cores and outcrops: American Association of Petroleum Geologists, Methods in Exploration Series, No. 7, 55 p.

Walker, D., Cudzil, M. R., Skelly, R. L., and Driese, S. G., 1988, The Chilhowee Group of east Tennessee: Sedimentology of the Cambrian fluvial-to-marine transition, in Driese, S. G., and Walker, D., eds., Depositional History of Paleozoic Sequences: Southern Appalachians, Midcontinent Section of Society of Economic Paleontologists and Mineralogists 6th annual fieldtrip guidebook: University of Tennessee Studies in Geology No. 19, p. 52.

——, and Driese, S. G., in press, Constraints on the position of the Precambrian–Cambrian boundary in the southern Appalachians: American Journal of Science.

Walker, D., Driese, S. G., and Hatcher, R. D., Jr., 1989, Paleotectonic significance of the quartzite of the Sauratown Mountains window, North Carolina: Geology, v. 17, p. 913–917.

——, 1990, The sedimentology and stratigraphy of the Chilhowee Group (uppermost Proterozoic to Lower Cambrian) of eastern Tennessee and western North Carolina: The evolution of the Laurentian-Iapetus margin: unpublished Ph.D. dissertation, University of Tennessee, Knoxville, 274 p.

Wehr, F., and Glover, L., III, 1985, Stratigraphy and tectonics of the Virginia–North Carolina Blue Ridge: Evolution of a late Proterozoic–early Paleozoic hinge zone: Geological Society of America Bulletin, v. 96, p. 285–295.

Whisonant, R. C., 1970, Paleocurrents in basal Cambrian rocks in eastern Tennessee: Geological Society of America Bulletin, v. 81, p. 2781–2786.

——, 1974, Petrology of the Chilhowee Group (Cambrian and Cambrian(?)) in central eastern and southern Tennessee: Journal of Sedimentary Petrology, v. 44, p. 228–241.

Wood, G. D., and Clendening, J. A., 1982, Acritarchs from the Lower Cambrian Murray Shale, Chilhowee Group, of Tennessee, U.S.A.: Palynology, v. 6, p. 255–265.

6
TERTIARY LITHOLOGY AND PALEONTOLOGY, CHESAPEAKE BAY REGION

Lauck W. Ward
Virginia Museum of Natural History
Martinsville, VA 24112

David S. Powars
U.S. Geological Survey
Reston, VA 22092

INTRODUCTION

Tertiary beds exposed along the Chesapeake Bay and Potomac River southeast of Washington, D.C., have been studied since the early 1800s. Because of their relative proximity to that city, where both the U.S. Geological Survey and the Smithsonian Institution are housed, many geologists and paleontologists have worked on the stratigraphy and fossils. Most of the earliest efforts were taxonomic ones, but Rogers (for a summary see Rogers, 1884) described the Eocene deposits of Virginia. Later work by Darton (1891) drew from these earlier investigations and he proposed the primary lithic division of Coastal Plain sediments in Virginia and Maryland. The term "Pamunkey Formation" was proposed for the glauconitic units and the overlying shelly sands and marls were named the "Chesapeake Formation." Since then, a steady refinement of the units has led to the present stratigraphic framework.

GEOLOGIC SETTING

Stratigraphic units exposed in the Chesapeake Bay area consist of Mesozoic and Cenozoic Coastal Plain beds deposited in a tectonic downwarp known as the Salisbury embayment (Figure 1). The Salisbury embayment includes parts of Virginia, Maryland, Delaware, and southern New Jersey and is bordered on the north and south by the South New Jersey arch and the Norfolk arch, respectively. Subsurface data indicates that these arches are characterized by stratigraphic thinning or truncation of Cretaceous and Tertiary age formations. The basement complex underlying the embayment includes Precambrian and Paleozoic age crystalline rocks and Mesozoic age rift-basin fill. The Salisbury embayment was the site of intermittent marine overlap and deposition during the Early and Late Cretaceous and most of the Tertiary. Beds are of fluvial, deltaic, and open-shelf origin and were deposited in a wedge-like configuration with their thin, westward edge overlapping the Piedmont. To the east, the Coastal Plain deposits thicken to several thousand feet.

The lithology, thickness, and dip of the various formations deposited in the Salisbury embayment are, to a great extent, structurally controlled. This tectonism occurred at several local and regional scales. Tectonism on a regional scale involved tilting of the entire Atlantic continental margin. Of lesser importance was the independent structural movement of the various basins, or depocenters, and the intervening arches, or high areas. These high and low areas moved independently of each other, creating a stratigraphic mosaic that is unique from basin to arch. Various tectonic models included block-faulting and possible movement of the landward. Parts of oceanic transform faults have been suggested as causes for the arch-basin configuration. Variations in the distribution and thickness of Cretaceous and Tertiary deposits also suggest the gradual migra-

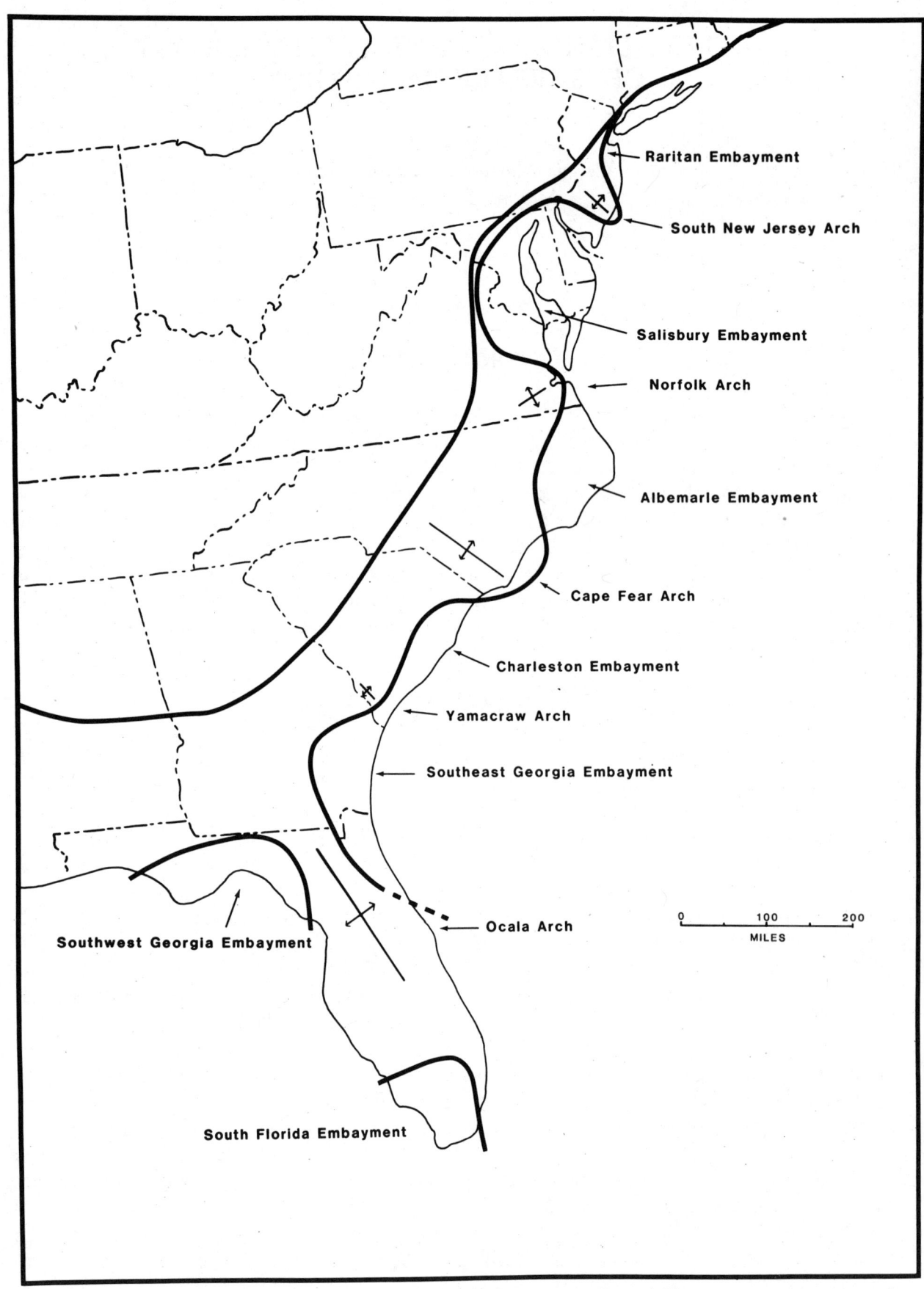

Figure 1. Map showing principal basins and arches of the Atlantic Coastal Plain.

tion of basins through time. Other structural deformation in the Salisbury embayment consists of localized, down-dropped grabens that occur along northeast-southwest trending lineaments. These grabens are related to early Mesozoic rifting and caused certain areas to be unstable. These areas were reactivated during the Cretaceous and Tertiary, possibly due to sediment loading. This resulted in structural highs behind which finer sediments accumulated. Thus, each of these various structural elements contributed to the overall depositional patterns on the Coastal Plain and in the Salisbury embayment.

Lower Tertiary deposits consist of glauconitic silty sands containing varying amounts of marine shells. The Tertiary beds are principally marine-shelf deposits. Fluvial, deltaic, and nearshore-shelf facies are generally lacking. The same is true for the upper Tertiary marine beds which consist of diatomaceous silts and silty and shelly sands. However, sands and gravels of fluvial and deltaic origin cap most of the higher interfluves in the Salisbury embayment area and are thought to be Miocene, Pliocene, and/or Pleistocene ages.

The Salisbury embayment had a warm-temperate to subtropical marine setting through much of its history. During the late Tertiary, a portion of the temperate molluscan fauna became endemic so that abrupt cooling in the late Pliocene caused a major local extinction involving taxa that had been successful since the Oligocene.

TERTIARY HISTORY OF THE SALISBURY EMBAYMENT

The Salisbury embayment and the entire Atlantic Coastal Plain have had a complex history. In contrast to "passive margin" descriptions, this was a structurally dynamic area whose sedimentary history clearly shows the effects of structural movement as well as of global sea-level events. To identify and eliminate local tectonic "noise" and detect actual global sea-level changes, one must compare the detailed stratigraphic records of several embayments. In Figure 2, the sea-level curves of three principal Atlantic Coastal Plain basins (Salisbury embayment, Albemarle embayment, and Charleston embayment) are summarized. A fourth curve for the Atlantic Coastal Plain combines the data obtained in the three basins and attempts to show the actual record of sea-level fluctuations. These curves are plotted against the cycles and super-cycles of Vail and Mitchum (1979). The curves are based on our interpretations of onshore outcrop and subsurface data. We have made no attempt to plot sea-level changes beyond the present coastline. The following trends occur and are based on the onlap relationships of formations in the three basins.

Paleocene

In the middle early Paleocene, there is agreement for a moderately strong marine pulse. This pulse is evidenced by the Brightseat Formation in the Salisbury embayment (Figures 3, 4A), the Jericho Run Member of the Beaufort Formation in the Albemarle embayment, and the Black Mingo Formation in the Charleston embayment. Another strong onlap sequence occurred during the late Paleocene and lasted almost that entire period. In the Salisbury Embayment, beds associated with the event are included in the Aquia Formation (Figure 4B). There are at least two recognizable sea-level pulses, represented by the Piscataway and Paspotansa Members, involved in that sequence. A final small transgression, probably only in the Salisbury embayment, resulted in the deposition of the Marlboro Clay (Figure 4C).

Eocene

During the early Eocene a moderately strong transgression occurred in the Salisbury embayment (Potapaco Member of the Nanjemoy Formation, Figure 5A). In the late early Eocene a second transgression occurred, which is reflected in the Salisbury embayment by the Woodstock Member of the Nanjemoy Formation (Figure 5B).

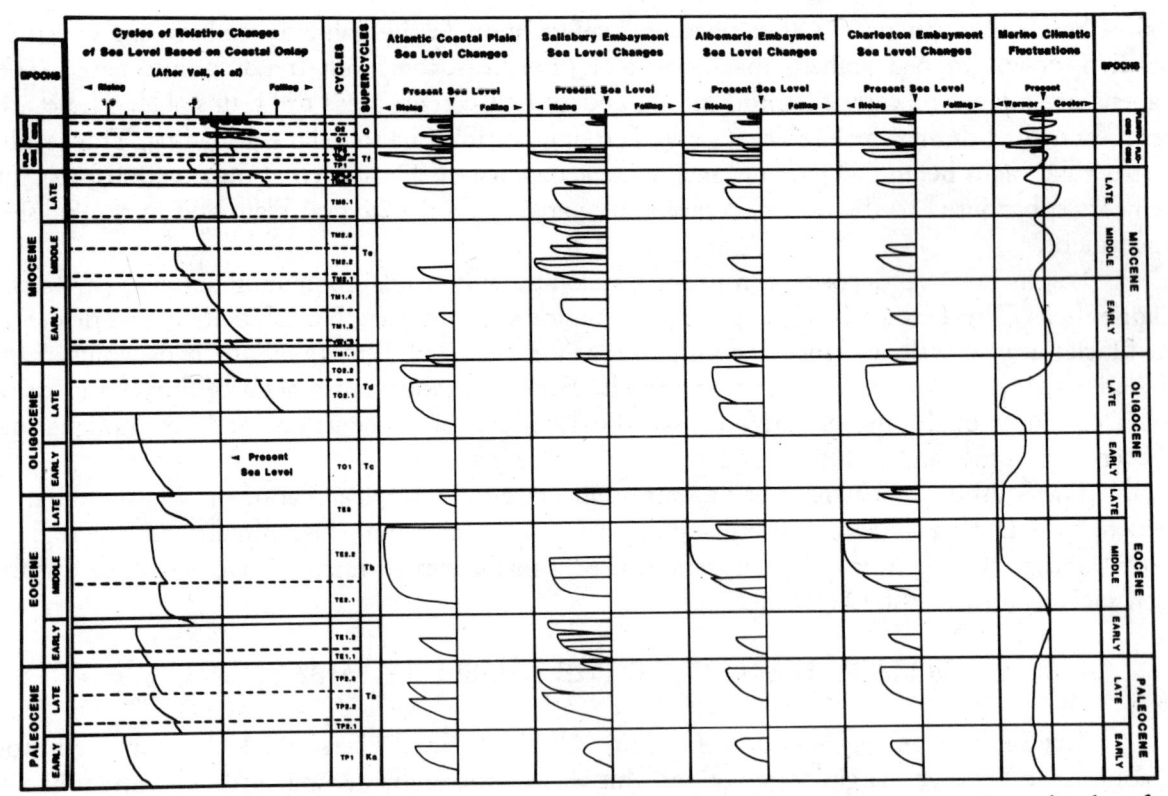

Figure 2. Onlap-offlap history of the Atlantic Coastal Plain, based on onshore outcrop and subsurface data. Sea-level fluctuations in the Salisbury, Albemarle, and Charleston embayments are plotted against a chart of cycles and supercycles by Vail and Mitchum (1979). Data from the basins are combined to approximate global sea-level events as seen along the Atlantic Coastal margin. The marine climate curve represents conditions in the Salisbury embayment and is based on data from fossil molluscan assemblages.

The most extensive transgression during the Tertiary occurred in the middle Eocene. In Virginia and Maryland it took place during the middle middle Eocene and resulted in the deposition of the Piney Point Formation (Figure 5C). To the south, this transgression consists of carbonate beds: Castle Hayne Formation in North Carolina, Moultrie Member of the Santee Limestone and McBean Formation in South Carolina and Georgia, and Lisbon Formation in Georgia and Alabama. Beds associated with this event are present in all areas of the Gulf Coastal Plain. It is clear that these deposits record a global sea-level rise. At least five small transgressions are reflected in the middle Eocene sequence, but they are plotted as a single event in Figure 2 because of the lack of correlative data. During the late Eocene, a small-scale transgression took place in Virginia (the Chickahominy Formation of Cushman and Cederstrom, 1945; Figure 5D). This thin unit contrasts with the thick stratigraphic sequence deposited in the Gulf area at that time. That record suggests a high sea-level stand, but the meager upper Eocene record in the Atlantic basins indicates a general sea-level lowering, unless most of that area was tectonically emergent.

Oligocene

During the early Oligocene, a thick sequence of beds was deposited in the Gulf, while in the Atlantic region there are only thin subsurface units of that age. In the late Oligocene, data indicate a relative high stand, which resulted in the deposition of beds in the Charleston embayment, Albemarle embayment, and the Gulf. During the very late Oligocene or very early Miocene a brief, small-scale, high stand left a sedimentary record in the Salisbury embayment (Old Church Formation, Figure 6A). In spite of the thinness of these deposits, their wide occurrence is good evidence for a global sea-level rise and the submergence of much of the Atlantic Coastal Plain (Ward, 1985).

Miocene

Following the Old Church transgression and a brief regression, onlap in the Salisbury embayment during the Miocene is characterized by nearly continuous sedimentation punctuated by short breaks, resulting in a series of thin, unconformity-bounded beds. Three of these transgressions produced the silty sands and diatomaceous clays of the Calvert Formation (Shattuck, 1902; 1904; Figure 6B). The diatom assemblages indicate the first and second transgressions occurred in the late early Miocene, and the third in the early middle Miocene (Abbott, 1978; Andrews, 1978). The axis of the depocenter was still to the northeast and it was apparently a restricted basin. Diatomaceous clays accumulated deep in the embayment while coarser-grained, sandy deposits predominate in a seaward direction. Small-scale marine pulses brought coarser sediments deep into the embayment and still-stands resulted in clay accumulations. This formed cyclic deposits of alternating thick beds of clay and sand. Each of the Calvert pulses was successively more extensive; the third pulse partially overlapped the Norfolk arch and extended into the Pungo River sea in the Albemarle embayment.

In the middle and late middle Miocene, the Salisbury embayment was again the site of two brief transgressions. Both were less extensive than the earlier Calvert seas and brought coarser sediments deeper into the embayment (Figure 6C). Beds of the first transgression, including the Drumcliff and St. Leonards Members (of Gernant, 1970) of the Choptank Formation, unconformably overlie the Calvert Formation. The second pulse of the Choptank, which corresponds to the Boston Cliffs Member of Gernant (1970), unconformably overlies beds of the first pulse. Molluscan assemblages indicate cool-temperature to warm-temperature, shallow-shelf, open-marine conditions.

In the early late Miocene another pair of marine transgressions occurred in the Virginia-Maryland area (Figure 6D). Predominantly clayey sands were deposited, with some beds containing a prolific and diverse molluscan assemblage. These beds, which have been assigned to the St. Marys Formation, conformably overlie the Choptank Formation and, in turn, are unconformably overlain by beds of the second pulse, which corresponds to Shattuck's (1904) zone 24. Both units contain abundant and diverse molluscan assemblages that indicate shallow-shelf, open-marine, warm-temperature to subtropical conditions. During the second pulse, the locus of marine deposition shifted substantially to the south. This shift indicates an end of the northeast-southwest depositional alignment that appeared to have dominated in the Salisbury embayment from the Paleocene to the middle Miocene. After the shift, the principal basinal area was centered in Virginia, while Maryland was largely emergent.

After a break of approximately 1.5–2.0 Ma, marine sedimentation resumed with a large-scale transgression in the late late Miocene (Figure 7A). It began with localized subsidence in central Virginia that caused the deposition of a thick sequence of inner-bay, shallow-shelf sediments, termed the Claremont Manor Member of the Eastover Formation (Ward and Blackwelder, 1980). The Claremont Manor Member is a poorly sorted mixture of clay and sand with the finer material concentrated in the westward portion of the basin. Toward the center, fine sands dominate and contain large concentrations of mollusks in the beds. Some of the nearshore clays deposited at that time contain appreciable concentrations of diatoms. Molluscan assemblages found in the Claremont Manor Member are less diverse than in either of the previous pulses in the St. Marys Formation and are less diverse than the subsequent Cobham Bay Member of the Eastover Formation. The composition of the fauna suggests cool to mild temperature conditions in a somewhat protected and restricted embayment.

After a brief low stand, a high sea-level pulse in the late Miocene resulted in a very thin, but

SERIES	STAGE	NEW JERSEY	DELAWARE	MARYLAND	VIRGINIA	NORTH CAROLINA	SOUTH CAROLINA	GEORGIA	ALABAMA
PLEISTOCENE			Columbia Formation	Columbia Formation	Norfolk Formation	Norfolk Formation	Socastee Formation / Canepatch Formation / Waccamaw Formation		
PLIOCENE UPPER / LOWER						Chowan River Formation / Yorktown Formation (Moore House Mbr., Morgarts Beach Mbr., Rushmere Mbr., Sunken Meadow Mbr.)	James City Formation / Chowan River Formation / Yorktown Formation (Moore House Mbr., Morgarts Beach Mbr., Rushmere Mbr., Sunken Meadow Mbr.)	Bear Bluff Formation / Goose Creek Limestone / Raysor Formation	Raysor Formation	Citronelle Formation
MIOCENE UPPER	TORTONIAN (MESSINIAN)	Cohansey Formation / Kirkwood Formation	Kirkwood Formation	Eastover Formation (Cobham Bay Member / Claremont Manor Member)	Eastover Formation (Cobham Bay Member / Claremont Manor Member)	Eastover Formation (Cobham Bay Member / Claremont Manor Member)			
MIOCENE MIDDLE	SERRAVALLIAN / LANGHIAN		Choptank Formation	Choptank Formation / St. Marys Formation ("Windmill Point Mbr.", "Little Cove Point Mbr.", Conoy Member, Boston Cliffs Mbr., St. Leonard Mbr., Drumcliff Mbr.)	St. Marys Formation / Choptank Formation		Coosawhatchie Clay Member / Hawthorn Formation	Coosawhatchie Clay / Marks Head Formation / Hawthorn Formation	
MIOCENE LOWER	BURDIGALIAN / AQUITANIAN	Kirkwood Formation	Calvert Formation / Kirkwood Formation	Calvert Formation (Calvert Beach Mbr., Plum Point Member, Fairhaven Member)	Calvert Formation	Pungo River Formation		Tampa Formation	
OLIGOCENE UPPER	CHATTIAN (CHICKASAWHAN)	Old Church(?) Formation	Old Church(?) Formation	Old Church Formation	Old Church Formation	Belgrade Formation (Pollocksville Mbr. / Haywood Landing Mbr.) / River Bend Formation	Cooper Marl / Edisto Formation / Ashley Member	Suwannee Limestone	Chickasawhay Limestone
OLIGOCENE LOWER	RUPELIAN (VICKSBURGIAN)							Bridgeboro Limestone	Byram Formation / Marianna Limestone

166

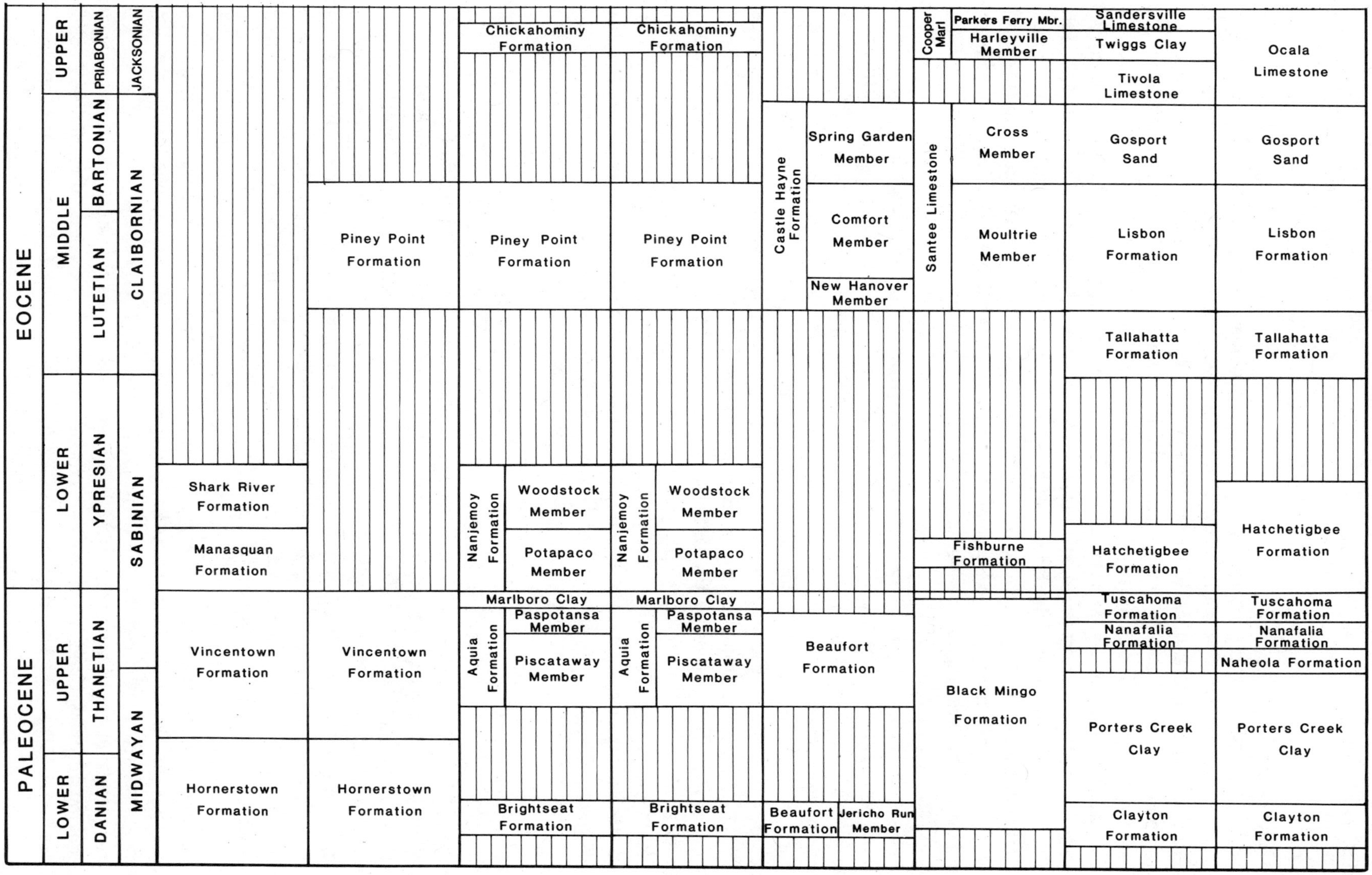

Figure 3. Correlation chart showing Tertiary units from New Jersey to Alabama.

widespread, marine deposit termed the Cobham Bay Member of the Eastover Formation (Ward and Blackwelder, 1980; Figure 7B).

STRATIGRAPHY

Brightseat Formation

The Brightseat Formation, named by Bennett and Collins (1952), consists of olive-black (5 Y 2/1), micaceous, clayey and silty sands. Ward (1985) reported that the Brightseat Formation crops out as far south as the Rappahannock River in Virginia. The Brightseat sea occupied principally the northeastern portion of the Salisbury embayment and was separated from the Albemarle embayment to the south by the Norfolk Arch (Figure 4A). In its type area, 1.0 miles west-southwest of Brightseat, Prince Georges County, Maryland, mollusks are abundant, but only the calcitic forms are well preserved. Away from the type area, the macrofossils are leached, leaving only molds and casts. In the Prince Georges County area, the Brightseat Formation unconformably overlies marine deposits of the Severn Formation (Upper Cretaceous). To the south, on the Potomac and Rappahannock Rivers, it overlies fluvial deposits of the Potomac Group (Lower Cretaceous). Beds now placed in the Brightseat Formation were originally assigned, with some reservations, to "Zone 1" of the Aquia Formation (Clark and Martin, 1901).

On the right bank of Aquia Creek (Stop 1), the Brightseat Formation unconformably overlies the Patapsco Formation of the Potomac Group (see Figure 8 for locality map). The Brightseat Formation is, in turn, unconformably overlain by the Aquia Formation. Macrofossils at the locality are leached and are present only as rare molds and casts, but the micaceous, silty sand, nearly devoid of glauconite, distinguishes the unit. The Brightseat Formation is not known in outcrop or in the subsurface to the south of the Rappahannock River exposures; however, it has recently been identified in cores from the Dismal Swamp area near Norfolk.

Hazel (1968, 1969), studying the ostracodes of the Brightseat Formation in the type area, found the unit to be equivalent to the upper part of the Clayton Formation in Alabama, and placed it in the *Globoconusa daubjergensis–Globorotalia trinidadensis* zone on the basis of planktic foraminifers. He showed the Brightseat Formation to be early Paleocene in age and placed it in the upper part of the Danian Stage. According to Gibson and Bybell (1984), calcareous nannofossils present in the Brightseat Formation indicate its placement in nannoplankton zone NP 3 (of Martini, 1971).

Aquia Formation

The use of the term "Aquia" as a stratigraphic unit was first introduced by Clark (1896). He gave the name "Aquia Creek Stage" to beds that crop out in the vicinity of Aquia Creek, Stafford County, Virginia. The concept of the unit was soon revised, and it was renamed the Aquia Formation by Clark and Martin (1901). Two members, the Piscataway and Paspotansa, were recognized by us. Bennett and Collins (1952) restricted Clark and Martin's (1901) earlier definition of the Aquia Formation when beds placed in Zone 1 by Clark were designated the Brightseat Formation. It is the Aquia Formation, in this restricted sense, that unconformably overlies the Brightseat Formation in the northeastern area of its range and unconformably overlies Lower Cretaceous deposits south of the Rappahannock River. The Aquia Formation consists of clayey, silty, very shelly, glauconitic sand. It crops out in a continuous arc from the upper Chesapeake Bay to the area around Hopewell on the James River, Virginia. Both members of the Aquia Formation are recognized along the Potomac, Rappahannock, Mattaponi, Pamunkey, and James Rivers, and both are extremely fossiliferous (see Figure 4B for the basinal outline of the Aquia).

Macrofossils in the Aquia Formation locally are well preserved but more commonly are leached, making recovery difficult. Microfossil groups consist of ostracodes, foraminifers, pollen, dinoflagellates, and calcareous nannofossils. Microfossil work has indicated placement of the Aquia

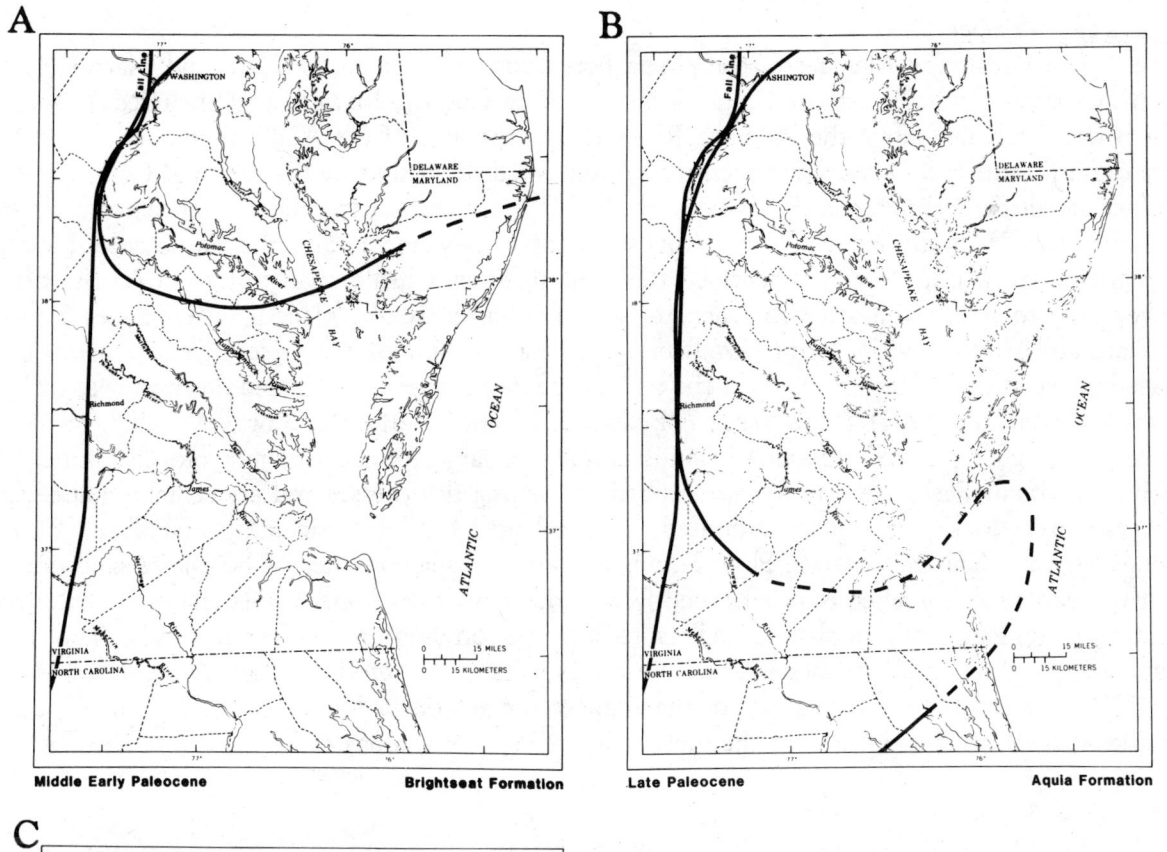

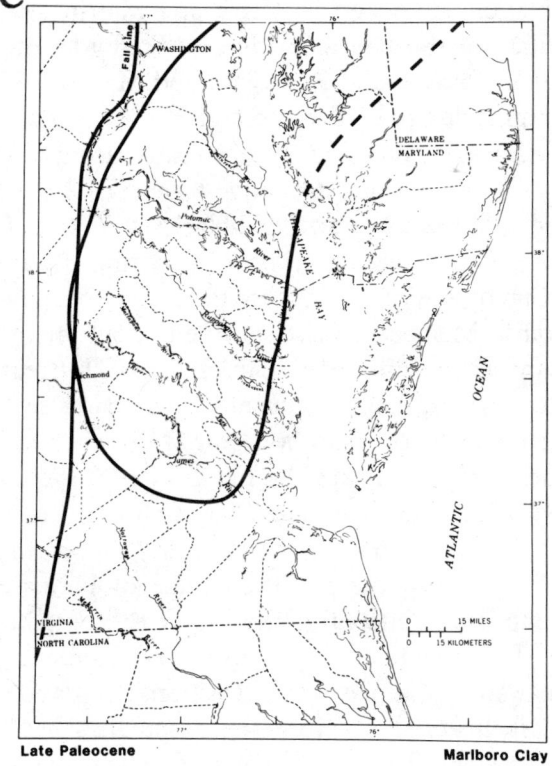

Figure 4. Maps showing depositional basins during the Paleocene. Dashed lines indicate areas where boundary data are lacking.

169

Formation in the upper Paleocene. Gibson and Bybell (1984), on the basis of calcareous nannofossils, placed the Aquia Formation in zones NP 5-9.

Piscataway Member

The Piscataway Member of the Aquia Formation was named by Clark and Martin (1901) from exposures along Piscataway Creek, Prince Georges County, Maryland. It included 7 "zones," which were traceable along the Potomac River in the type area of the Aquia Formation. Zone 1 of Clark and Martin (1901) has since been recognized as a distinct unit by Bennett and Collins (1952) and was termed the Brightseat Formation. Study of the lectostratotype section (principal reference section; Stop 2) of the Aquia Formation, the Piscataway Member, and the Paspotansa Member (designated by Ward, 1985) has revealed the most significant lithic change, from a poorly sorted, clayey sand to a very well-sorted, micaceous, silty, fine sand, to take place between Zones 5 and 6 of Clark and Martin (1901). Ward (1985) proposed that the boundary between the two members be placed between those "Zones" and that the base of the Paspotansa be extended downward to include Clark and Martin's (1901) Zones 6 and 7. Beds 2–5, assigned to the Piscataway Member, consist of clayey, silty, poorly sorted glauconitic sands containing large numbers of macrofossils, principally mollusks. The mollusks are concentrated in beds of varying thicknesses and are cemented at several intervals into locally traceable indurated ledges. Large bivalves, including *Cucullaea*, *Ostrea*, *Dosiniopsis*, and *Crassatellites* (see Plate 1), are the most conspicuous taxa. The quartz sand present in the Piscataway Member is usually poorly sorted, angular, and clear. Glauconite is extremely abundant, ranging from sand-sized pellets to coatings on and in molluscan fossils. The sand, glauconite, and mollusks are interspersed in a clayey, silty matrix producing a very tough, olive-gray (5 Y 4/1), calcareous marl. Glauconite percentages range from a low of 20 percent in far updip localities to 70 percent or more in the more seaward parts of the basin.

Paspotansa Member

The Paspotansa Member of the Aquia Formation, described by Clark and Martin (1901), received its name from Passapatanzy Creek, a tributary of the Potomac River in Stafford County, Virginia. As originally defined, the Paspotansa included Zones 8 and 9 of Clark and Martin (1901). However, as previously discussed, Ward (1985) recommended that "Zones 6 and 7" also be included in the Paspotansa Member. "Zone 6" is an olive-gray (5 Y 4/1), very fine, micaceous, glauconitic sand containing large numbers of *Turritella mortoni*. "Zones 7 and 8" consist of olive-black (5Y 2/1), fine, glauconitic sand, with scattered, thin *Turritella* beds. "Zone 9" is an olive-black (5 Y 2/1), fine, glauconitic sand containing large numbers of closely packed *Turritella* in beds of varying thickness. The thicknesses of the units, as well as their fossil content, vary from locality to locality, but several characteristics are internally consistent. The Paspotansa Member consists of fine to very fine, silty, well-sorted, micaceous, glauconitic and quartzose sand in massive or very thick beds. This texture is strikingly different from the underlying poorly-sorted, clayey, shelly, glauconitic and quartzose sand of the Piscataway Member. The Paspotansa Member is usually overlain by a gray (N 7, when fresh), tough clay termed the Marlboro Clay. This bed, where present, makes the recognition of the upper boundary of the Paspotansa Member relatively easy. Where the Marlboro Clay is absent, the well-sorted, fine sands of the Paspotansa Member may be distinguished from the overlying, clayey, highly bioturbated, poorly sorted glauconitic sands of the Potapaco Member of the Nanjemoy Formation.

As in all formations in the Pamunkey Group, the glauconite content of the Paspotansa Member varies with proximity to the paleo-shoreline. Percentages are much lower near the perimeter of the basin, in some areas less than 10 percent. Seaward, in an eastward direction, the glauconite content may reach 90 percent. The nature of the shell deposits within the Paspotansa Member further serves to distinguish that unit from the underlying Piscataway Member and the overlying

Nanjemoy Formation. Massive glauconitic sands containing considerable numbers of large *Turritella* in thin beds or lenses characterize the Paspotansa Member, in striking contrast to the very shelly, silty sands of the Piscataway Member, which are usually dominated by closely packed, medium-sized to large bivalves. Also notable is the massive nature of the Paspotansa Member, often exposed in the high vertical walls of bluffs along the Potomac, Rappahannock, and Pamunkey Rivers. Where very fresh the unit is dark olive-black (5 Y 2/1). Where partially weathered it is grayish-orange (10 YR 7/4), and where very weathered it appears yellowish-orange (10 YR 7/6), because of the oxidation of iron in the glauconite.

The Paspotansa Member apparently disconformably overlies the Piscataway Member, but the nature of the contact is commonly obscured. Clark and Martin (1901) described the nature of the contact between "Zones" 6 and 8 along the bluffs below Aquia Creek, but the most notable lithic change occurs at the contact between their "Zones" 5 and 6. On the Rappahannock and Mattaponi Rivers, the contact between the two members is obscured by slumping and poor outcrops. Along the Potomac River, however, the contact between the two members is sharp and undulating. No phosphate accumulations or burrows are present, indicating only a brief period of nondeposition.

The Paspotansa Member crops out in a broad arc from the Eastern Shore of Maryland to the James River in Virginia (see Figure 4B). Clark and Martin (1901, p. 73) described the Paspotansa Member from the Chester River in Kent County, Maryland, and their descriptions of the sections on the Severn and South Rivers in Anne Arundel County indicate the presence of the unit there. Additional sections are described in the Upper Marlboro area of Prince Georges County, where the Paspotansa Member includes a heavy concentration of bryozoans. The most extensive outcrops of the member extend along the Virginia shore of the Potomac River from the mouth of Aquia Creek to near Fairview Beach. Between Potomac Creek and Aquia Creek (Stop 2), 61.5 feet (18.8 m) of the Paspotansa Member occurs in steep, almost vertical bluffs, which have been weathered to a reddish-orange color. This section was designated the principal reference section (lectostratotype locality) of the Paspotansa Member by Ward (1985). Several distinct shell beds are present, as well as several discontinuous ledges of boulder-sized concretions. Other shell concentrations, consisting principally of large, current-oriented *Turritella mortoni* (see Plate 2), occur in lens-shaped masses.

Marlboro Clay

Clark and Martin (1901, p. 65) first applied the term "Marlboro clay" to sediments included in "Zone 10" of Clark (1896, p. 42). The name was derived from exposures of that unit near Upper Marlboro, Prince Georges County, Maryland. Clark and Martin (1901) considered this unit to be the basal unit of the Potapaco Member of the Nanjemoy Formation. Clark and Miller (1906) briefly described the outcrop area of the "Marlboro clay" across Maryland and Virginia and included it in the basal bed of the Nanjemoy Formation. Clark and Miller (1912) again included the pink clay as the basal bed in the Nanjemoy Formation. However, at only one locality, below Hopewell on the James River (Clark and Miller, 1912, p. 115), is a specific outcrop section described. Darton (1948), in a short note, described the areal extent of the Marlboro Clay and referred to that unit as the basal member of the Nanjemoy Formation. This had the effect of formalizing the name. A more detailed study, including a detailed geologic map by Darton (1951), continued the placement of the Marlboro Clay as the basal bed of the Nanjemoy Formation. Glaser (1971), however, was the first to formally propose the elevation of the Marlboro Clay to formational rank. This restricted the original concept of the Nanjemoy Formation and, more specifically, of the Potapaco Member. This treatment of the Marlboro Clay, as a separate formation, was continued by Reinhardt et al. (1980).

Glaser (1971, p. 14) characterized it as "a silvery-gray to pale-red plastic clay interbedded with much subordinate yellowish-gray to reddish silt." Glaser noted that both the lower and upper contacts of the Marlboro Clay were sharp and nongradational and probably represented at least a brief hiatus between the underlying and overlying units. A more recent study by Reinhardt et al. (1980) on a core from Westmoreland County, Virginia, concluded that the Aquia-Marlboro contact was somewhat gradational, whereas the upper Marlboro-Nanjemoy contact was sharp and was marked by burrows into the underlying Marlboro Clay.

The areal distribution of the Marlboro Clay was mapped by Darton (1951) and schematically shown by Glaser (1971), but no detailed study of the formation has been made in much of the Virginia Coastal Plain. Outcrops of the Marlboro Clay examined by us have been limited to the Potomac, Mattaponi, Pamunkey, and James Rivers. Outcrop patterns indicate a spotty, though widespread, occurrence of the unit (see Figure 6).

Because of the lack of calcareous fossils, the age of the Marlboro Clay has been assigned principally on the fact that it occurs between the Aquia and Nanjemoy Formations. This brackets its age but does not afford primary evidence. A detailed study of a core from Oak Grove, Westmoreland County, Virginia, by Gibson et al. (1980) and Frederiksen (1979) afforded the best paleontologic evidence of its age. The consensus of pollen and dinoflagellate data suggested an age of very late Paleocene and possibly very early Eocene. Mixing of the two floral assemblages may have occurred through reworking and bioturbation, or the unit may contain the Paleocene-Eocene boundary.

Nanjemoy Formation

Beds now included in the Nanjemoy Formation were first studied in detail by Clark (1896), who divided them, along with those now included in the Aquia Formation, into "zones." Those "zones" above Aquia Creek Stage of Clark (1896) were numbered 10 through 17. "Zone" 17 was described as the Woodstock Stage. Clark and Martin (1901) revised this terminology and placed their "Zones" 10 through 17 in the "Nanjemoy Formation or Stage." The Nanjemoy Formation was divided into the "Potapaco member or substage," including "Zones" 16 and 17. In this same publication Clark and Martin (1901, p. 65) introduced the term "Marlboro clay," informally, for part of their Zone 10. In a brief, preliminary report on the stratigraphy of the Virginia Coastal Plain, Clark and Miller (1906) dropped the stage and substage terminology and referred only to the Aquia and Nanjemoy Formations. Clark and Miller (1912) continued this usage and retained both in formational status. Again, the "Marlboro clay" was briefly mentioned (Clark and Miller, 1912, p. 103) but was specifically reported from only one locality (Clark and Miller, 1912, p. 155). The zonation of beds and their breakdown into members was retained only for those well-studied exposures along the Potomac River. South of the Potomac, only assignment to formation was attempted. The term "Marlboro Clay" was finally formalized by Darton (1948), but the unit was retained as a basal member of the Nanjemoy. This removed the Marlboro Clay ("Zone 10," in part, of Clark) from beds previously included in the Potapaco Member. The Marlboro Clay was retained as a member of the Nanjemoy until Glaser (1971) elevated it to formational rank. This, in effect, restricted the original concept of the Nanjemoy Formation, and only part of "Zone 10" and "Zones 11–17" remained in that formation. Beds younger than "Zone 17" were not included in the original description or sections of the Nanjemoy Formation but were later lumped under the term "Nanjemoy Formation" by Clark and Miller (1912).

Potapaco Member

The Potapaco Member of the Nanjemoy Formation was described by Clark and Martin (1901, p. 65) and received its name from ". . . the word Potapaco found on the (Captn. John) Smith and others early maps . . . ." The Potapaco Member included Clark and Martin's (1901) Zones 10–15. Part of their "Zone" 10 included the Marlboro Clay. Clark (1896), Clark and Martin (1901), and Clark and Miller (1912) described the beds ("Zones") found in the Potapaco Member at sections

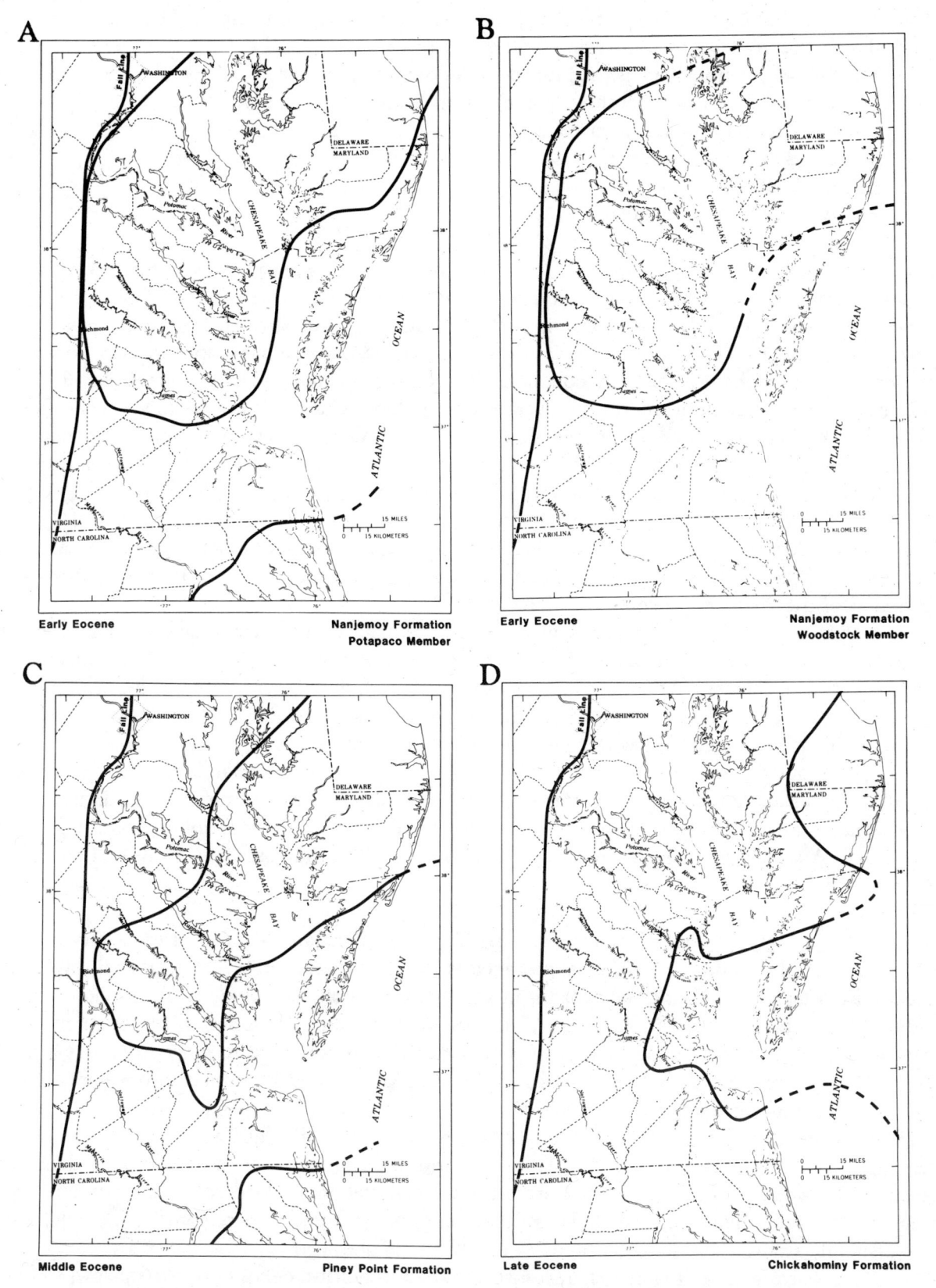

Figure 5. Maps showing depositional basins during the Eocene. Dashed lines indicate areas where boundary data are lacking.

upriver from Popes Creek on the left bank of the Potomac River, Charles County, Maryland. The section described by Clark and Martin (1901, p. 70, Section VIII) was designated the principal reference section (lectostratotype) of the Nanjemoy Formation and the Potapaco Member by Ward (1985). The exposure is just downstream of Stop 6 in this report.

The following terminology is used for the series of four beds that have been recognized in the Potapaco Member (lettered from bottom to top) (from Ward, 1984; 1985):

Bed D - Concretion-bed Potapaco
Bed C - Burrowed Potapaco
Bed B - Bedded Potapaco
Bed A - Non-bedded Potapaco

Bed A - Non-bedded Potapaco. Bed A is found on the Potomac, Rappahannock, Mattaponi, and Pamunkey Rivers. It consists of a clayey, silty, fine sand containing scattered, small mollusks including *Venericardia potapacoensis* Clark and Martin, 1901 (see Plate 3). Glauconite occurs in relatively small amounts in the sand-sized fraction in updip areas, but glauconite percentages increase in a seaward direction. Small phosphate pebbles are common. The bed is estimated to be 15–20 feet (4.6–6.1 m) thick and, in most places, unconformably overlies the Marlboro Clay. Bed A is distinguishable from Bed B by its darker color, lack of bedding, and less clayey texture. Calcareous fossils are generally leached, leaving only molds and casts. The unit is present on the right bank of the Potomac River 1.75 miles (2.8 km) below Fairview Beach in King George County, Virginia (Stop 4).

The lithic and faunal makeup of Bed A suggests an initial marine pulse and basal transgression following the quiet, protected embayment indicated by the Marlboro Clay. Physical and paleontologic evidence indicates that little time occurred in the break between the Marlboro Clay and Bed A of the Potapaco Member.

Low molluscan diversity and small glauconite percentages suggest restricted conditions during the deposition of Bed A. In spite of this evidence, renewed marine influence is apparent. Dinoflagellate assemblages are marked by reduced diversity; the flora is dominated by a single taxon, indicating restricted marine conditions (L. E. Edwards, personal communication). Mollusks, in general, are poorly preserved but where present are low in diversity.

Calcareous nannofossils found in the Oak Grove core in Westmoreland County (Gibson et al., 1980), from the interval just above the Marlboro Clay, probably come from Bed A and indicate the placement of that bed in calcareous nannoplankton zone NP 10. This zonation indicates an early Eocene age. Assemblages of pollen, dinoflagellates, foraminifers, and ostracodes substantiate this placement.

Bed B - Bedded Potapaco. Bed B, the most striking unit in the Potapaco Member, is easily recognized by its thinly bedded appearance. This appearance is due to the accumulation of a small bivalve, *Venericardia potapacoensis* Clark and Martin, in vast numbers along numerous, discontinuous, thin bedding planes. Bed B varies in thickness from locality to locality. Its exact thickness in surface exposures is difficult to determine because of poor outcrops. It is estimated to range from only a few feet (about 1 m) to more than 15 feet (4.6 m). The sediment in Bed B consists of olive-gray (5 Y 4/1), very clayey, glauconitic sand to sandy clay. The clay, when fresh, appears grayish-orange pink (5 YR 7/2) and contains varying amounts of fine to medium-sized glauconite and quartz sand. Glauconite content ranges from less than 10 percent in the westernmost exposures to more than 75 percent with increasing distance eastward from the paleo-shoreline. Bedded concretions ranging up to boulder size are common in Bed B. These nodules, although sometimes regionally traceable, are not sufficiently stratigraphically continuous to be used as marker beds.

Sedimentological and faunal characteristics of Bed B indicate deposition in a shallow, somewhat restricted embayment. Glauconite grains, which appear to be concentrated in burrows, are common. The burrows and the concentrations of abraded bivalves along bedding planes suggest shallow depths, probably not below wave base. Glauconite-coated, worn, disarticulated valves of *Venericardia* indicate periods of slow sediment accumulation. Bivalves may be concentrated along those winnowed zones because of intermittently favorable bottom conditions or storms. Elsewhere in the section, where soft clays inhabited by burrowing organisms predominated, the bottom may not have been suitable for the settlement of bivalve larvae. The molluscan assemblage of Bed B is dominated (up to 95 percent of the assemblage) by *Venericardia potapacoensis* Clark and Martin, 1901 (Plate 3).

Bed C - Burrowed Potapaco. Above the thin-bedded clayey sand of Bed B is a series of sandy clays to clayey sands that are easily recognizable by their intensely burrowed appearance. Bedding, if it was ever present, has been obscured by bioturbation except along a few very thin planes. Along those surfaces sedimentation appears to have been interrupted and is marked by local diastems, by a concentration of glauconitic sand, and by glauconite-filled burrows extending down into the underlying sediment. These stratigraphic breaks, if that is what they are, have not been traced over a wide area and may be only local, possibly current-scoured surfaces. The dominant lithic characteristic of Bed C is its very clayey texture, with interspersed grains of fine to medium sand-sized glauconite in a grayish-orange-pink (5 YR 7/2) clay matrix. In some areas, the concentration of glauconite is such that the lithology is best described as a clayey sand. This very clayey, glauconitic texture is typical of the entire Potapaco Member, but the intensely burrowed nature of the unit is unique to Bed C. The macrofossil component of Bed C consists principally of small or broken, poorly preserved mollusks that are concentrated in burrows and make up a small percentage of the bed. Thicknesses of as much as 20 feet (6.1 m) of Bed C have been observed in outcrop.

Bed C overlies Bed B with no distinct contact between the two, suggesting a gradation from one environmental regime to another. In most of its outcrop area, Bed C is overlain by the Woodstock Member of the Nanjemoy Formation. On the Pamunkey River, at least, Bed C is overlain by a thin bed, 1.5–3.0 feet (0.5–0.9 m) thick, of clayey, glauconitic sand marked by a series of boulder-sized indurated blocks. The contact between Bed C and the younger unit, Bed D, is abrupt and is marked by a sharp but burrowed contact indicating a probable diastem. Elsewhere, where Bed D is missing, the Bed C–Woodstock boundary is disconformable and is marked by an abrupt change in lithology, a lag deposit of phosphate, bone, pebbles, and wood in the lower Woodstock Formation, and burrows containing Woodstock sediment several feet into the underlying Bed C. The olive-black (5 Y 2/1), very fine, well-sorted, micaceous, glauconitic sand of the Woodstock Member is easily distinguishable from the very clayey, burrowed sand of Bed C. This contact has been observed on the Patuxent, Potomac, Mattaponi, and Pamunkey Rivers. Upriver of Popes Creek on the left bank of the Potomac River, Charles County, Maryland, the area in which Bed C should descend to water level is slumped and obscured by weathering of the cliff face.

Dinoflagellate assemblages indicate near-shore or high-energy conditions with an abundant source of nutrients (L. E. Edwards, personal communication). On the basis of the dinoflagellate flora, Bed C may be correlated with calcareous nannoplankton Zone NP 10 or 11.

Bed D - Concretion-bed Potapaco. Bed D crops out only along the Pamunkey River below the mouth of Totopotomoy Creek in Hanover County, Virginia, and therefore is discussed only briefly here. Bed D, in its small outcrop area, consists of 1.5–2.0 feet (0.5–0.6 m) of clayey, very glauconitic sand and has sharp upper and lower contacts. Both contacts are marked by abrupt changes in lithology and color and contain concentrations of quartz and phosphate pebbles, and wood. Burrows at both contacts extend down into the underlying beds. The high glauconite content of Bed D makes it easily differentiated from the lighter-colored clays of Bed C and the less glauconitic silty sand of the basal portion of the overlying Woodstock Member. The bed is marked by a line of cobble to boulder-sized concretions, which occur in the middle of the unit.

Woodstock Member

The Woodstock Member of the Nanjemoy Formation was first proposed by Clark and Martin (1901, p. 66) for beds of glauconitic sand exposed near Woodstock, "an old estate situated a short distance above Mathias Point," King George County, Virginia. The term "Woodstock" had previously been used by Clark (1896) to describe the Woodstock Stage, a unit defined principally on its fauna. Clark and Martin (1901) described the Woodstock Member as consisting of their "Zones" 16 and 17. The bluff described by Clark (1896, p. 40, Pl. IV, Section III) and Clark and Martin (1901, p. 70, Section IX) exhibits both the Potapaco and Woodstock Members. Stop 5 of this guidebook was designated the lectostratotype section by Ward (1985).

The Woodstock Member consists of olive-black (5 Y 2/1), very fine, well-sorted, silty, glauconitic sands. The glauconite content increases markedly from a low of 10–15 percent in its most inland outcrops to 70–80 percent in its most seaward exposures. Carbonaceous material in the form of logs, branches, and nuts is abundant. The Woodstock Member unconformably overlies the Potapaco Member and is unconformably overlain either by the Piney Point Formation in the James and Pamunkey River areas or by younger beds in the northern section. The lower contact with the Potapaco Member is a constant and easily recognized feature that may be seen from the Patuxent River to the Pamunkey River. There is a significant faunal and floral change at this boundary, though it marks only a relatively brief hiatus. The Woodstock Member may be distinguished from the underlying Potapaco Member by its fine-textured, micaceous, massive appearance, while it lacks the very clayey, poorly sorted, bioturbated texture of the underlying Potapaco Member.

Molluscan assemblages in the Woodstock Member are diverse and abundant and are scattered throughout the fine matrix (Plate 4). Large valves of *Venericardia ascia* Rogers and Rogers are concentrated along bedding planes in some areas but are easily distinguished from the much smaller *Venericardia potapacoensis* found in the Potapaco Member. Along the Potomac River above Mathias Point on the right bank (Stop 5) and below Popes Creek on the left bank (Stop 6), the Woodstock Member is overlain by transgressive sediments of the Calvert Formation that range from early to middle Miocene in age. The contact is marked by a basal lag concentration of phosphate and quartz pebbles, a burrowed surface, and an abrupt lithic change from glauconitic sand to olive-brown, clayey, phosphatic sand. At the end of the bluffs, downriver from Popes Creek (below Stop 6), a very thin tongue of burrowed gray clay and a bed of glauconitic sand occur between typical, easily recognized Woodstock Member and the Calvert Formation. These two beds thicken downstream but are beveled off upstream south of Popes Creek. Macrofossils are leached from the beds, but dinoflagellates indicate that they are early Eocene in age (L. E. Edwards, personal communication). Therefore, we include them in the Woodstock Member in spite of their very different lithologies. We believe that these units are represented in the Oak Grove core by the clay and sand beds shown as occurring in the upper Nanjemoy by Reinhardt et al. (1980, fig. 1).

The Woodstock sea occupied a broad embayment reaching from at least the Patuxent River in Maryland to a short distance south of the James River in Virginia (Figure 5B). The locus of the embayment was somewhat south of the Potomac River.

The Woodstock Member contains a moderately diverse molluscan fauna (Plate 4). Many of the taxa are new and their stratigraphic significance is, at present, poorly understood. A number of species were listed by Clark and Martin (1901) as being present in the Woodstock Member, but the list is in serious need of updating.

Best evidence, at this time, of the age and correlation of the Woodstock Member is found in the dinoflagellate and calcareous nannofossil assemblages. Calcareous nannofossils in the Putney Mill core, New Kent County, Virginia, indicate an approximate equivalence with nannofossil zone NP 12 (L. M. Bybell, personal communication). This zone was also reported in the Oak Grove core (Gibson et al., 1980) in the interval between 227.0 and 269.4 feet (69.2 and 82.1 m). L. M. Bybell (personal communication) now believes that only the 69.2-meter interval in the Oak Grove core contains calcareous nannofossils indicative of NP 12.

CHESAPEAKE GROUP

The term "Chesapeake Formation" was introduced by Darton (1891, p. 433) for a series of beds in southeastern Maryland and Virginia that consist of sands, clays, marls, diatomaceous clays, and shell fragments. Dall and Harris (1892) elevated the unit to group status and included all stratigraphically equivalent beds at the same horizon from Delaware to Florida. Shattuck (1902) subdivided the Chesapeake Group in Maryland into (in ascending order) the Calvert Formation, Choptank Formation, and St. Marys Formation. Shattuck (1904) greatly expanded this work and described the units, and their contained molluscan fauna, in detail. Clark and Miller (1906) expanded the definition of the Chesapeake Group by including the Yorktown Formation in Virginia. Clark and Miller (1912) included beds along the Chowan River in Bertie County, North Carolina, in the Yorktown Formation. Mansfield (1944) also included the Chowan River beds in the Yorktown. Blackwelder (1981) named those beds the Chowan River Formation; he split the unit into two members, the Edenhouse Member (lower) and the Colerain Beach Member (upper), and included the new formation in the Chesapeake Group.

Ward (1985) recommended that a new unit, the Old Church Formation, be included in the Chesapeake Group. The Old Church Formation is a calcareous, shelly sand containing only small amounts of glauconite, which unconformably underlies the Calvert Formation and unconformably overlies the Piney Point Formation. It is easily differentiated from the underlying, very glauconitic beds of the Pamunkey Group. It is unclear whether Darton (1891) or Clark and Miller (1912) actually observed the bed that Ward (1985) termed the Old Church. Therefore, where they would have placed it is conjectural.

The following units constitute the Chesapeake Group:

Chowan River Formation	Colerain Beach Member	upper Pliocene
	Edenhouse Member	upper Pliocene
Yorktown Formation	Moore House Member	upper Pliocene
	Morgarts Beach Member	upper Pliocene
	Rushmere Member	upper Pliocene
	Sunken Meadow Member	lower Pliocene
Eastover Formation	Cobham Bay Member	upper Miocene
	Claremont Manor Member	upper Miocene
St. Marys Formation	Windmill Points beds*	upper Miocene
	Little Cove Point beds*	upper Miocene
	Conoy Member	lower upper Miocene
Choptank Formation	Boston Cliffs Member	upper middle Miocene
	St. Leonard Member	middle middle Miocene
	Drumcliff Member	middle middle Miocene
Calvert Formation	Calvert Beach Member	lower middle Miocene
	Plum Point Marl Member	lower middle Miocene
	Fairhaven Member	lower and lower middle Miocene
Old Church Formation		upper Oligocene and lower Miocene

*Informally named units of Ward (1984)

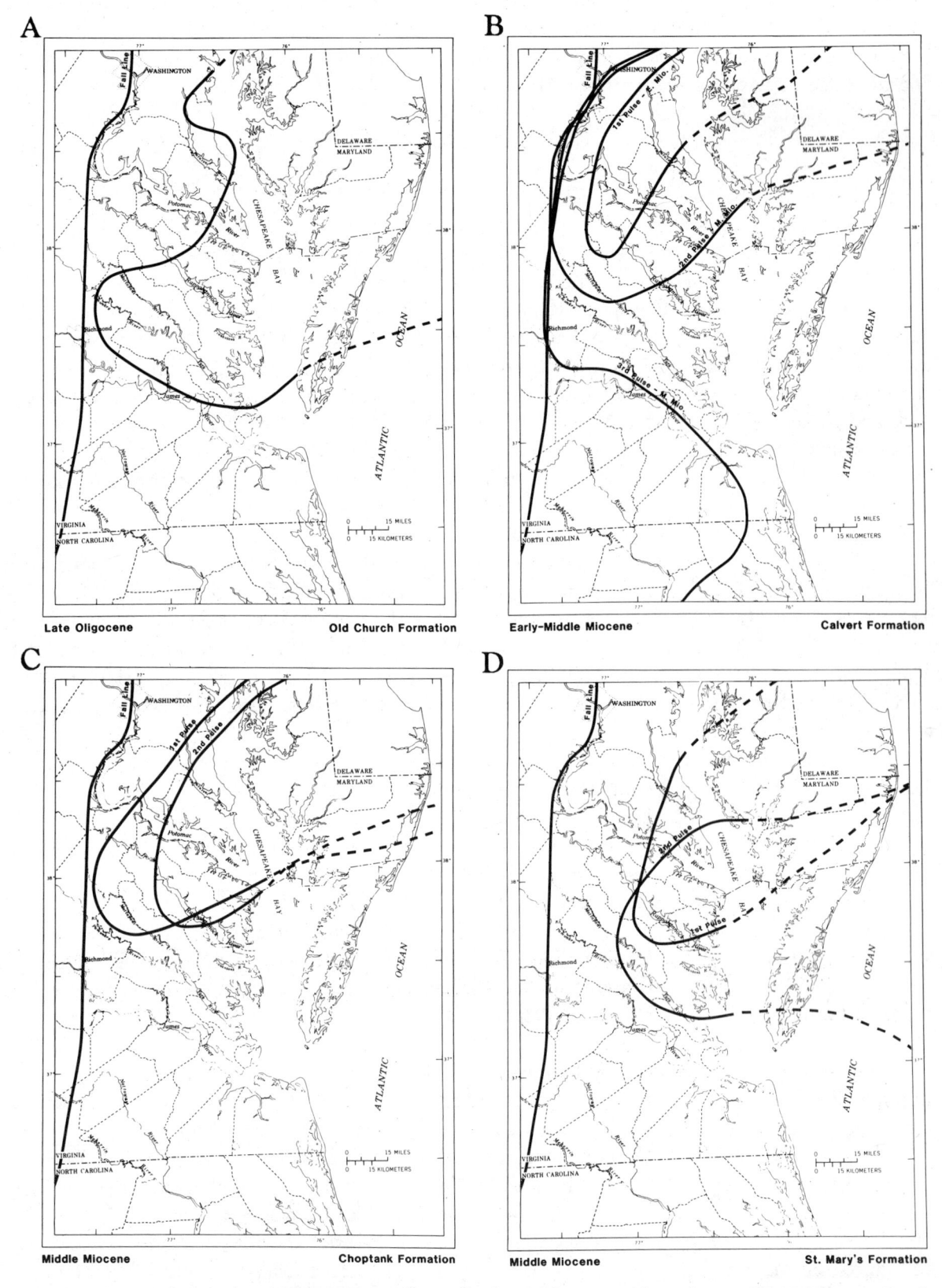

Figure 6. Maps showing depositional basins from the late Oligocene thru the middle Miocene. Dashed lines indicate areas where boundary data are lacking.

Only beds of the Calvert, Choptank, St. Marys, and Eastover Formations crop out in the field trip area and are described here.

Calvert Formation

The Calvert Formation was named and described by Shattuck (1902; 1904) for Miocene beds exposed along the Calvert Cliffs in Calvert County, Maryland. Sections for the Calvert Formation were given by that author, principally along the Chesapeake Bay, but he described a few other localities in scattered areas of Maryland. The Calvert Formation in Virginia was first mentioned by Clark and Miller (1906) and it was soon thereafter mapped in the Richmond area by Darton (1911). Clark and Miller (1912) documented, rather completely, the extent of the Calvert and other Coastal Plain units; no such exhaustive treatment has since been attempted. More recent descriptions have been in guidebooks, treating exposures described by Clark and Miller (1912) (see Stephenson et al., 1933; Ruhle, 1962).

Shattuck (1904) described 15 "Zones" or beds in the Calvert Formation. The Fairhaven Member of Shattuck (1904) included "Zones, 1, 2, and 3." "Zone 2" is merely the basal trangressive sand that accumulated during the first Calvert pulse or sea-level rise (see Figure 6B). "Zone 3," a massive series of diatomaceous clays, includes most of the thickness of the Fairhaven Member. "Zone 3" contains two distinct marine pulses involving basal trangressive lags and fining-upward sequences. Exposures at the Kaylorite pit on Ferry Road, Calvert County, Maryland, and the lower 10 feet (3.0 m) of the Fairhaven Member below Popes Creek, Charles County, Maryland, contain beds associated with the first pulse of the Calvert. This bed is separated from the remaining, upper portion of the Fairhaven Member by a phosphate pebble lag indicating an unconformity or at least a diastem. This lower unit was named the Popes Creek Sand Member by Gibson (1982; 1983) and was excluded from the Fairhaven. We feel that it is very important to recognize unconformities and the sea-level changes that they represent. However, the lithologies on either side of that contact are very similar, consisting of silty diatomaceous clays, which cannot be separated lithologically. For this reason, we prefer to discard the new name and include both beds in the Fairhaven Member. Beds associated with the first transgression can be found as far south as the area of Wilmont on the Rappahannock River, Westmoreland County, Virginia. Beds of the second pulse are known as far south as the vicinity of Reedy Mill on the Mattaponi River, Caroline County, Virginia.

Plum Point Marl Member

The Fairhaven Member is overlain, unconformably, by a series of shelly sands interbedded with diatomaceous clays grouped under the term Plum Point Marl Member. This series contains a number of identifiable pulses: the first pulse includes "Zones 4–9" (of Shattuck, 1904), the second pulse includes "Zones 10–11," the third pulse includes "Zones 12–13," and the fourth pulse includes "Zones 14–16." The pulses are included in the area plotted as the third pulse on Figure 6B. "Zone 16," as exposed at Calvert Beach, Calvert County, Maryland, was originally included in the Choptank Formation even though it is the equivalent of "Zone 15" of the Calvert at Scientists Cliffs, Calvert County, Maryland (Stop 10). This miscorrelation and the fact that "Zone 16" contains "Choptank fossils" led to its inclusion in that unit despite its very different lithology and a striking unconformity separating the two. Mollusks that characterize "Zones 17 and 19" of the Choptank Formation make their first appearance at least as far down in the sequence as "Zone 14," and some taxa may be present in "Zone 12." It was recommended that "Zones 14, 15, and 16" be included in the Calvert Beach Member (Ward, 1984). The Plum Point Marl Member, as a lithic entity, is recognizable only as far south as the Westmoreland bluffs in Westmoreland County, Virginia. Farther to the southeast, beds equivalent to the Plum Point Marl Member grade into silty, diatomaceous clays.

Mollusks are common in the Calvert Formation but are well preserved only in beds along the

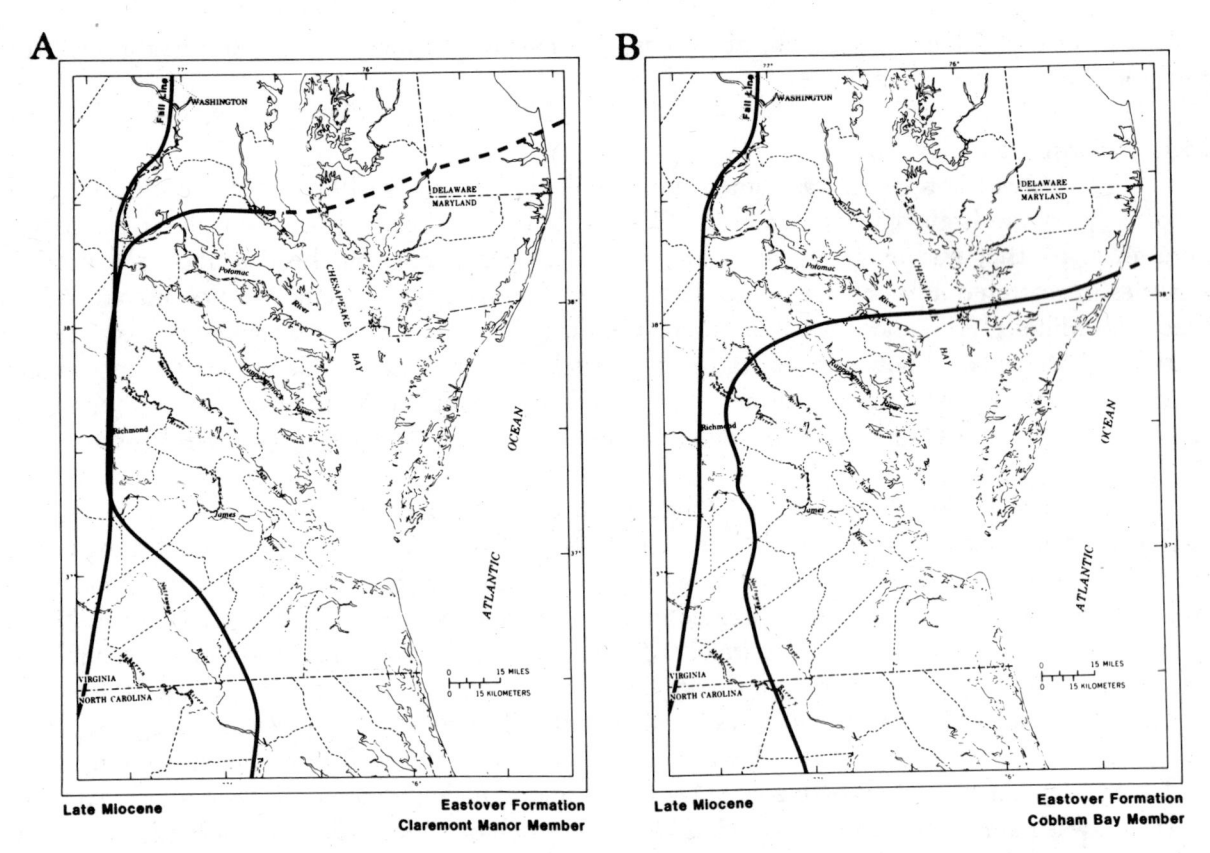

Figure 7. Maps showing depositional basins during the late Miocene. Dashed lines indicate areas where boundary data are lacking.

Chesapeake Bay in Calvert County, Maryland. Some of the common forms are (Plate 5):

Marvacrassatella melinus (Conrad, 1832)
Astarte sp.
Mercenaria sp.
Cyclocardia sp.
Pecten humphreysii Conrad, 1842

Melosia staminea (Conrad, 1839)
Bicorbula idonea (Conrad, 1833)
Lucinoma contracta (Say, 1824)
Lirophora latilirata (Conrad, 1841)
Ecphora tricostata pamlico Wilson, 1987

Choptank Formation

The Choptank Formation was named and described by Shattuck (1902; 1904) for the shelly, sandy Miocene beds exposed along the Choptank River, Talbot County, and in the Calvert Cliffs in Calvert County, Maryland. The Choptank was originally composed of "Zones 16 through 20" of Shattuck. Ward (1984) recommended the placement of "Zones 14, 15, and 16" in the Calvert Beach Member, as defined by Gernant (1970), and expanded by Ward (1984). Blackwelder and Ward (1976) recommended "Zone 20," or the Conoy Member of Gernant (1970), removed from the Choptank Formation and placed in the St. Marys Formation. Distribution of the Choptank beds is shown in Figure 6C.

The Choptank Formation consists of three members (in ascending order), the Drumcliff, St. Leonard, and Boston Cliffs. Mollusks common in the Choptank Formation are shown on Plate 6.

St. Marys Formation

The St. Marys Formation was named and described by Shattuck (1902; 1904) for Miocene beds exposed along the Calvert Cliffs in Calvert County, Maryland, and along the St. Marys River,

St. Marys County, Maryland. The St. Marys was divided into three units by Ward (1984): Conoy Member ("Zone" 20 of Shattuck, 1904), Little Cove Point Member ("Zones" 21–23 of Shattuck, 1904), Windmill Point Member ("Zone" 24 of Shattuck, 1904). Common mollusks are shown on Plate 7. The distribution of the St. Marys beds is shown in Figure 6D.

Eastover Formation

Claremont Manor Member

Two facies of the Claremont Manor Member are very evident: a sandy, basal trangressive portion which grades into a silty clay containing numerous diatoms and an overlying clayey sand. Mollusks in the Claremont Manor Member are low in diversity and are usually poorly preserved (Plate 8). The distribution of the Claremont Manor Member is shown on Figure 7A.

Cobham Bay Member

The Cobham Bay Member consists of very shelly, well-sorted sand and unconformably overlies the Claremont Manor Member. The unit is quite thin and is approximately 12.0 feet (3.6 m) thick at Cobham Wharf, Surry County, Virginia, the type area. The distribution of the Cobham Bay Member is shown in Figure 7B.

Mollusks in the Cobham Bay Member are much more diverse than the Claremont Manor Member and probably represent open-shelf, more normally saline, warm conditions.

POTOMAC RIVER SECTIONS

STOP 1. AQUIA CREEK 0.5 miles (0.8 km) above Thorney Point.

	Ft	m
Sloped and covered by vegetation	3.0	0.9
Paleocene		
Aquia Formation		
Piscataway Member		
Sand, grayish-orange (10 YR 7/4), silty, fine, very glauconitic, poorly sorted, weathered, and leached; some molds and casts	9.0	2.0
-Unconformity-		
Brightseat Formation		
Sand, dark-olive-black (5 Y 2/1), micaceous, clayey, silty, very fine, well-sorted in the lower half, weathered to grayish-orange in the upper half; an 8 inch (20 cm) indurated capping present at the lower end of the exposure but beveled off at the upper end of the exposure	7.0	2.1
-Unconformity-		
Lower Cretaceous		
Patapsco Formation		
Sand, well-consolidated, clayey, silty light blue-gray (5 B 7/1); very irregular, burrowed and eroded upper surface	0–1.0	0–0.3
-Sea Level-		

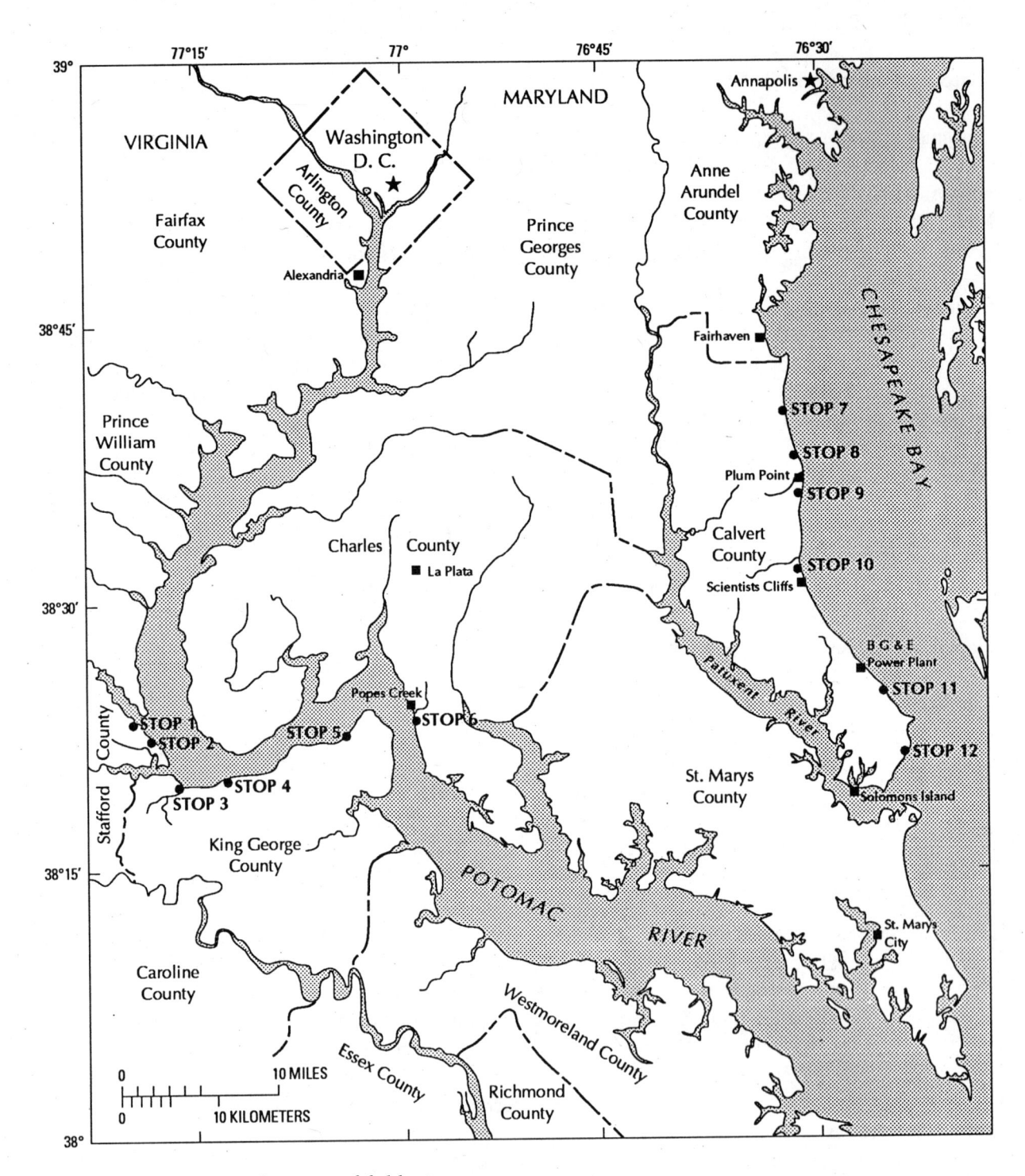

Figure 8. Map showing location of field trip stops.

Below the mouth of Aquia Creek, most of the good exposures are on the Virginia shore for the next 5.0 miles (8.0 km). The bluffs immediately downriver of the mouth of Aquia Creek are the site of the lectostratotype section of the upper Paleocene Aquia Formation. Ward (1985, p. 62) described the section as follows:

STOP 2. 1.5 miles (2.4 km) below the mouth of Aquia Creek.

	Ft	m
Covered	5.0	1.5

Eocene
  Nanjemoy Formation
    Sand, yellowish-gray (5 Y 8/1), weathered, moderately glauconitic, fine
    — 5.0 / 1.5

Paleocene
  Marlboro Clay
    Clay, light gray (N 8), weathered; where this bed is absent there is a distinct line between the Aquia and the overlying bed
    — 0.0–.75 / 0.0–0.23

  Aquia Formation
    Paspotansa Member
      Sand, weathered, grayish-orange (10 YR 7/4), glauconitic, fine; contains large number of *Turritella* in lenses, bands, and large indurated masses
      — 35.0 / 10.7

      Sand, olive-black (5 Y 2/1), fine, well-sorted, silty; scattered poorly preserved *Turritella*
      — 25.0 / 7.6

      Sand, olive-black (5 Y 2/1), glauconitic, very-fine, well sorted; many *Turritella* ("zone" 6 of Clark and Martin, 1901)
      — 1.5 / 0.5

                -Unconformity-

  Aquia Formation
    Piscataway Member
      Sandstone, light olive-gray (5 Y 6/1), indurated, glauconitic: many molds and casts, some siliceous replacements ("zone" 5 of Clark and Martin, 1901)
      — 2.0 / 0.6

      Sand, olive-gray (5 Y 4/1), very glauconitic, silty, clayey, very shelly poorly sorted; packed with large bivalves and *Turritella*. Appears light olive-gray (5 Y 6/1) from a distance due to large numbers of mollusks present; preservation poor to moderate; irregularly indurated in beds where *Ostrea* are concentrated
      — 12.0 / 3.7

                -Sea Level-

STOP 3. Belvedere Beach

Right bank of the Potomac River, 0.3 mile (0.5 km) above Belvedere Beach, King George County, Virginia.

	Ft	m
Covered	5.0	1.5

Paleocene
- Aquia Formation
  - Paspotansa Member
    - Sand, olive-black (5 Y 2/1), fine, well-sorted, silty, micaceous, glauconitic; numerous *Turritella*, scattered as well as in distinct bands, common *Ostrea sinuosa*, moderate molluscan diversity — 12.0 / 3.7

-Sea Level-

STOP 4. Bluffs Below Chatterton (Ward, 1985, p. 64)

Right bank of the Potomac River, 1.75 miles (2.8 km) below the large wharf at Fairview Beach, King George County, Virginia.

	Ft	m
Sloped and covered	5.0	1.5

Eocene
- Nanjemoy Formation
  - Potapaco Member (Bed A)
    - Sand, grayish-yellow, weathered, clayey, fine, poorly sorted, glauconitic — 6.0 / 1.8

-Unconformity-

Paleocene
- Marlboro Clay
  - Clay, light gray (N 7), somewhat weathered, blocky — 6.0 / 1.8
- Aquia Formation
  - Paspotansa Member
    - Sand, grayish-yellow, silty, very weathered; molds of *Turritella mortoni* — 2.5 / 0.7

-Sea Level-

STOP 5. Woodstock Member Lectostratotype

Right bank of Potomac River, high bluffs 2.0 miles (3.2 km) above Mathias Point.

	Ft	m

Pleistocene
- Sand, orange, coarse — 5.0 / 1.5

        Conglomerate, sand, gravel, cobbles, boulders

                                                      3.0   0.9

                      -Unconformity-

Miocene

    Calvert Formation

        Fairhaven Member

            Clay, yellowish-gray (5 Y 8/1), silty, somewhat sandy near base, weathered, blocky, diatomaceous; phosphate and sand along contact

                                              17.0   5.2

                      -Disconformity-

Eocene

    Nanjemoy Formation

        Woodstock Member

            Sand, pale greenish-yellow (10 Y 8/2), weathered, fine, micaceous; molds and casts; upper surface very eroded and burrowed with as much as 3 feet (1.0 m) of relief

                                            25.0   7.6

            Sand, olive-black (5 Y 2/1), silty, very fine, micaceous, glauconitic; small mollusks with moderate preservation

                                            15.0   4.6

                      -Sea Level-

STOP 6. Popes Creek, 0.95 miles (1.5 km) below.

                                                            Ft   m

Pleistocene

    Conglomerate, yellowish-orange, weathered; gravel, sand, cobbles, boulders

                                                25.0   7.6

Miocene

    Calvert Formation

        Fairhaven Member

            Clay, light yellowish-gray (5 Y 9/1), blocky, diatomaceous

                                            10.0   3.0

            Sand, yellowish-gray (5 Y 8/1), silty

                                              2.0   0.6

                      -Disconformity-

            Clay, light yellowish-gray (5 y 9/1), blocky

                                              0.5   0.2

            Sand, yellowish-gray (5 Y 8/1), silty

                                              5.0   1.5

            Sand, olive-brown (5 Y 4/4), silty, phosphatic, pebbles

                                              1.5   0.5

            Sand, yellowish-gray (5 Y 7/2), silty

                                              2.5   0.8

            Sand, olive-brown (5 Y 4/4), silty phosphatic, pebbles

                                              2.0   0.6

                      -Unconformity-

Eocene
> Nanjemoy Formation (?) The following unit is provisionally referred to the Woodstock Member.
>> Sand, olive-gray; (5 Y 4/1), medium, very glauconitic; many molds and casts; unit becoming thicker downstream
>> 0.75   0.23

>> -Disconformity-

Nanjemoy Formation
> Woodstock Member
>> Sand, olive-black (5 Y 2/1), very fine, micaceous, silty, glauconitic; many small mollusks, poorly preserved
>> 5.0   1.5

>> Concretions, olive-gray (5 Y 4/1), calcareous, sandy, glauconitic, rounded
>> 5.0   1.5

>> Sand, olive-black (5 Y 2/1), very fine, micaceous, silty, glauconitic; many small mollusks, moderate preservation
>> 5.0   1.5

- Sea Level -

CHESAPEAKE BAY SECTION

STOP 7. Randle Cliffs (Northern End), high bluff just south of Chesapeake Beach.

	Ft	m

Soil
  2.0   0.6

-Unconformity-

Miocene
> Choptank Formation
>> Boston Cliffs Member
>>> Silt, sandy, blocky (Bed 19)
>>> 7.0   2.1

>> -Unconformity-
>> Drumcliff–St Leonard Member
>>> Silt, sandy, blocky (Bed 17-18)
>>> 10.0   3.0

> Calvert Formation
>> Calvert Beach Member
>>> Silt, clayey, blocky (Bed 15)
>>> 6.0   1.8

>> Sand, silty, shelly (Bed 14)
>> 17.0   5.1

>> Plum Point Member
>>> Silt, clayey, blocky (Bed 13)
>>> 6.0   1.8

Sand, silty, poorly preserved shell (Bed 12)

    3.0   0.9

Silt, clayey, blocky (Bed 11)

    3.0   0.9

Sand, silty, very shelly (Bed 10)

    10.0   3.0

-Unconformity-

Sand, silty, with numerous Corbulids concentrated in bands (Bed 4-9)

    35.0   10.6

Fairhaven Member

  Silt, clayey, blocky, burrowed (Bed 3)

    2.0   0.6

-Sea Level-

STOP 8. Camp Roosevelt.

    Ft   m

Soil

    2.0   0.6

Miocene

  Choptank Formation

    Boston Cliffs Member

      Silt, sandy, fine with sand near base (Bed 19)

    14.0   4.2

    Drumcliff Member

      Sand, silty, clayey, poorly-preserved shell (Bed 17)

    1.0   0.3

-Unconformity-

Calvert Formation

  Calvert Beach Member

    Silt, clayey, blocky (Bed 15)

    10.0   3.0

    Sand, silty, clayey, poorly-preserved shells (Bed 14)

    10.0   3.0

  Plum Point Member

    Silt, clayey, blocky (Bed 13)

    7.0   2.1

    Sand, silty, shelly (Bed 12)

    4.0   1.2

-Unconformity-

Silt, clayey, blocky (Bed 11)

    4.0   1.2

Sand, silty, very shelly (Bed 10)

    12.0   3.6

-Unconformity-

Sand, silty, numerous small *Varicorbula* concentrated in several distinct bands (Bed 4-9)

    25.0   7.6

STOP 9. Plum Point, 1.0 miles (1.6 km) south.

	Ft	m
Miocene		
Choptank Formation		
Boston Cliffs Member		
Silt, clayey, block (Bed 19)	18.0	5.4
St. Leonards Member		
Silt, clayey, block (Bed 18)	9.0	2.7
Drum Cliff Member		
Sand, silty, clayey, some preserved shells (Bed 17)	4.0	1.2
Calvert Formation		
Calvert Beach Member		
Silt, clayey, blocky, laminated (Bed 15)	10.0	3.0
Sand, silty, clayey, moderately shelly (Bed 14)	15.0	4.5
Plum Point Member		
Silt, clayey, blocky (Bed 13)	13.5	4.1
Sand, silty, clayey, poorly preserved mollusks (Bed 12)	2.5	0.7
Silt, clayey, blocky (Bed 11)	11.0	3.3
Sand, silty, very shelly (Bed 10)	9.0	2.7

-Sea Level-

STOP 10. Parkers Creek, Just Above Scientists Cliffs.

	Ft	m
Covered with vegetation		
Miocene		
St. Marys Formation		
Clay, silty	4.9	1.5
Choptank Formation		
Boston Cliffs Member		
Sand, silty, fine, with many mollusks	13.1	4.0
St. Leonard Member		
Sand, clayey, silty, well-burrowed, some molds of mollusks	14.7	4.5
Drumcliff Member		
Sand, very shelly, fine, many large mollusks, well-		

        preserved

                                                    13.1   4.0

Calvert Formation
    Calvert Beach Member
        Sand, silty, fine, scattered, small, poorly preserved mollusks

                                                    6.5   2.0

        Sand, shelly, silty, many mollusks, especially *Glossus*

                                                    4.9   1.5

    Plum Point Member
        Clay, blocky, silty

                                                    9.8   3.0

        Sand, shelly, silty mollusks numerous but poorly preserved

                                                    0.9   0.3

        Clay, blocky, silty

                                                    4.2   1.3

                      -Sea Level-

STOP 11. Baltimore Gas and Electric Power Plant.

The section given below is from the bluff just upbay from the power plant, which is now inaccessible. A similar section, near Rocky Point below the plant site, will be visited.

                                                      Ft   m

Pleistocene (?)
Miocene
    St. Marys Formation
        Little Cove Point Member
            Soil

                                                    1.6   0.5

            Sand, pebbly, coarse

                                                    10.0   3.0

            Sand, silty fine, molluscan molds

                                                    14.7   4.5

            Sand, fine, burrowed, clean, well-sorted

                                                    3.2   1.0

            Sand, medium, well-sorted

                                                    3.2   1.0

            Shell hash, clayey, sandy, very worn mollusks

                                                    1.9   0.6

            Clay, sandy, scattered, small fragmented mollusks

                                                    4.9   1.5

            Clay, sandy, scattered, small poorly preserved mollusks

                                                    1.3   0.4

            Clay, sandy scattered, small shells

                                                    4.9   1.5

            Sand, shelly, fine

                                                    1.6   0.5

                    Clay, blocky, molluscan molds
                                                                    1.9   0.6
                Sand, clayey, fine
                                                                    5.1   4.0
                Clay, blocky, molluscan molds abundant along thin
                horizontal planes
                                                                    5.1   4.0
Choptank Formation
    Boston Cliffs Member
        Sand, shelly, fine, abundant large mollusks, upper 1.0
        m indurated
                                                                    5.1   4.0
    St. Leonard Member
        Sand, silty, fine, very burrowed mollusks scarce,
        scattered, poorly preserved
                                                                    5.1   4.0
    Drumcliff Member
        Sand, shelly, silty, fine, abundant mollusks, cetacean
        remains common
                                                                   10.4   3.2

STOP 12. Little Cove Point.
Bluff 0.6 miles (0.9 km) downbay from Little Cove Point.

                                                                    Ft   m
Pliocene (?)
    Sand, orange-gray, interbedded with thin clay layers, flaser-bedded, ripple-marked
                                                                   30.0   9.1
    Sand, reddish-orange (10 YR 5/6) medium to coarse, burrowed, x-bedded, with
    pebbles and cobbles at base
                                                                    5.0   1.5

Miocene
    St. Marys Formation
        Little Cove Point Member
            Sand, yellow-orange (10 YR 5/6), poorly sorted,
            burrowed
                                                                   13.0   3.9
            Sand, olive-gray (5 Y 4/1), fine, silty, interbedded with
            silty-clay
                                                                   15.0   4.5
            Sand, olive-gray (5 Y 4/1), silty, fine, molluscan molds
            only
                                                                   11.0   3.3
            Sand, olive-gray (5 Y 4/1), silty fine, glauconitic;
            abundant mollusks
                                                                    5.0   1.5
            Sand, olive-gray (5 Y 4/1), silty fine, few mollusks
                                                                    6.0   1.8
            Sand, olive-gray (5 Y 4/1), fine, very shelly; mollusks

dominated by *Turritella*, many worn

                                                            1.0   0.3

Sand, grayish-olive-gray (5 G 4/1), silty, fine, burrowed, with small, fragile, mollusks

                                                            3.0   0.9

REFERENCES CITED

Abbott, W. H., 1978, Correlation and zonation of Miocene strata along the Atlantic margin of North America using diatoms and silicoflagellates: Marine Micropaleontology, v. 3, p. 15–34.

Andrews, G. W., 1978, Marine diatom sequence in Miocene strata of the Chesapeake Bay region, Maryland: Micropaleontology, 24, p. 371–406.

Bennett, R. R., and Collins, G. G., 1952, Brightseat Formation, a new name for sediments of Paleocene age in Maryland: Journal of the Washington Academy of Science, v. 42, no. 4, p. 114–116.

Blackwelder, B. W., 1981, Stratigraphy of upper Pliocene and lower Pleistocene marine and estuarine deposits of northeastern North Carolina and southeastern Virginia: U.S. Geological Survey Bulletin 1502-B, 16 p.

———, and Ward, L. W., 1976, Stratigraphy of the Chesapeake Group of Maryland and Virginia: Geological Society of America, NE-SE Meeting, Arlington, VA, Guidebook for Fieldtrip 7B, p. 1–55.

Clark, W. B., 1896, The Eocene deposits of the middle Atlantic Coastal Plain: American Journal of Sciences, 4th Ser., v. 40, p. 499–506.

———, and Martin, G. C., 1901, The Eocene deposits of Maryland: Maryland Geological Survey, Eocene volume, pp. 1–91, 122–204, 258–259.

———, and Miller, B. L., 1906, A brief summary of the geology of the Virginia Coastal Plain: Virginia Geological Survey Bulletin 2, p. 11–24.

———, and ———, 1912, The physiography and geology of the Coastal Plain Province of Virginia: Virginia Geological Survey Bulletin 4, p. 1–58, 88–222.

Cushman, J. A., and Cederstrom, D. J., 1945, An upper Eocene foraminiferal faunal from deep wells in York County, Virginia: Virginia Geological Survey, Bulletin 67, 58 pp.

Dall, W. H., and Harris, G. D., 1892, Correlation papers, Neogene: U.S. Geological Survey Bulletin 84, 349 pp.

Darton, N. H., 1891, Mesozoic and Cenozoic formations of eastern Virginia and Maryland: Bulletin Geological Society of America, v. 2, p. 431–450.

———, 1911, Economic geology of Richmond, Virginia, and vicinity: U.S. Geological Survey Bulletin 483, 48 pp.

———, 1948, The Marlboro clay: Economic Geology, v. XLIII, no. 2, p. 154–155.

──, 1951, Structural relations of Cretaceous and Tertiary formations in parts of Maryland and Virginia: Bulletin Geological Society of America, v. 62, p. 745–780.

Frederiksen, N. O., 1979, Sporomorph biostratigraphy, northeastern Virginia: Palynology, v. 3, p. 129–167.

Gernant, R. E., 1970, Paleoecology of the Choptank Formation (Miocene) of Maryland and Virginia: Maryland Geological Survey Report of Investigations no. 12, 90 pp.

Gibson, T. G., 1982, Depositional framework and paleoenvironments of Miocene strata from North Carolina to Maryland, in Scott, J. M., and Upchurch, S. B., eds., Miocene of the southeastern United States: Florida Bureau of Geology, Special Publication 25, p. 1–22.

──, 1983, Stratigraphy of Miocene through lower Pleistocene strata of the United States central Atlantic Coastal Plain, in Ray, C. E., ed., Geology and paleontology of the Lee Creek mine, North Carolina, I: Smithsonian Contributions to Paleontology, 53, p. 35–80.

──, Andrews, G. W., Bybell, L. M., Frederiksen, N. O., Hansen, T., Hazel, J. E., McLean, D. M., Wilmer, R. J., and Van Nieuwenhuise, D. S., 1980, Geology of the Oak Grove Core: Part 2, Biostratigraphy of the Tertiary strata of the core: Virginia Division of Mineral Resources Publication 20, p. 14–30.

──, and Bybell, L. M., 1984, Foraminifers and calcareous nannofossils of tertiary strata in Maryland and Virginia: A summary, in Frederiksen, N. O., and Krafft, K., eds., American Association of Stratigraphic Palynologists Field Trip Volume and Guidebook, p. 181–189.

Glaser, J. D., 1971, Geology of mineral resources of southern Maryland: Maryland Geological Survey Report of Investigations 15, 84 pp.

Hazel, J. E., 1968, Ostracodes from the Brightseat Formation (Danian) of Maryland: Journal of Paleontology, v. 42, p. 100–142.

──, 1969, Faunal evidence for an unconformity between the Paleocene Brightseat and Aquia Formations (Maryland and Virginia): U.S. Geological Survey Professional Paper 650-C, p. 58–65.

Mansfield, W. C., 1944, Stratigraphy of the Miocene of Virginia and the Miocene and Pliocene of North Carolina: Pt. 1, Pelecypoda: U.S. Geological Survey Professional Paper 199-A, p. 1–19.

Martini, E., 1971, Standard Tertiary and Quarternary calcareous nannoplankton zonation, in Farinacci, A., ed., Proceedings of the Second Planktonic Conference: Roma, Edizioni Tecnoscienza, v. 2, p. 739–785.

Reinhardt, J., Newell, W. L., and Mixon, R. B., 1980, Geology of the Oak Grove core, Part 1, Tertiary lithostratigraphy of the core: Virginia Division of Mineral Resources, Publication 20, p. 1–13.

Rogers, W. B., 1884, A Reprint of Annual Reports and Other Papers on the Geology of the Virginias: New York, D. A. Appleton and Company, 832 p.

Ruhle, J. L., 1962, Guidebook to the Coastal Plain of Virginia north of the James River: Virginia Division of Mineral Resources, Information Circular 6, 46 p.

Shattuck, G. B., 1902, The Miocene Formations of Maryland: Science, v. XV, no. 388, p. 906.

———, 1904, Geological and paleontological relations, with a review of earlier investigations: Maryland Geological Survey, Miocene Volume, p. 33–94.

Stephenson, L. W., Cooke, C. W., and Mansfield, W. C., 1933, Chesapeake Bay Region, International Geological Congress, XVI Session, United States, 1933, Guidebook 5, Excursion A-5, 49 p.

Vail, P. R., and Mitchum, R. M., Jr., 1979, Global cycles of relative changes of sea level from seismic stratigraphy, in Geological and geophysical investigations of continental margins: American Association of Petroleum Geologists, Memoir 29, p. 469–472.

Ward, L. W., 1984, Stratigraphy and paleontology of the outcropping Tertiary beds along the Pamunkey River–central Virginia Coastal Plain, in Ward, L. W., and Krafft, K., eds., Stratigraphy and paleontology of the outcropping Tertiary beds in the Pamunkey River Region, central Virginia Coastal Plain: Atlantic Coastal Plain Geological Association 1984 Field Trip Guidebook, p. 11–17, 240–280.

———, 1985, Stratigraphy and characteristic mollusks of the Pamunkey Group (lower Tertiary) and the Old Church Formation of the Chesapeake Group–Virginia Coastal Plain: U.S. Geological Survey Professional Paper 1346, 78 p.

———, and Blackwelder, B. W., 1980, Stratigraphic revision of upper Miocene and lower Pliocene beds of Chesapeake Group, middle Atlantic Coastal Plain: U.S. Geological Survey Bulletin 1482-D, 61 p.

PLATE 1. Mollusks common in the Piscataway Member of the Aquia Formation.

All specimens were collected from the Pamunkey River 0.5 mi (0.3 km) east of Wickham Crossing, Hanover County, Va (USGS Locality 26337).

Figure 1. *Ostrea alepidota* Dall, 1898. Left valve of specimen (USNM 366570); length 73.5 mm, height 85.1 mm.
2. *Ostrea alepidota* Dall, 1898. Right valve of specimen (USNM 366470); length 65.4 mm, height 82.6 mm.
3. *Turritella humerosa* Conrad, 1835. Apertural view of an incomplete specimen (USNM 366471); height 32.5 mm.
4. *Ostrea alepidota* Dall, 1898. Left valve of specimen (USNM 366472); length 58.9 mm, height 74.3 mm.
5. *Ostrea alepidota* Dall, 1898. Left valve of specimen (USNM 366473); length 36.3 mm, height 44.6 mm.
6. *Turritella mortoni* Conrad, 1830. Apertural view of incomplete specimen (USNM 366474); height 35.8 mm.
7. *Cucullaea gigantea* Conrad, 1830. Left valve of specimen (USNM 366475); length 42.4 mm, height 27.4 mm.
8. *Pitar pyga* Conrad, 1845. Left valve of a double-valved specimen (USNM 366476); length 46.3 mm, height 38.9 mm.
9. *Crassatellites capricranium* (Rogers, 1839). Left valve of specimen (USNM 366477); length 59.1 mm, height 38.6 mm.
10. *Dosiniopsis lenticularis* (Rogers, 1839). Left valve of specimen (USNM 366478); length 47.1 mm, height 44.5 mm.

PLATE 2. Mollusks common in the Paspotansa Member of the Aquia Formation.

All except Figure 5 were collected from the Potomac River, 0.3 mi (0.5 km) above Belvedere Beach, King George County, Va.

Figure 1. *Cucullaea gigantea* Conrad, 1830. Left valve of specimen (USNM 366479); length 83.3 mm, height 71.8 mm.

2. *Pycnodonte* sp. Left valve of specimen (USNM 366480); length 74.2 mm, height 74.6 mm.

3. *Turritella mortoni* Conrad, 1830. Apertural view of specimen (USNM 366481); height 105.0 mm.

4. *Ostrea sinuosa* Rogers and Rogers, 1837. Left valve of specimen (USNM 366482); length 152.8 mm, height 134.3 mm.

5. *Venericardia regia* Conrad, 1865. Left valve of specimen (USNM 366483) from the Potomac River, 0.1 miles (0.3 km) below the mouth of Passapatanzy Creek, King George County, Va. (USGS Locality 26341); length 75.4 mm, height 71.7 mm.

6. *Turritella mortoni* Conrad, 1830. Apertural view of nearly complete specimen (USNM 366484); height 92.3 mm.

7. *Crassatellites alaeformis* Conrad, 1830. Right valve of specimen (USNM 366485); length 51.9 mm, height 22.7 mm.

8. *Ostrea sinuosa* Rogers and Rogers, 1837. Right valve of specimen (USNM 366482); length 98.8 mm, height 118.3 mm.

9. *Pitar pyga* Conrad, 1845. Right valve of specimen (USNM 366486); length 33.6 mm, height 28.1 mm.

10. *Turritella humerosa* Conrad, 1835. Apertural view of an incomplete specimen (USNM 366487); height 49.8 mm.

PLATE 3. Mollusks common in the Potapaco Member of the Nanjemoy Formation.

Figures 1, 3, and 5 from the Pamunkey River, 0.8 mi (1.3 km) below Hanover-town, Hanover County, Va. (USGS Locality 263777). Figures 2, 4, and 6 from the Pamunkey River, 0.45 mi (0.72 km) above the mouth of Millpond Creek on the right bank, Hanover County, Va. (USGS Locality 26424). Figures 7-9, 11, and 12 from the Potomac River, 2.3 mi (3.9 km) above Popes Creek, Charles County, Md. (USGS Locality 26425). Figure 10 from the Rappahannock River, opposite Goat Island, Caroline County, Va. (USGS Locality 26360).

Figure
1. *Cubitostrea* sp. Left valve of specimen (USNM 366488); length 48.5 mm, height 51.6 mm.
2. *Cubitostrea* sp. Left valve of specimen (USNM 366489); length 35.7 mm, height 50.4 mm.
3. *Cubitostrea* sp. Left valve of specimen (USNM 366490); length 37.7 mm, height 58.5 mm.
4. *Cubitostrea* sp. Left valve of specimen (USNM 366491); length 18.7 mm, height 27.6 mm.
5. *Cubitostrea* sp. Right valve of specimen (USNM 366492); length 47.9 mm, height 56.2 mm.
6. *Cubitostrea* sp. Right valve of specimen (USNM 366493); length 18.8 mm, height 21.4 mm.
7. *Venericardia potapacoensis* Clark and Martin, 1901. Left valve of specimen (USNM 366494); length 29.0 mm, height 24.1 mm.
8. *Nuculana parva* (Rogers, 1837). Right valve of specimen (USNM 366495); length 4.62 mm, height 2.85 mm.
9. *Vokesula* sp. Right valve of specimen (USNM 366496); length 4.48 mm, height 4.20 mm.
10. *Lucina* sp. Right valve of specimen (USNM 366497); length 5.14 mm, height 4.66 mm.
11. *Vokesula* sp. Right valve of specimen (USNM 366498); length 4.29 mm, height 3.65 mm.
12. *Cadulus* sp. Lateral view of incomplete specimen (USNM 366499); length 3.45 mm.

PLATE 4. Mollusks common in the Woodstock Member of the Nanjemoy Formation.

Figures 5, 6, and 10–14 from the Pamunkey River, in a small ravine, 0.63 mi (1.01 km) south-southeast of the mouth of Totopotomoy Creek, Hanover County, Va. (USGS Locality 26403). Other localities are as described below.

Figure 1. *Glycymeris* sp. Right valve of specimen (USNM 366500) from the Pamunkey River at the termination of Rte. 732, Hanover County, Va. (USGS Locality 26393); length 22.0 mm, height 21.0 mm.

2–4. Specimens from the Potomac River, 0.95 miles (1.53 km) below the mouth of Popes Creek, Charles County, Md. (Stop 6)(USGS Locality 26397).

2. *Cubitostrea* sp. Left valve of specimen (USNM 366501); length 13.8 mm, height 20.2 mm.

3. *Cubitostrea* sp. Left valve of specimen (USNM 366502); length 11.8 mm, height 17.0 mm.

4. *Cubitostrea* sp. Right valve of specimen (USNM 366503); length 12.0 mm, height 22.0 mm.

5. *Nuculana* sp. Right valve of specimen (USNM 366504); length 4.14 mm, height 2.56 mm.

6. *Macrocallista subimpressa* (Conrad, 1848). Left valve of specimen (USNM 366505); length 21.3 mm, height 13.9 mm.

7–9. Specimens from the Pamunkey River, just upstream of the old Newcastle Bridge, Hanover County, Va. (USGS Locality 26405).

7. *Corbula aldrichi* Meyer, 1885. Right valve of specimen (USNM 366506); length 10.0 mm, height 7.5 mm.

8. *Venericardia ascia* Rogers and Rogers, 1839. Right valve of specimen (USNM 366507); length 41.4 mm, height 37.5 mm.

9. *Venericardia ascia* Rogers and Rogers, 1839. Right valve of specimen (USNM 366508); length 62.6 mm, height 54.1 mm.

10. *Lucina dartoni* Clark, 1895. Left valve of specimen (USNM 366509), length 6.1 mm, height 5.2 mm.

11. *Lunatia* sp. Apertural view of specimen (USNM 366510); height 14.2 mm.

12. *Turritella* sp. Apertural view of incomplete specimen (USNM 366511); height 12.2 mm.

13. *Turritella* sp. Apertural view of incomplete specimen (USNM 366512); height 14.4 mm.

14. *Cadulus* sp. Lateral view of nearly complete specimen (USNM 366513); height 4.93 mm.

PLATE 5. Mollusks common in the Calvert Formation.

Figures 1, 3, 4, 9 are from Plum Point, on the Chesapeake Bay, Calvert County, Md. (Stop 9). Figures 2, 5, 6, 8 are from Camp Roosevelt on the Chesapeake Bay, Calvert County, Md. (Stop 8). Figure 7 is from Mrs. Anderson's Cottages, near Plum Point, on the Chesapeake Bay, Calvert County, Md.

Figure 1. *Pecten humphreysii* Conrad, 1842. Right valve of specimen (USNM 380693); length 52.2 mm, height 47.6 mm.

2. *Bicorbula idonea* Conrad, 1833. Right valve of specimen (USNM 380694); length 28.0 mm, height 24.9 mm.

3. *Pecten humphreysii* Conrad, 1842. Left valve of specimen (USNM 380695); length 48.4 mm, height 35.5 mm.

4. *Lirophora latilirata* (Conrad, 1841). Right valve of specimen (USNM 380696); length 19.9 mm, height 15.6 mm.

5. *Melosia staminea* (Conrad, 1839). Right valve of specimen (USNM 380697); length 30.0 mm, height 28.0 mm.

6. *Astarte cuneiformis* Conrad, 1840. Left valve of specimen (USNM 380698); length 35.6 mm, height 25.0 mm.

7. *Mercenaria* sp. Right valve of specimen (USNM 380699); length 92.5 mm, height 91.4 mm.

8. *Marvacrassatella melinus* Conrad, 1832. Left valve of specimen (USNM 280700); length 80.9 mm, height 51.8 mm.

9. *Ecphora tricostata pamlico* Wilson, 1987. Apertural view of specimen (USNM 280701); height 73.6 mm.

PLATE 6. Mollusks common in the Choptank Formation.

Figure 1. *Stewartia anodonta* (Say, 1824). Right valve of specimen (USNM 405221) from Drumcliff (Jones Wharf), on the Patuxent River, St. Marys County, Md. (USGS Locality 26557); length 46.2 mm, height 49.0 mm.

2. *Lucinoma contracta* (Say, 1824). Right valve of specimen (USNM 405225) from Calvert Beach, on the Chesapeake Bay, Calvert County, Md.; length 35.6 mm, height 32.1 mm.

3. *Timothynus subvexa* (Conrad, 1838). Right valve of specimen (USNM 405227) from Calvert Beach, on the Chesapeake Bay, Calvert County, Md.; length 35.6 mm, height 32.1 mm.

4. *Astarte thisphila* Glenn, 1904. Left valve of holotype specimen (USNM 405239) from Drumcliff (Jones Wharf), on the Patuxent River, St. Marys County, Md. (USGS Locality 26557); length 27.0 mm, height 25.3 mm.

5. *Isognomon* (*Hippochaeta*) sp. Left valve of nearly complete specimen (USNM 405192) from Long Beach, on the Chesapeake Bay, Calvert County, Md.; length 105.0 mm.

6, 7. "*Arca*" *carolinensis* Dall, 1898.

6. Apertural view of holotype (USNM 405320) from Drumcliff (Jones Wharf), on the Patuxent River, St. Marys County, Md. (USGS Locality 26557); height 64.6 mm.

7. Posterior view of the same specimen.

8. *Turritella* sp. Apertural view of specimen (USNM 405306) from Drumcliff (Jones Wharf), on the Patuxent River, St. Marys County, Md (USGS Locality 26557); height 21.4 mm.

9. *Scaphella virginiana* Dall, 1890. Apertural view of neotype (USNM 405342) from Drumcliff (Jones Wharf), on the Patuxent River, St. Marys County, Md (USGS Locality

PLATE 7. Mollusks common in the St. Marys Formation.

Figure 1. *Conus deluvianus* Green, 1830. Apertural view of neotype (USNM 405343) from above Windmill Point, St. Marys River, St. Marys County, Md. (USGS Locality 26554); height 55.3 mm.

2. *Urosalpinx subrusticus* (d'Orbigny, 1852). Apertural view of specimen (USNM 405326); height 27.7 mm.

3. *Nassarius (Tritiaria) peralta* (Conrad, 1868). Apertural view of specimen (USNM 405335) from above Windmill Point, St. Marys River, St. Marys County, Md. (USGS Locality 26554); height 14.9 mm.

4, 6. *Mactrodesma subponderosa* (d'Orbigny, 1852).

4. Exterior of right valve (USNM 405256) from above Windmill Point, St. Marys River, St. Marys County, Md. (USGS Locality 26554); length 103.8 mm, height 78.2 mm.

6. Interior view of left valve (USNM 405257) from the same locality; length 104.2 mm, height 80.9 mm.

5. *Buccinofusus parilis* (Conrad, 1832). Apertural view of specimen (USNM 405339) from above Windmill Point, St. Marys River, St. Marys County, Md. (USGS Locality 26554); height 101.5 mm.

PLATE 8. Mollusks common in the Claremont Manor Member of the Eastover Formation.

Figures 1, 4, and 6 from Cobham Wharf, Va. (USGS Locality 26052). Figures 2, 3, 5, 10, and 12 from just above Sunken Meadow Creek, Surry County, Va. (USGS Locality 26041). Figures 7–9 and 11 from just below the mouth of Upper Chippokes Creek on the James River, Surry County, Va. (USGS Locality 26042).

Figure 1. *Isognomon* sp. Right valve of nearly complete specimen (USNM 258347); length 150.3 mm.

2. *Ecphora gardnerae whiteoakensis* Ward and Gilinsky, 1988. Apertural view of specimen (USNM 258348); height 74.4 mm.

3. *Mercenaria* sp. Left valve of specimen (USNM 258349); length 93.5 mm, height 75.9 mm.

4. *Glycymeris virginiae* Dall, 1898. Left valve of specimen (USNM 258350); length 60.5 mm, height 65.1 mm.

5. "*Arca*" *virginiae* Dall, 1898. Right valve of specimen (USNM 258351); length 82.2 mm, height 57.3 mm.

6. *Lirophora dalli* Olsson, 1914. Right valve of specimen (USNM 258352); length 19.3 mm, height 17.2 mm.

7. "*Arca*" *carolinensis clisea* Dall, 1898. Left valve of specimen (USNM 258353); length 42.7 mm, height 44.1 mm.

8. *Glossus fraterna* (Say, 1824). Left valve of incomplete specimen (USNM 258354); length 82.3 mm, height 80.2 mm.

9. *Euloxa latisulcata* (Conrad, 1839). Left valve of an incomplete specimen (USNM 258355); approx. length 15 mm, height 13.1 mm.

10. *Ostrea* sp. Left valve of specimen (USNM 258356); length 97.7 mm, height 111.9 mm.

11. *Turritella plebeia* ssp. Apertural view of incomplete specimen (USNM 258357); height 98.8 mm.

12. *Chesapecten middlesexensis* (Mansfield, 1936). Right valve of specimen (USNM 258358); length 98.8 mm, height 91.4 mm.

201

7
BOTTOM SEDIMENTS OF THE CHESAPEAKE BAY: PHYSICAL AND GEOCHEMICAL CHARACTERISTICS

Jeffrey P. Halka and James M. Hill
Maryland Geological Survey
Baltimore, MD 21218

INTRODUCTION

Research efforts in the Coastal and Estuarine Geology Program of the Maryland Geological Survey have focused on understanding the recent biogeochemical history of sediments accumulating in the Chesapeake Bay, and the record of sediment infilling which has occurred in the Quaternary. Determining the details of this history requires an integration of sediment chemistry, early stages of diagenesis, sediment textural characteristics, and activities of the benthic infauna.

Participants in this field trip will board the Maryland Geological Survey's Research Vessel *Discovery* at Sandy Point State Park located south of Baltimore (Figure 1). The vessel will travel 35 km south into the middle mesohaline portion of the Bay across from the mouth of the Choptank River. We will deploy acoustic reflection profiling equipment in this area and run a transect across the Bay. The 5.0 kHz component of this system provides a subbottom record showing variations in the thickness and character of Quaternary sediments across the estuary. The higher frequency 300 kHz portion provides an indication of the density structure of the water column, and processes related to that structure. To the north of this transect, off of Kent Island, we will occupy two stations where fine-grained bottom sediments will be collected using a gravity corer. One station will be located in the deep axial channel in 30.5 m water depth and the second on the shallower western flank in 16 m water depth. At each station the bottom core will be extruded on deck and examined for physical and biogenic characteristics, and electrode measurements made to determine the basic geochemical characteristics of the sediment. The information collected will be related to more detailed analyses which have been performed in the lab on cores taken previously at these stations. These studies include detailed grain size and bulk properties, X-Ray analyses, and thorough geochemical work-ups.

GENERAL SETTING

The Chesapeake Bay, the largest estuary in the United States, is a classic coastal plain estuary located on a trailing edge continental margin. It is nearly 300 km long and ranges from 8 to 48 km in width, having a surface area of nearly 6000 km^2. The modern Bay formed in response to sea level changes during and following the last major continental glaciation. At the height of the glacial advance, the Susquehanna River, debouching from the Piedmont physiographic province at the head of the Bay, eroded the gently dipping Tertiary age strata of the Coastal Plain. The resulting river valley was drowned by the rising sea level as the glaciers waned. This history of formation is reflected in the bathymetry, which is characterized by a deep axial channel flanked by broad shallow benches, and in the deeply dissected shoreline. Water depths exceed 30 m in the deep channel, but overall the Bay is quite shallow with an average depth of only 8 m.

The drainage basin encompasses 167,000 km^2 with the Susquehanna River, discharging at the head of the Bay, providing the primary source of fresh water to the system. The Bay is a microtidal

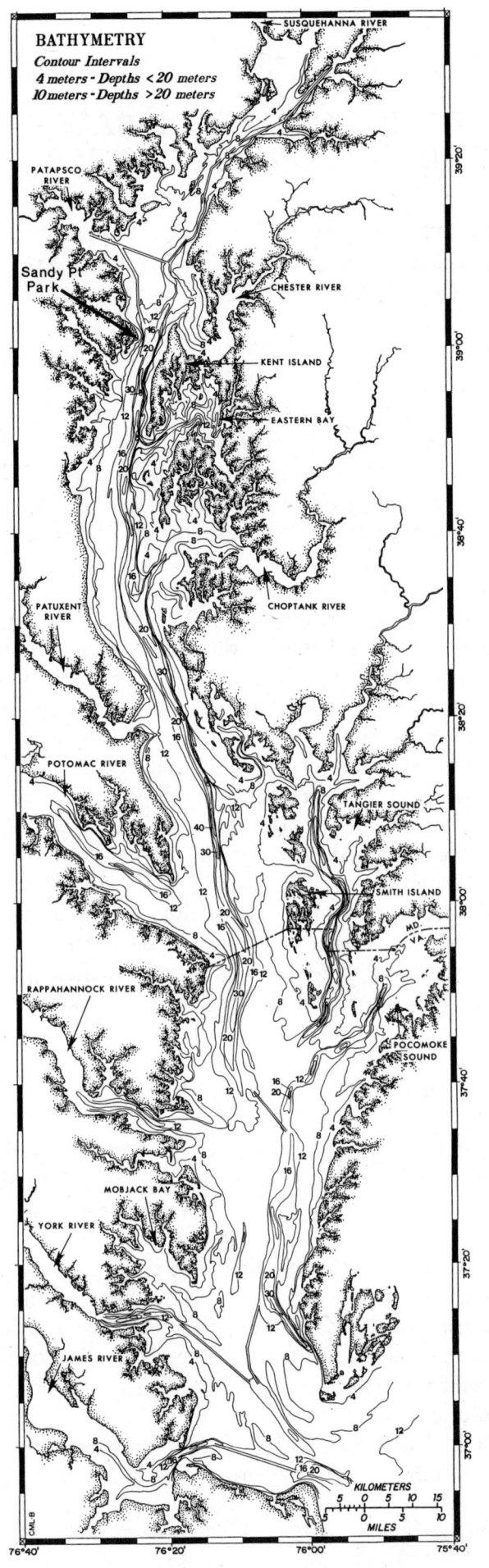

Figure 1. Bathymetric Map of the Chesapeake Bay showing major tributaries and Sandy Point State Park from which the vessel will depart (from Hill and Halka, 1989).

system with a tide range of one meter at the mouth, decreasing progressively up-bay to a minimum of 30 cm in the vicinity of Baltimore, and then increasing to nearly 60 cm at the more constricted head of the Bay. The tidal influence is, however, sufficiently large relative to the freshwater input to produce a partially mixed estuary. Mixing between the relatively fresh surface water and the denser, more saline bottom water occurs throughout the system. The Susquehanna River provides 60 percent of the total freshwater input to the Bay, with the Potomac and the James Rivers ranking second and third.

The water circulation within the Bay is controlled by freshwater inflow, density stratification, basin morphology, tidal forcing, and the Coriolis force. Southward-flowing freshwater travels predominantly along the western side of the basin. Saline water entrained in this southward flow, by mixing, is removed from the Bay to the Atlantic Ocean. The lost volume is replaced by the inflow of seawater, which enters the system as bottom water at the mouth of the Bay. The more saline water travels northward along the eastern side of the basin, where the flow eventually becomes trapped and directed by the main axial channel. The turbidity maximum, located slightly downstream of the landward limit of saline water intrusion, is generally found 30 km from the mouth of the Susquehanna River, although its location varies with river discharge.

The salinity of the Bay varies seasonally in response to the freshwater inputs. Typically, the lowest salinity occurs in the spring and the highest in the fall. Along the length of the Bay the salinity varies from less than 2 ppt at the head to greater than 30 ppt at the mouth, seasonally averaged. The dissolved load is a source of chemical species to the sediments. Of particular interest are dissolved oxidants (such as O_2, NO_3^-, and $SO_4^=$) which affect the geochemical environments of the sediment. Non-conservative processes strongly influence the geochemical environment of the water column and, in turn, the sediment. The most significant of these is the occurrence of hypoxic/anoxic water. These events are confined to the warm summer months when fresh water inflow is low, and are confined primarily to water depths in excess of 10 m. During severe anoxic events, sulfide species are measurable in the water column. A detailed description of this trip and a summary of our most recent research may be found in Hill, J. M., and Halka, J. P., 1989, Seismic and Geochemical Research in Chesapeake Bay, Maryland: American Geophysical Union, Field Trip Guidebook T231, 28th International Geological Congress, 28 p.

8
SEDIMENTOLOGY AND SEQUENCE STRATIGRAPHIC FRAMEWORK OF THE MIDDLE DEVONIAN MAHANTANGO FORMATION IN CENTRAL PENNSYLVANIA

Anthony R. Prave
Department of Earth and Planetary Sciences
The City College of The City University of New York
New York, NY 10031

William L. Duke
Department of Geosciences
Pennsylvania State University
University Park, PA 16802

INTRODUCTION

Rocks of the Middle Devonian Mahantango Formation are some of the earliest coarse deposits of the west-northwestward prograding clastic wedge shed from the Acadian orogenic highlands into the central Appalachian foreland basin (e.g., Faill, 1985). They consist of shallow-marine fossiliferous mudstone, sandstone, and quartz-pebble conglomerate arranged in coarsening upward cycles several meters to a few tens of meters in thickness. Contacts between superposed cycles generally are sharp and erosional, and are commonly overlain by a thin accumulation of poorly sorted coarse clastic fragments and fossiliferous debris. Although a principally shallow marine origin for the Mahantango Formation is accepted by most workers, the mechanism producing depositional cyclicity is not. Consequently, differing sedimentologic interpretations exist for the Mahantango Formation. We propose a variation and expansion of a previous interpretation. We interpret the depositional cyclicity in the Mahantango Formation as having resulted from episodic progradation of a mostly straight tide-dominated shoreline into a storm-dominated marine basin (Duke and Prave, 1989; Prave and Duke, 1989).

The Mahantango Formation forms the bulk of the Givetian-aged strata of the Hamilton Group in Pennsylvania (Figures 1 and 2). These strata exhibit pronounced lithologic variations over relatively short distances, but generally coarse lithologies (the Montebello Member) occur in central Pennsylvania and vertical alternation of shale, siltstone, and, in places, limestone occur elsewhere. This lithological variation is reflected in complicated stratigraphic subdivisions (Figure 2) which have been subject to continuous revision and redefinition (Faill et al., 1978). In our depositional framework for the Mahantango Formation we have downplayed the formal member subdivisions and relied more on what we infer are genetically significant depositional cycles.

Our objectives for this trip are to examine the facies we have defined from observations of eleven carefully measured sections in the central Pennsylvania fold and thrust belt (Figure 1). This data will be used to support our sedimentological interpretations and depositional model for what we believe is a genetically significant depositional, i.e., sequence stratigraphic, framework for the Mahantango Formation. Hopefully, this trip will provide the data necessary to evaluate and discri-

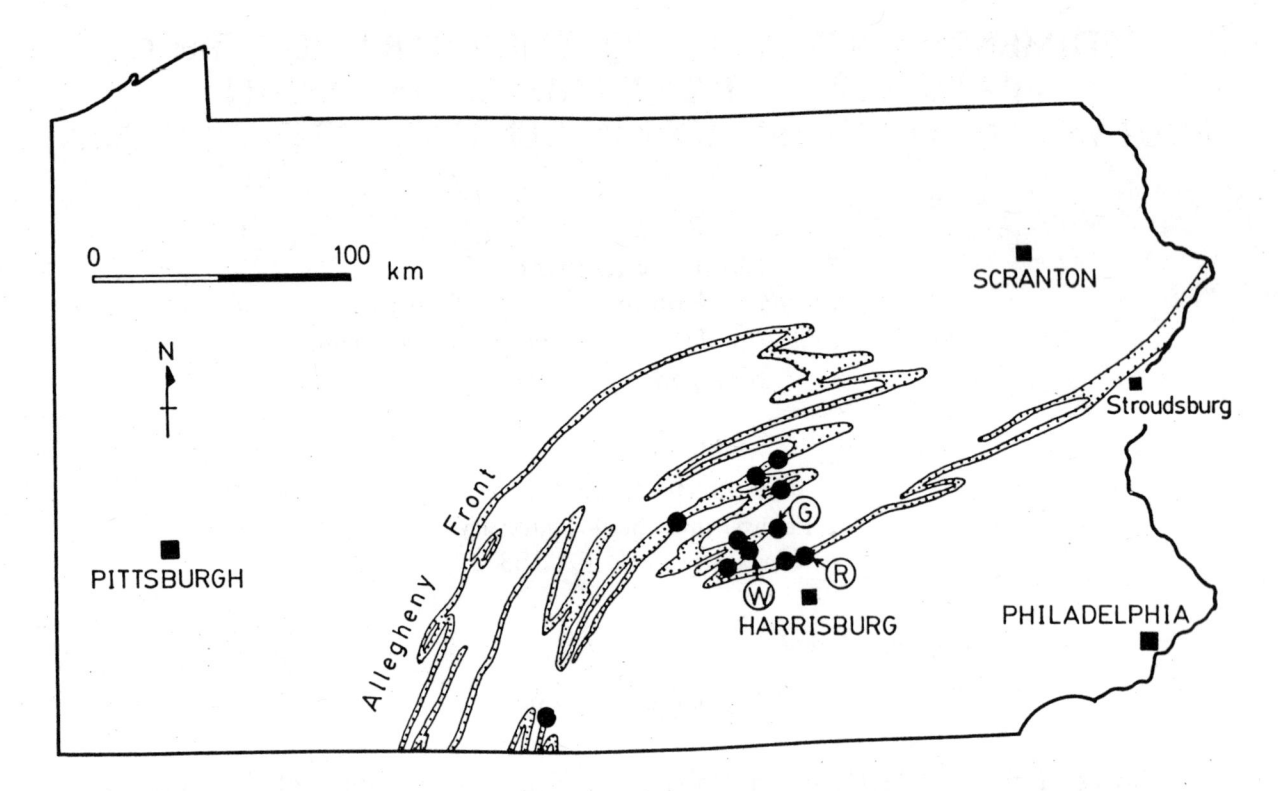

Figure 1. Location of measured sections along the outcrop belt (stippled) of Middle Devonian strata in Pennsylvania. Sections examined during this field trip are indicated by circled letters: G-Girty's Notch; R-Rockville; W-Watts.

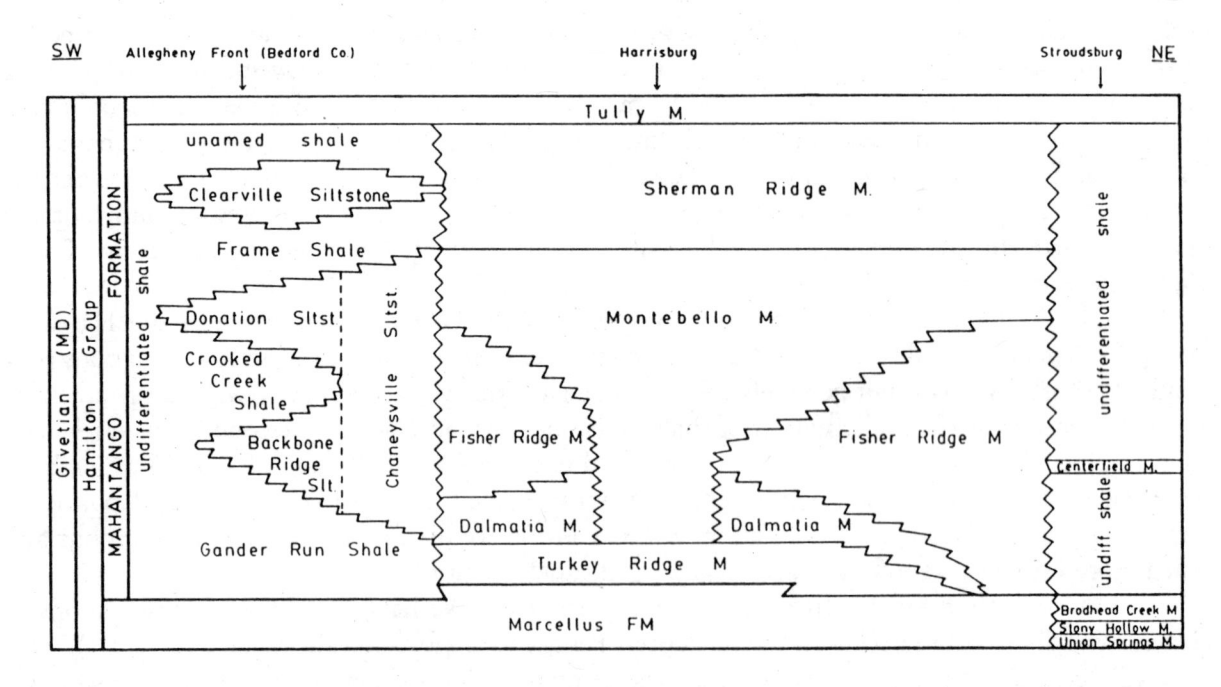

Figure 2. Lithostratigraphic subdivision of the Middle Devonian strata in Pennsylvania. Compiled from numerous sources, notably Faill et al. (1978) and Dennison and Hasson (1976).

minate between our interpretation and previous ones.

During this trip we will examine three of our measured sections (Figures 1, 3–5). These were chosen because they are typical of the vertical facies development and lateral facies variation exhibited by the Mahantango Formation throughout the study area. To date, our work has focused primarily in those areas dominated by sandstone. Clearly the interpretations presented herein must be tested in the laterally adjacent regions dominated by mudstone. Thus, this trip is a progress report on our larger and ongoing investigation of the validity and applicability of sequence stratigraphic methodology to clastic rocks deposited in the Acadian foreland basin.

PREVIOUS WORK

Investigations of sedimentology and stratigraphy of the Mahantango Formation have produced a spectrum of paleoenvironmental interpretations. These include a fluvial-dominated delta (Willard, 1935; Ellison, 1965; Kaiser, 1972; Faill et al., 1978), a storm-dominated prograding shoreline (Goldring and Bridges, 1973), and a system of storm-generated shelf sand ridges (Sarwar and Smoot, 1983; Sarwar, 1984).

The well-developed coarsening-upward cyclicity exhibited by the Mahantango Formation has been recognized and discussed by numerous workers, principally Kaiser (1972), Goldring and Bridges (1973), Faill and Wells (1974), and Hoskins (1976). The hierarchical arrangement of small- and large-scale cycles was described as "nested" cyclicity by Faill et al. (1978), and the straightforward nature of large-scale cycle correlations (in contrast to the complicated member correlations) was demonstrated by Faill and Wells (1974) in their mapping of the Millersburg 15-minute quadrangle.

FACIES DESCRIPTION AND INTERPRETATION

We have defined 19 facies from our descriptive data base collected from careful measurement of eleven sections (Figure 1); we have combined these into seven facies associations (summarized in Table 1).

Facies Assoc. 1: Shale and Mudstone with Thin Beds of Siltstone and Sandstone

Occurrences of Facies Association 1 range in thickness from several decimeters to a few tens of meters and consist mostly of grayish (or rarely red) colored shale or mudstone. These fine clastic strata are fossiliferous, containing a variety of brachiopods, crinoids, and rugose and tabulate corals and less commonly bryozoans, gastropods, pelecypods, and cephalopds. A *Cruziana* ichnofacies occurs locally and includes *Teichichnus, Chondrites, Paleophycus,* and *Planolites.* In places, intervals have been completely homogenized by bioturbation, resulting in a muddy sandstone with a ropey texture of anastomosed, poorly defined burrows.

Thin siltstone and fine sandstone beds are, in many places, interbedded with the mudstone and shale; they locally can comprise up to 40 percent of this facies. These beds typically are a few centimeters thick but can be as much as 20 cm thick. Most are laterally continuous across the outcrop, although some are lenses. Beds have sharp, flat bases and internally are flat to wavy laminated, ripple cross-laminated, or structureless. Some of the thicker beds are small-scale hummocky cross-stratified. Many beds are graded, with upper contacts gradational with overlying mudstone, and resemble B- or BC-type Bouma sequences. Others display sharp upper contacts that locally exhibit small current, symmetrical vortex, or combined-flow ripples.

We interpret the mudstone and shale as fair-weather, offshore marine hemipelagic deposits. The thin siltstone and sandstone beds are interpreted as distal storm deposits, as suggested by their

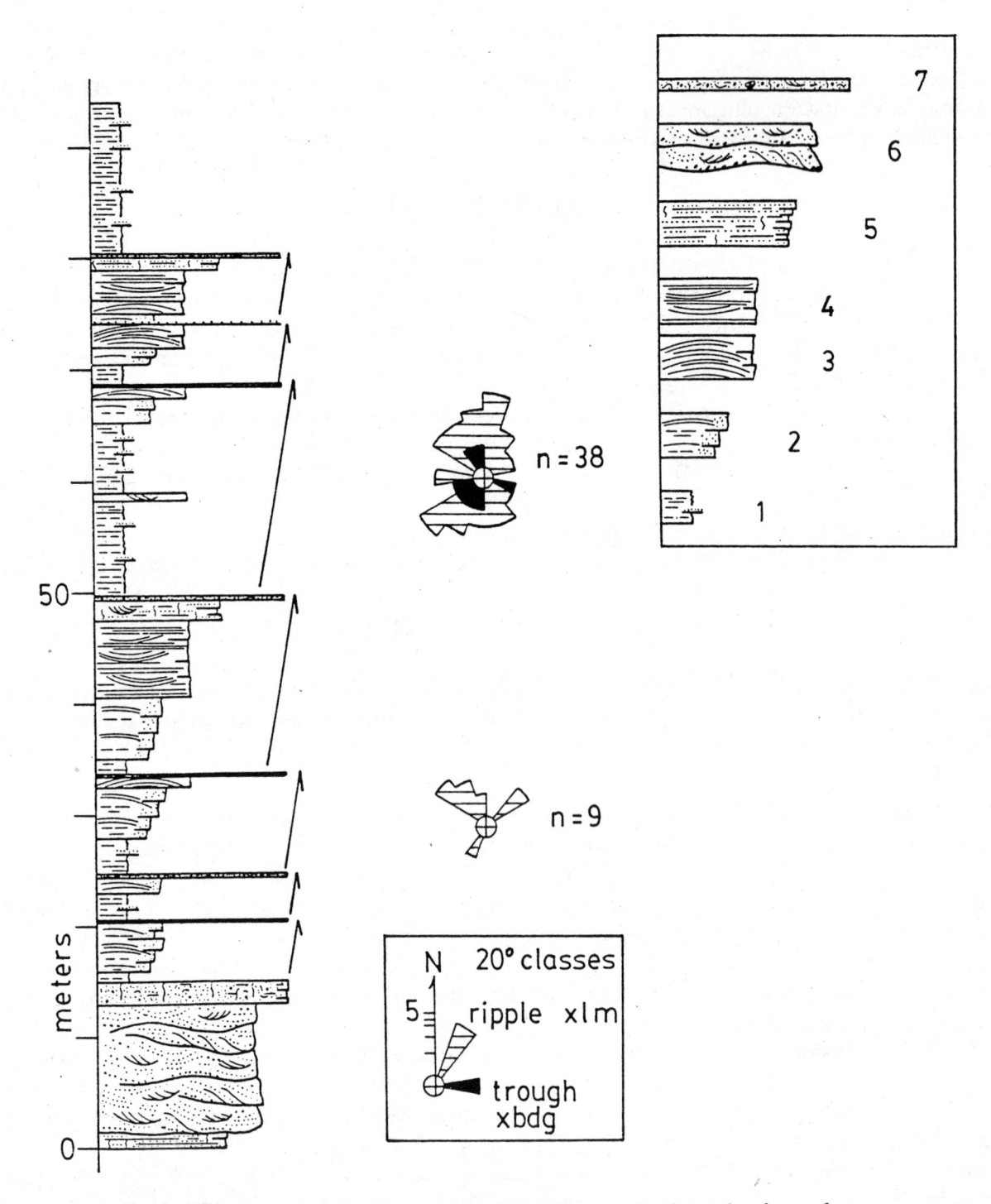

Figure 3. Vertical graphic log of the Watts exposure. Numbers in the legend refer to facies associations defined and described in the text.

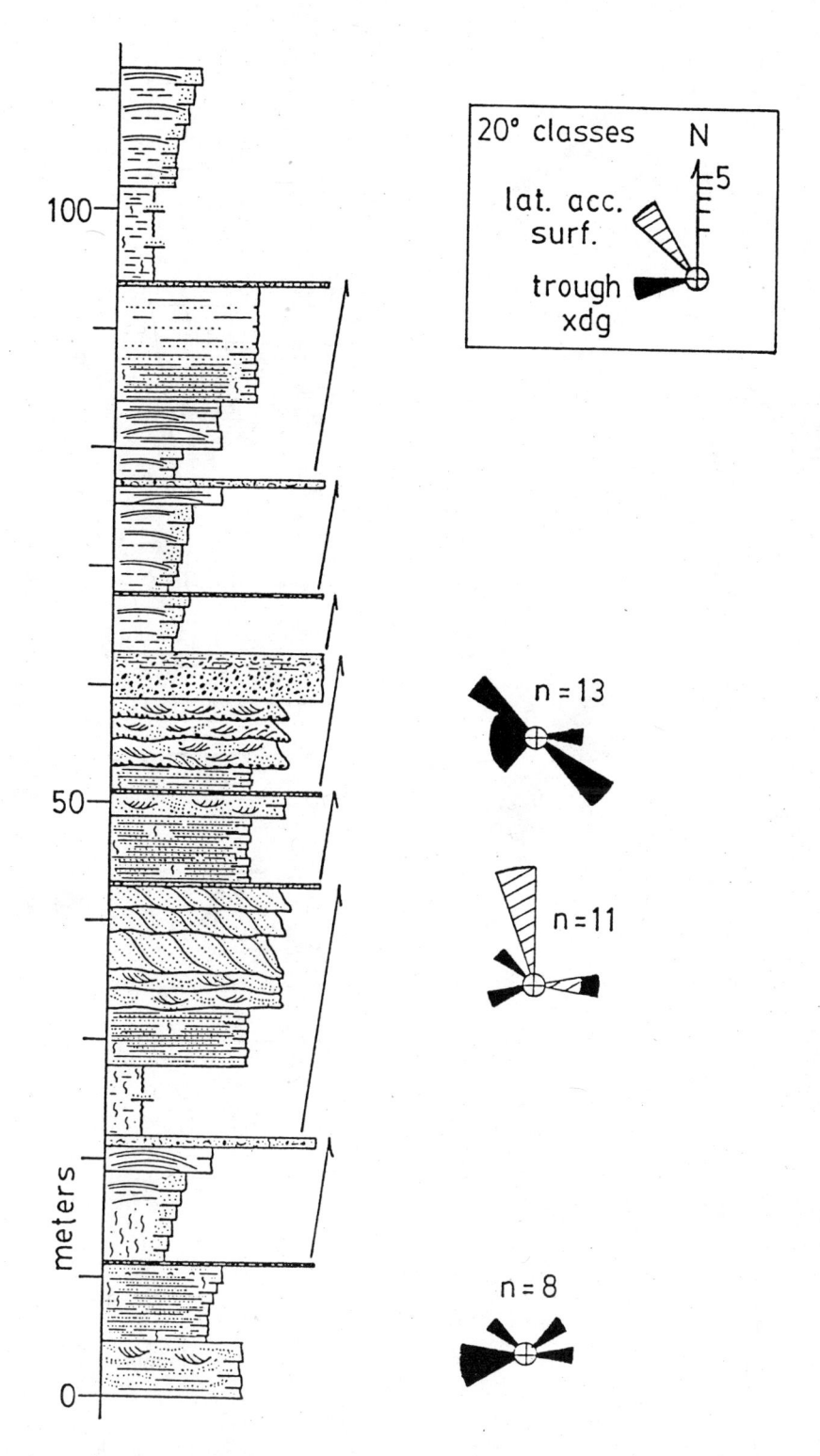

Figure 4. Vertical graphic log of the Girty's Notch exposure. See Figure 3 for key to lithological symbols.

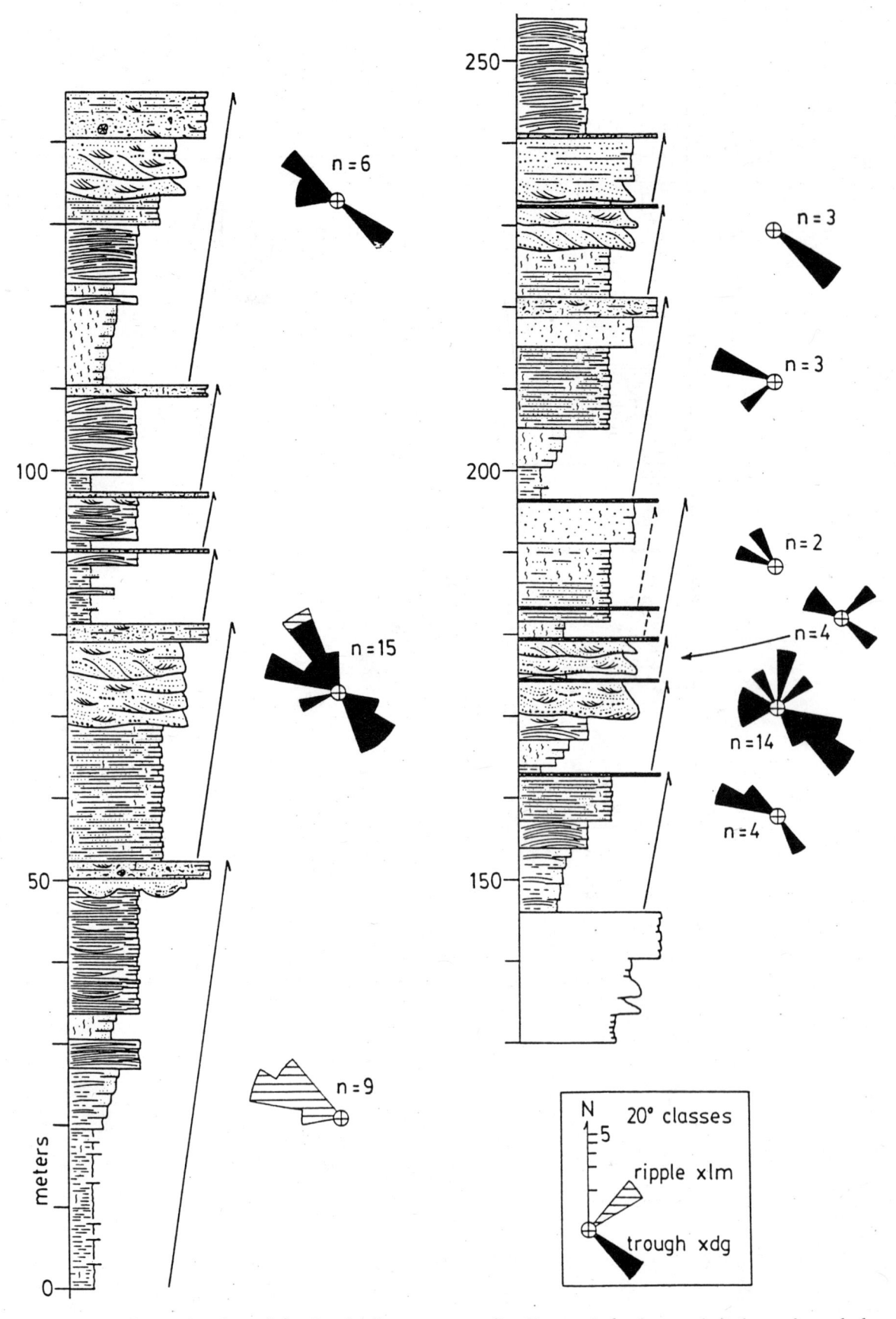

Figure 5. Vertical graphic log of the Rockville exposure. See Figure 3 for key to lithological symbols.

sharp bases, their graded nature, and the local occurrence of vortex ripples and small-scale hummocky cross-stratification.

Facies Assoc. 2: Hummocky Cross-stratified Beds Interbedded with Mudstone

Facies Association 2 strata consist of sharp-based, many decimeters thick (20–150 cm), hummocky cross-stratified siltstone and fine sandstone interbedded with shale or mudstone similar to those of Facies Association 1. This facies association typically occurs in intervals a few meters to over 10 m thick, which display well-developed thickening- and coarsening-upward trends. Hummocky beds have sharp, flat bases locally displaying oriented sole marks and sharp to gradational upper surfaces. Locally, beds are capped with symmetrical vortex ripples, current ripples, or combined-flow ripples. Hummock spacings typically are 1 to 2 m and many of the thicker beds exhibit a pinching and swelling over several meters laterally. In places, flat lamination is common, and, rarely, medium sandstone beds included in this facies exhibit trough cross-bedding.

A wide variety of trace fossils from the *Cruziana* and *Skolithos* ichnofacies are contained within this facies. In places, decimeters-thick intensely bioturbated intervals are present. These commonly exhibit isolated remnants of hummocky cross-stratified and flat-laminated beds. Disarticulated and reworked brachiopod and other faunal fragments are present in many of the sandstones and locally form basal lags.

We interpret the sharp-based hummocky beds as discrete storm deposits emplaced in a proximal offshore setting. The finer clastic deposits represent hemipelagic deposition during fairweather conditions. The intensely bioturbated intervals record the opportunistic invasion by burrowing organisms onto tempestites in proximal offshore settings.

Facies Assoc. 3: Amalgamated Hummocky Cross-stratified Sandstone Bodies

This facies association consists of numerous, discrete fine sandstone beds a few to several decimeters in thickness. They are characterized by undulatory, erosional bases which cause lateral pinching and swelling and bed amalgamation; this results in a tabular sandstone body a few to several meters thick. In places, thin, discontinuous layers or partings of shale separate beds. Internally, beds are either massive or hummocky cross-stratified and flat laminated. Symmetric vortex ripples and other sedimentary structures are rare. In places, small, rounded quartz pebbles and shale rip-up clasts are present and occur either dispersed along laminae or concentrated into thin, discontinuous layers above basal scour surfaces.

Disarticulated and reworked thick-shelled brachiopods and other shallow marine body fossils are present in low abundance, although "fossil hash" layers of brachiopod shells are present locally. Trace fossils, principally of the *Skolithos* ichnofacies, occur in varying densities.

We interpret the amalgamated, hummocky cross-stratified sandstone bodies as storm-dominated deposits of the lower shoreface. As suggested by the rare preservation of fines, water depths were shallow enough for storms to scour and erode most "fairweather" hemipelagic deposits.

Facies Assoc. 4: Swaley Facies Sandstone Bodies

This facies association consists of fine, "clean" sandstone in beds many decimeters thick which form tabular, laterally continuous sandstone bodies up to 5 m thick. Laminae are difficult to distinguish but where discernible are either flat or define broad, shallow concave-up scours or "swales" several meters in width and are draped by flattening-upward laminae. "Hummocks" are absent or rare and are subordinant in scale to the "swales." In places, medium sandstone layers are present and display isolated sets of trough cross-bedding. Thick-shelled brachiopod valves and rounded quartz pebbles are concentrated above the nadirs of some swales. Other biogenic or sedimentary structures were not observed.

Table 1. Interpretive summary of facies defined in this study.

FACIES ASSOC. 1: distal offshore deposits of mainly hemipelagic mudstones and shales with some thin, distal tempestites.

1SH-hemipelagic shale deposited far offshore.
1BM-bioturbated mudstone deposited far offshore.
1MS-offshore mudstone with thin distal tempestites.

FACIES ASSOC. 2: proximal offshore mudstone interbedded with tempestites.

2H1-proximal offshore mudstone with thin hummocky cross-strat. storm beds.
2H2-proximal offshore mudstone with thick hummocky cross-strat. storm beds.
2BS-bioturbated argillaceous sandstone from a proximal offshore setting.

FACIES ASSOC. 3: fine sandstone from the storm-dominated lower shoreface.

3HB-"hummocky to burrowed" sandstone couplets from the lower shoreface.
3HS-amalgamated hummocky cross-strat. beds from the lower shoreface.
3TB-trough cross-bedded sandstone from the lower-upper shoreface.

FACIES ASSOC. 4: swaley facies sandstones from the storm-dominated upper shoreface.

4SW-swaley facies fine sandstone bodies from the upper shoreface.

FACIES ASSOC. 5: tide-dominated channel-mouth shoals (ebb-tidal deltas?) from the lower-upper shoreface.

5VL-tabular, thin beds of fine to medium sandstone with variable lamination types; beds commonly separated by shale partings.
5BB-same as above but bioturbated by traces of the *Skolithos* ichnofacies.

FACIES ASSOC. 6: sandy subtidal (to intertidal?) flats and tidal channels incised to subtidal depths on the shoreface.

6T1-trough cross-bedded medium to pebbly sandstones; lateral accretion surfaces may be developed locally.
6T2-trough cross-bedded to cross-laminated medium sandstones; lateral accretion surfaces may be present locally.
6VB-largely structureless medium to coarse sandstones.

FACIES ASSOC. 7: reworked lag deposits of argillaceous sandstone to conglomerate, forming a system of transgressive, storm-generated shallow marine sand ridges.

7SL-erosive-based single bed lag (inter-ridge facies).
7ML-multi-bed lags interbedded with thin mudstones (inter-ridge facies).
7BL-intensely bioturbated lag (inter-ridge facies).
7CL-low-angle, cross-bedded thick lag deposits (ridge facies).

We interpret these facies as deposits of the upper shoreface or energetic nearshore zone. The absence of biogenic structures, finer clastic deposits, and other sedimentary structures suggests the substrate was under continuous, high-energy agitation.

Facies Assoc. 5: Thin, Tabular Sandstone Beds Separated by Argillaceous Partings

Occurrences of this facies association range from 1 to 20 m in thickness and consist of laterally continuous, many centimeters thick, fine to medium sandstone beds with thin argillaceous partings at bedding-plane contacts. The bases and tops of beds are sharp, flat, and mostly nonerosive; bed tops locally display both current and symmetrical vortex ripples. Internally, beds are ripple cross-laminated, flat to wavy-parallel laminated, trough cross-bedded, small-scale hummocky cross-stratified, and, in places, climbing ripple cross-laminated. Some beds are structureless. The most common vertical transition within any given bed is from flat lamination upward into ripple cross-lamination. Trough cross-bedded bedsets are limited to the coarser sandstone, which generally occurs in the upper parts of this facies. Texturally, sandstone in this association contains minor amounts of argillaceous material and thus is less well sorted than sandstone in adjacent facies associations. Coarsening- or thickening-upward trends are *not* well developed.

A relatively diverse trace fossil assemblage dominated by forms from the *Skolithos* ichnofacies is present in varying densities in the sandstone. The most abundant traces are *Skolithos*, *Rosselia*, and *Teichichnus* burrows a few to several centimeters in length. *Planolites* is present on many bedding planes. Thin layers of disarticulated and reworked brachiopod valves are common. In rare cases, thin brachiopod coquinite lenses, in which many of the valves are articulated, are present, suggesting these are death assemblages. A few sandstones contain abundant ostracods.

The variable stratification types, lateral persistence and thinness of beds, and presence of abundant shale partings suggest the magnitude of the depositing flows varied rapidly and were separated by intervening periods of slack water of relatively short duration. This interpretation also is suggested by the thinness of the shale layers and lack of a "quiet-water" trace fossil assemblage. The tabular beds, their nonerosive nature, local development of climbing ripples, and presence of argillaceous material in the sandstone suggest deposition in a region of pronounced flow expansion and rapid sediment fallout. The presence of wave- and current-formed sedimentary structures, thick-shelled brachiopods, and a *Skolithos* ichnofacies suggest an open, energetic, nearshore marine setting. The presence of ostracods are consistent with a shallow, nearshore marine setting and may reflect a brackish water source. Additionally, the brachiopod death assemblages may reflect salinity changes associated with proximity to the shore.

This data leads us to conclude that the combination of rapid deposition followed by brief slack-water intervals indicates a tidal influence. The bimodality of paleocurrent data strongly supports this conclusion (Figure 6). Based on the stratigraphic context of this facies association, we interpret these deposits as having accumulated in a subtidal setting within or seaward of the mouths of tidal channels incised into the shoreface. Tidal currents appear to have been dominant, but the presence of wave ripples and hummocky cross-strata records the occasional overprinting by oscillatory and combined flows. Consequently, we interpret these facies as channel-mouth shoals or ebb-tidal deltas exposed to the open ocean.

Facies Assoc. 6: Thick, Medium to Pebbly Sandstones

Relatively thick bodies (1 to 10 m) of poorly to moderately sorted, medium to pebbly sandstone compose this facies association. Pebbles are mostly moderately to well-rounded quartz. Sandstones display abundant sets of decimeters-thick trough cross-bedding; paleocurrent bimodality from cross-bed data is common (Figure 6). Many of the boundaries between cross-bed sets are characterized by shale breaks and/or small current ripples, some yielding paleocurrent directions

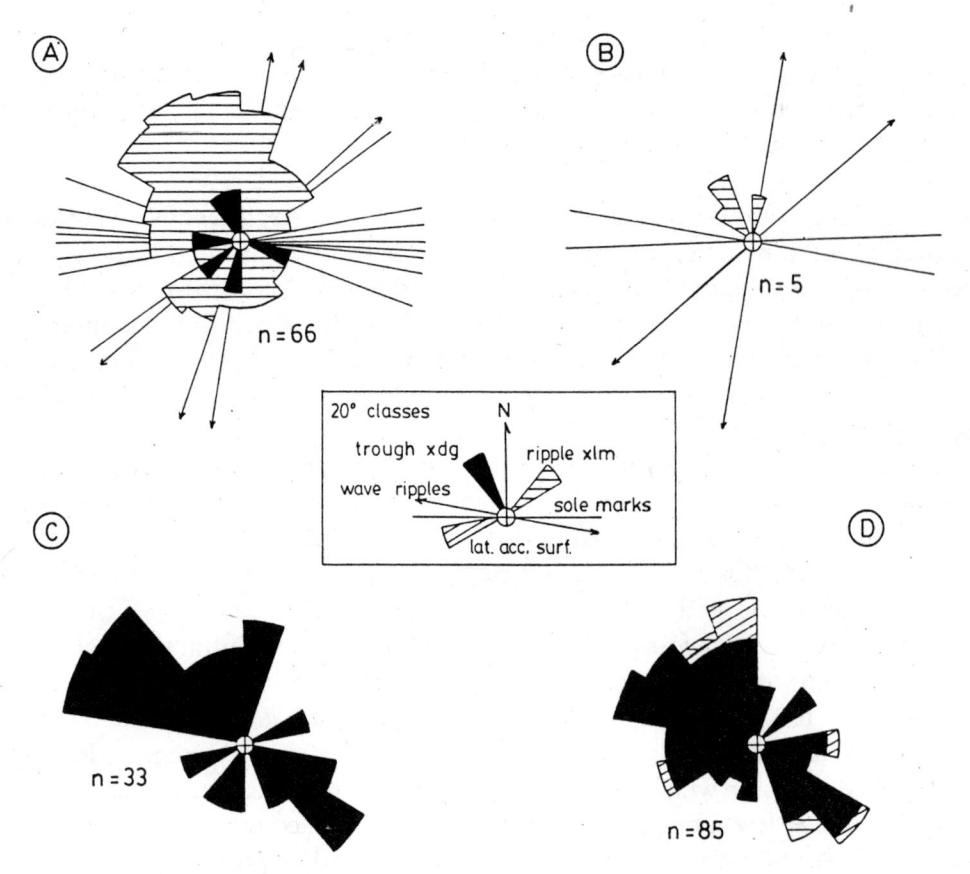

Figure 6. Summary of paleocurrent data measured throughout the study area. Rosettes are equal area. A, Data for offshore storm deposits (Facies Assoc. 1 and 2). B, Data for shoreface sandstones (Facies Assoc. 3 and 4). C, Data for tidal channel mouth shoals/ebb-tidal deltas (Facies Assoc. 5). D, Data for tidal channel deposits (Facies Assoc. 6).

opposite that of the underlying cross-bedding. Mud drapes and reactivation surfaces occur on some bedsets. Layers of shale, mudstone, and fine sandstone are present locally, as are shale rip-up clasts. Some examples of this facies association contain thick intervals (many decimeters up to a few meters) of apparently structureless coarse to pebbly sandstone which locally exhibit faint trough cross-bedding and/or flat lamination.

Basal contacts of many of the sandstones in this facies association are sharply erosional, in places displaying decimeter-scale relief. Pebbly lags are common above these erosional bases and fining-upward trends are well developed. In some of the larger channels lateral accretion bedding is present in erosive-based sets 2 to 10 m thick; restored dips of these range from 5° to 15°.

Large, vertical to oblique burrows are common and locally abundant. Many of the burrows are ornamented; *Rosselia* has been observed, but most remain unidentified. There also are unidentified sub-horizontal, cylindrical burrows with diameters of many millimeters and up to 10 cm in length; their fill is nearly identical to the surrounding sandstone. Thick-shelled, mostly disarticulated and reworked brachiopod valves are common.

We interpret the sandstones of Facies Association 6 as the deposits of a sandy subtidal (possibly intertidal?) flat reworked by ebb and flood tidal currents. The bimodality of trough cross-bedding, development of smaller current ripples with asymmetries opposed to the cross-beds, and presence of thin, shaley breaks support our interpretation of periodic flow reversals separated by slack-water intervals. The lateral accretion beds and fining-upward trends were formed by point-bar migration and lateral abandonment in tidal channel systems incised into the sandy tidal flat. The thickness of some of the larger channels (up to 10 m) indicate they probably were incised to subtidal depths. The coarseness of the channel-fill deposits suggests a strong fluvial influence; thus, many of the channels probably were tide-dominated distributaries.

Facies Assoc. 7: Cycle Bounding Lag Deposits

This facies association consists of laterally extensive, coarse "lags" of poorly sorted, argillaceous, medium to granular sandstone and/or argillaceous to sandy quartz-pebble conglomerate (either matrix- or clast-supported). These deposits commonly are fossiliferous (in places coquinite), containing a variety of reworked and abraded brachiopod valves and other faunal fragments, notably rugose and tabulate corals (some have diameters as large as 12 cm). Many are bioturbated to varying degrees of intensity. "Lags" containing abundant argillaceous material are both well-cemented and standing in relief and others are poorly cemented and weather recessive. In the latter, the exposure must be examined with great care in order not to mistake the deposit for similarly colored mudstone. Some of the deposits in this facies are ferruginous, locally becoming ironstones (Kaiser, 1972; Prave and Duke, 1990).

In places, rocks of Facies Association 7 are only a thin (one to a few grain diameters thick), laterally discontinuous veneer of coarse clasts. Elsewhere they are laterally continuous, many decimeters to a few meters thick, and composed of several distinct beds. We have recognized four basic types of "lags": 1) Single-bed lags that consist of a single, erosive-based, centimeter- to decimeter-thick deposit of argillaceous, relatively coarse sandstone, pebbly sandstone, and/or fossil-hash debris of mostly abraded brachiopod valves; vertical burrows are present locally but bioturbation typically is not intense. 2) Multi-bed lags in which two or more "single-bed lags" are separated by centimeters-thick beds of nonfossiliferous shale or mudstone; intervals containing multi-bed lags generally are many decimeters to several meters thick. 3) Bioturbated lags that consist of beds of poorly sorted, argillaceous sandstone or sandy mudstone many decimeters thick; intense bioturbation has resulted in a ropey or knotted texture. Coarse grains span a wide range of sizes, and many occurrences contain widely disseminated quartzose pebbles. Disarticulated brachiopod valves, coralline fragments, and other faunal remains are common. 4) Cross-bedded lags occur in intervals several meters thick and consist of poorly sorted, coarse to pebbly sandstones or sandy conglomerates exhibiting crude cross-bedding. Abraded brachiopod valves vary in abundance from bed to bed and shale rip-up clasts are present locally. In many examples of this facies, groups of beds exhibit low angular relationships with subjacent deposits suggesting sedimentation along gently inclined depositional surfaces.

All these deposits occupy stratigraphically distinct positions: they only occur at or just above coarsening-upward cycle boundaries (i.e., the transition from relatively coarser intervals below to relatively finer intervals above). Most boundaries, but not all, between coarsening-upward cycles (both large and small scale) are marked by these deposits. In nearly all instances, the "lags" are sharp-based and erosional and clearly are hydrodynamic "concentrates" of the coarsest material eroded from underlying lithologies. Others contain clasts much coarser than any present immediately below and indicate transport over some distance prior to deposition.

We interpret occurrences of Facies Association 7 as reworked, transgressive lag deposits formed during erosional shoreface retreat associated with rises of relative sea level. This interpretation is based on both the internal characteristics of these deposits and their relationships with overlying and underlying facies in the coarsening-upward cycles. Below we present a complete discussion of vertical facies successions.

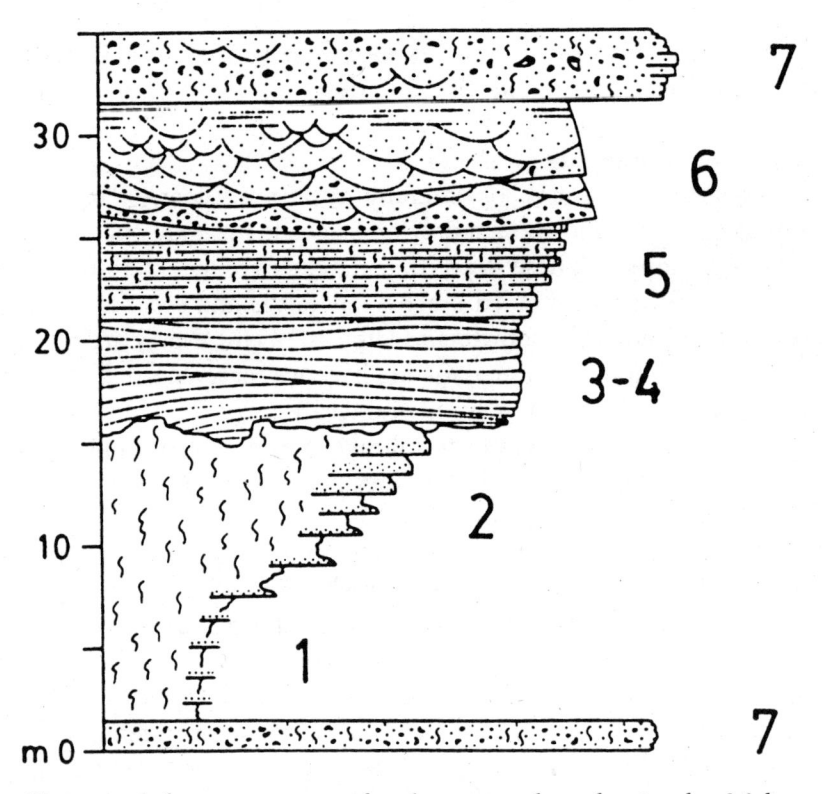

Figure 7. "Idealized" vertical facies transition for depositional cycles in the Mahantango Formation. Numbers refer to facies associations described in text. See Figure 3 for key to lithologic symbols.

VERTICAL FACIES SUCCESSION AND THE "IDEALIZED" MAHANTANGO DEPOSITIONAL CYCLE

The nature and scales of coarsening-upward cycles typical of the Mahantango Formation can be recognized by simple examination of the three measured sections presented herein (Figures 3–5). Figure 7 is our "idealized" vertical facies succession for depositional cycles in the Mahantango Formation based on our observations from these and the seven other measured sections. Thicknesses of individual facies in Figure 7 are approximate mean values of all occurrences of that facies underlying the next facies in the normative succession. The resulting "idealized" cycle is approximately 30 m thick. The reduced thickness of the "ideal" facies succession relative to some of the thicker cycles reflects the non-representation of superimposed small-scale cyclicity.

Based on our facies interpretations, the idealized cycle represents a shoaling succession produced by the progradation of a tide-influenced shoreline into a storm-influenced marine basin. Offshore mudstone with distal tempestites (Facies Assoc. 1) passes upward into more proximal offshore tempestites interbedded with mudstone (Facies Assoc. 2), which in turn are overlain by storm emplaced amalgamated hummocky cross-stratified sandstone of the lower shoreface (Facies Assoc. 3). Swaley sandstone of the upper shoreface (Facies Assoc. 4) is preserved only locally, probably due to the strong tidal overprinting in nearshore settings. At this point in the vertical transition, tide-dominated deposition replaces storm-dominated deposition. Tidal channel-mouth shoals (Facies Assoc. 5) or possibly ebb-tidal deltas are the lowest, clearly tide-dominated deposits. These are immediately overlain by the sandy subtidal to intertidal(?) flat deposits dissected by fluvially influenced tidal channel/distributary deposits (Facies Assoc. 6). A relative rise in sea level indicated by a relatively thin transgressively reworked lag deposit (Facies Assoc. 7) terminates the shoaling succession. This is overlain by distal offshore deposits heralding the beginning of a new depositional shoaling cycle.

The coarse nature of the tidal channel fill indicates it must have been supplied from a fluvial source (the much finer shoreface deposits cannot be a source); this suggests these channels were connected to rivers through a distributary system maintained by strong tidal currents during progradation. Exposures of the fluvial systems are not present in central Pennsylvania, but the coarseness of the tidal channel fill indicates they probably were high gradient, braided(?) rivers which debouched directly onto the shoreline.

There is no evidence for barrier islands or associated inlets in the depositional system inferred by us for the Mahantango Formation. The absence of barriers along modern shorelines broadly coincides with high mesotidal or macrotidal coasts. However, modern coasts generally are experiencing transgression. The Mahantango shoreline was progradational and the absence of barriers most likely is a consequence of that. We see no need to invoke a macrotidal range for this ancient shoreline system. However, given the thickness of some of the lateral accretion sets, tidal channels must have been incised locally to over 10 m deep. This depth of incision suggests to us the Mahantango shoreline in central Pennsylvania experienced a probable high mesotidal range.

PALEOFLOW DATA AND MAHANTANGO PALEOGEOGRAPHY

Paleoflow data obtained from our eleven measured sections strongly support the above interpretations. These are summarized on Figure 6.

Data from offshore storm deposits (Figure 6A) indicate a more-or-less NE-SW oriented shoreline. Wave-formed vortex-ripple crests average NE-SW, whereas small paleocurrent indicators (ripple cross-lamination and sole marks) are broadly NW-directed. These latter indicators are inferred by us to be oriented generally shore-normal. They most likely reflect the direction of peak instantaneous shear stress beneath storm-generated, oscillatory-dominant combined flows, resulting from superimposition of strong, shore-normal storm-wave orbital motions on weaker, geostrophically balanced steady bottom currents (see Duke, 1990, for detailed discussion of such storm deposits). The Mahantango shoreline was situated at low southern latitudes (Scotese et al., 1985); thus the storms probably were hurricanes (Duke, 1985) and the geostrophic component of such circulation, although weak, would have been alongshore to the W-SW, an orientation suggested by the sole marks. The few paleoflow data obtained from Facies Associations 3 and 4 support this model (Figure 6B).

Paleoflow data from Facies Associations 5 and 6 (Figure 6C, D) mainly reflect reversing onshore-offshore tidal currents associated with the shoals and channels. The dispersion of data and the varied restored dip directions of the lateral accretion surfaces (evident on Figure 4) suggest a somewhat sinuous geometry for the tidal channels. Ebb-tide dominance is indicated by the greater abundance of offshore-directed cross-strata relative to onshore-directed data. We suggest this reflects a strong fluvial influence on tidal currents within the channels, rather than a true tidal asymmetry.

Paleoflow data from the transgressive lag deposits are sparse and difficult to obtain. Nonetheless, our qualitative observations indicate the restored dip directions, from a variety of scales of cross-bedding, are westerly. We attribute this to the seaward migration of low-amplitude transgressive shelf ridges formed by downwelling geostrophic circulation during storms. We are now investigating the nature and origin of these deposits and will deal with them at greater length in a forthcoming paper.

Our reconstruction of the shoreline system for the Mahantango Formation in central Pennsylvania is shown on Figure 8. The paleoflow data from section to section as well as for any given section (see Figures 3–5) display no major variation in orientation. This suggests that our reconstruction is valid for any individual cycle in the formation, that the configuration of the shoreline remained relatively straight during progradational events, and that there was no major "swithching" of point source locations over time. There is no evidence to suggest a "lobate" shoreline configuration. We believe that a possible partial modern analog for this ancient system is the barrierless portion of the German Bay of the North Sea, near the Weser and Elbe estuaries (Reineck and Singh, 1980, p. 388, figure 535).

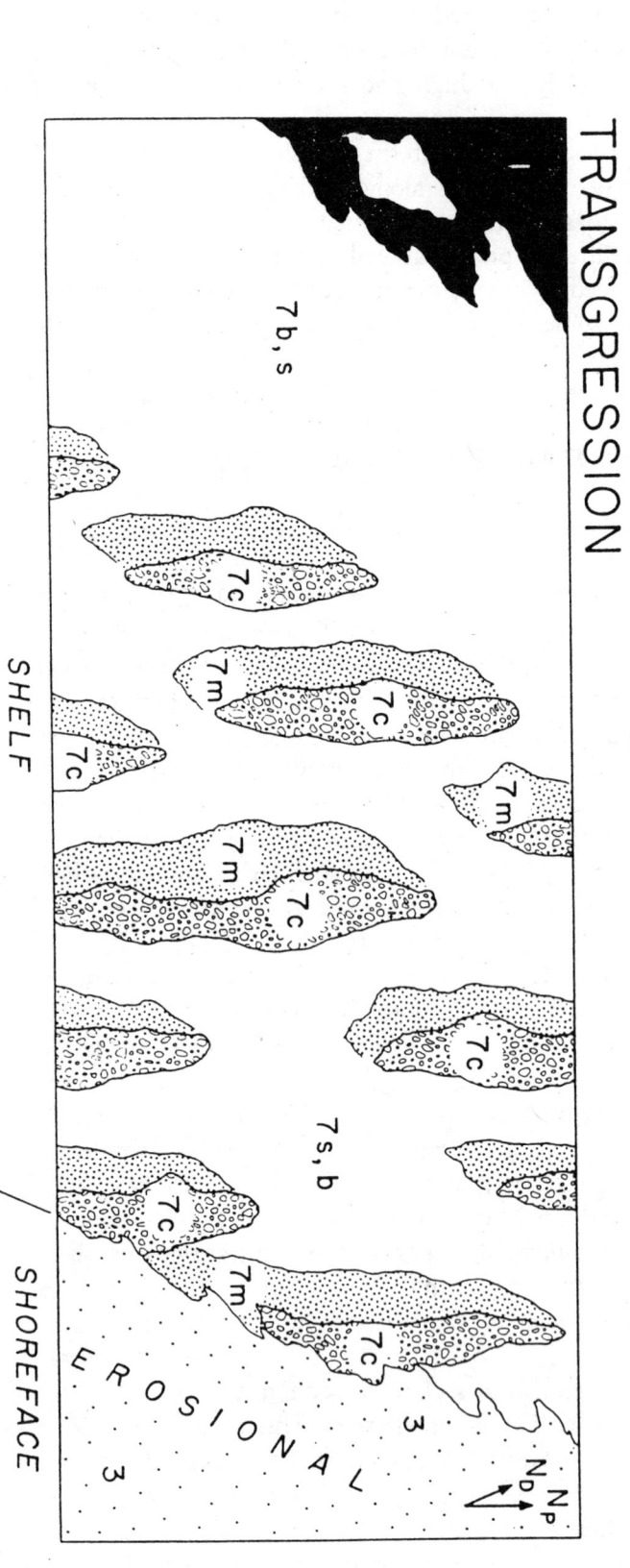

Figure 8. Conceptual paleogeographic reconstruction of the Mahantango shoreline. Numbers refer to facies associations defined and described in text.

Discussion

Goldring and Bridges (1973) attributed the large-scale coarsening-upward cycles in the Mahantango Formation to a prograding storm-dominated shoreline. An essentially similar origin involving multiple shoreline systems is implicit in the deltaic interpretations of Kaiser (1972) and Faill et al. (1978). We largely agree with these interpretations, with two exceptions. Our work indicates that tide-influenced sedimentation dominated the nearshore settings, and progradation into the Mahantango depocenter was characterized by a regionally straight, rather than lobate, shoreline.

In contrast, Sarwar and Smoot (1983) and Sarwar (1984) attribute the genesis of large-scale shoaling cycles to the lateral migration of storm-generated shallow marine sand ridges. We question this interpretation for the following reasons: 1) the large coarsening-upward cycles are too thick (some are over 50 m thick) to be reasonably attributed to shelf sand ridges; 2) the coarseness and thickness of the sandstones at the tops of cycles must reflect a proximity to a fluvial source—none of the underlying facies contain such coarse material to have been reworked into a shelf sand ridge system; 3) the persistent and systematic vertical paleobathymetric variation within cycles and from cycle to cycle is too consistent to be attributed to a random "autocyclic" process of shelf sand ridge migration; 4) the restored dip directions of the lateral accretion surfaces (between 5° and 15°) are too steep to be considered as the gentle, inclined margins of shelf sand ridges; and 5) the paleoflow data and relationships between various paleocurrent indicators, i.e., abundant reversing cross-beds, near orthogonal relationships between lateral accretion surfaces, tidal indicators, and the inferred shoreline, are difficult to resolve in a shelf ridge system. As we have shown, the above are readily explained in our depositional model.

Our preferred interpretation is that depositional cyclicity in the Mahantango Formation can be best attributed to the episodic progradation of a mostly straight, tide-dominated shoreline into a storm-dominated marine basin. Previous deltaic interpretations either explicitly or implicitly suggested the shoreline was lobate. This conclusion appears to have been based on the "lobate" geometry of isopachus maps constructed principally for the Montebello Member and positive correlation of the maps to the distribution of large clasts (e.g., Willard, 1935). We believe these are an "artifact" of miscorrelation. The lithostratigraphic, rather than allostratigraphic, basis for the definition of the Montebello Member has resulted in the inadvertent classification of genetically distinct depositional cycles. As previously stated, the consistent orientation of varied paleocurrent indicators from cycle to cycle and from place to place supports our inference of a relatively straight shoreline and does not support the alternative, a system of switching, lobate deltas.

Based on the known lateral variation of Middle Devonian strata in Pennsylvania and adjacent states, there can be little doubt that central Pennsylvania was a depocenter during that time. As early as the 1930s (Willard, 1935), this region was inferred to be the locus of a relatively long-lived fluvial input system(s) carrying siliciclastic detritus derived from the nearby Acadian orogenic highlands. That a regionally straight shoreline could be maintained over time, even during periods with rates of clastic influx high enough to force progradation, is remarkable. We speculatively suggest this reflects the combined interactions between strong tidal and wave currents coupled with relatively high rates of subsidence and clastic influx.

CYCLE CORRELATIONS, LATERAL VARIATIONS, AND GENESIS

Cycle Correlations and Lateral Variations Within Cycles

The three measured sections (Figures 3–5) exhibit several scales of coarsening-upward cyclicity which are representative of the cyclicity developed throughout the study area. At each outcrop, the larger-scale cycles can be subdivided into smaller superimposed cycles. In the source-proximal Rockville section (Figure 5), the vertical, paleobathymetric transition from storm-dominated shelf and lower shoreface deposits (Facies Assoc. 1 to 3) to tide-dominated shoal and channel deposits (Facies Assoc. 5 and 6) provides a record of the depth-controlled transition from one depositional regime to another. The more distal (basinward) correlative cycles at Girty's Notch (Figure 4) and Watts (Figure 3) similarly record shoaling, but only to depths as shallow as the shoreface.

The lateral facies variation within individually correlated depositional cycles provides the basis for interpreting sedimentological relationships of the shoreline system of the Mahantango Formation. Figure 9A is our correlation diagram for the large-scale cycles in the three measured sections. Correlations are made on the basis of the consistency of the internal facies successions, the magnitude of inferred paleobathymetric variation across cycles and within groupings of cycles, relative stratigraphic position, and hierarchical rank. Note the relatively straightforward nature of cycle correlation, which in many instances extends to the small-scale cycles. The datum on Figure 9A is the top of a thick, regionally traceable lag deposit. This lag separates depositional cycles developed in overall "shallower" paleoenvironments below from those in overall "deeper" paleoenvironments above.

The most insight into the nature of lateral facies transitions is gained from examination of the two uppermost large-scale cycles above the datum (Figure 9A). Both are present at the Rockville and Watts exposures; only the lower cycle is complete at the Girty's Notch exposure. Tidal deposits comprise the Rockville cycles, whereas storm deposits comprise the Watts cycles. The cycles at Girty's Notch display a "mixed" influence. We conclude that tide-dominated and storm-dominated shoaling successions can be laterally correlative (Figure 9B). The predominance of one over the other is solely a function of the original paleogeographic position within the basin.

Had we lacked the stratigraphic data provided by the source-proximal exposures, the environmental and hydrodynamic significance of the tidal shoal deposits (Facies Assoc. 5) might not have been recognized. In the more basinward cycles these could have been easily misinterpreted as wave-dominated nearshore sandstones. Consequently, studies of storm-dominated stratigraphic successions must consider this caveat: a possible strong tidal influence cannot be eliminated unless unequivocal beach and/or non-marine deposits are preserved.

Cycle Genesis

The first-order variables commonly invoked as controls on depositional cyclicity are climate, tectonism (subsidence), and eustacy. Because the Acadian foreland basin remained within low-latitudinal positions throughout the Middle and Late Devonian (e.g., Scotese et al., 1985), we have assumed that climatic variations, such as changes in precipitation, which could cause changes in clastic influx, can be assumed as minimal (relative to those possible in mid-latitudinal settings).

It is possible that the inferred tens of meters of relative sea level variation indicated by the shoaling cycles of the Mahantango Formation were eustatic in origin. The currently in vogue Milankovitch cyclicity is commonly cited as the preeminent causative factor for such higher-order eustatic fluctuations. In our opinion, such an assessment requires evidence from numerous measured sections of coeval strata from at least two tectonically disconnected sedimentary basins. Furthermore, there are two lines of strong circumstantial evidence suggesting the depositional cycles reflect regional tectonic events. First, the Acadian orogeny was at or near its peak during this time (Faill, 1985). Basinal development and deposition along the active margin of the foreland basin is known to have been strongly dependent on Acadian tectonism (Quinlan and Beaumont, 1984). Second, the coarse clastic deposits in many of the cycles suggest deposition near a region of active uplift. Consequently, we tentatively attribute the large-scale cycles in the Mahantango Formation to episodes of Acadian thrusting. In this model, thrust nappes load and depress a continental margin, causing a relative rise in sea level. Subsequent erosion of the nappes results in clastic detritus prograding into the basin. Repeated episodes of thrusting and erosion produce stacked progradational cycles. However, intra-

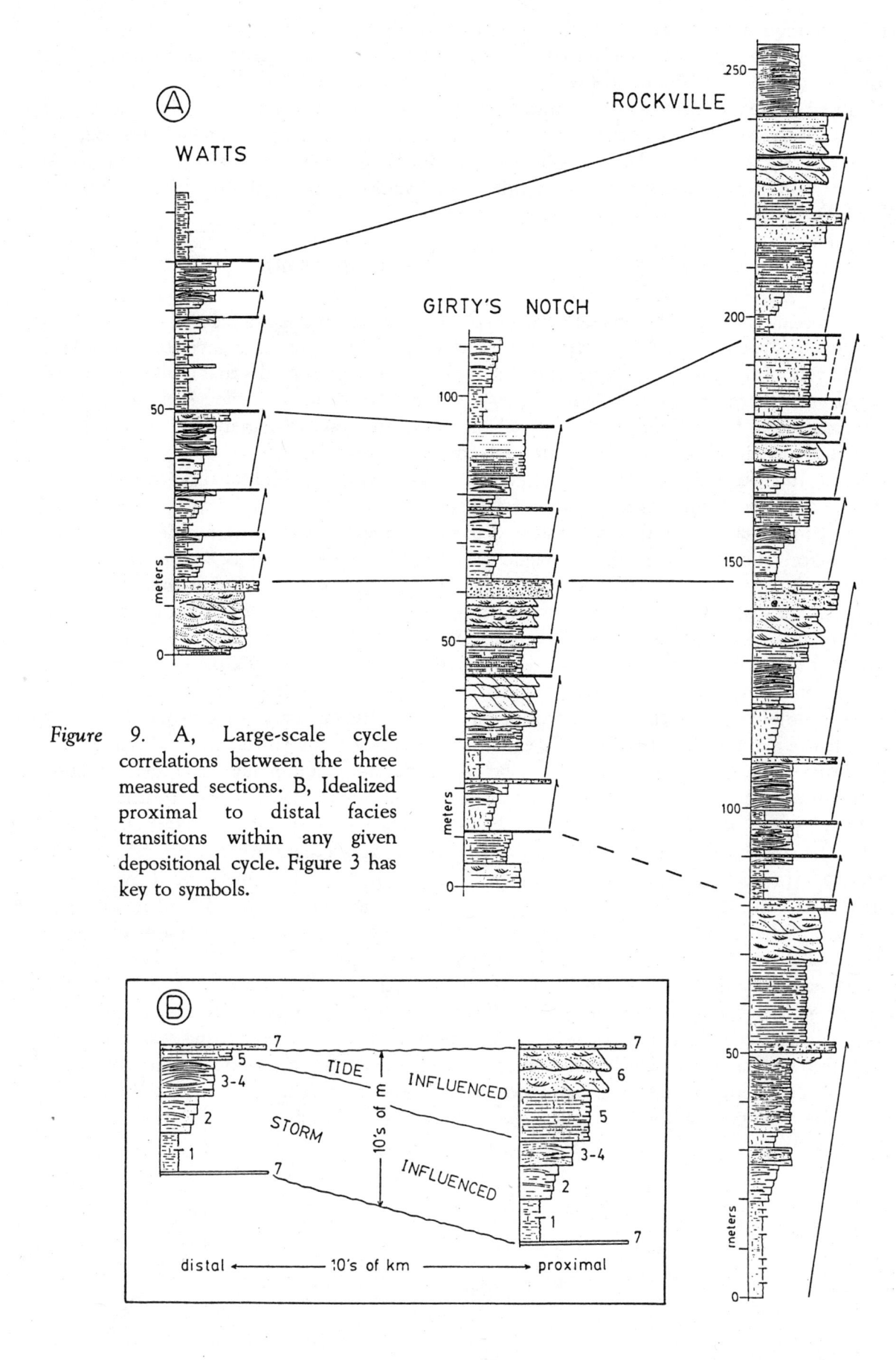

Figure 9. A, Large-scale cycle correlations between the three measured sections. B, Idealized proximal to distal facies transitions within any given depositional cycle. Figure 3 has key to symbols.

basinal relationships between Middle Devonian units perhaps may reflect a long-term eustatic rise. Such a rise is indicated in the sea-level curves of Dennison and Head (1975), Dennison (1985), Johnson et al. (1985), and Vail et al. (1977).

As best as we can determine, the smaller-scale cycles are much less laterally extensive. For these we do not necessarily invoke the large-scale tectonic mechanism suggested above. Instead, the smaller-scale cycles appear to be best attributed to "autocyclic" processes such as short-term spatial and temporal variation in the advancing shoreline, local variation in the position of the locus of the "point" source, or short-lived fluctuations in clastic influx rate.

SEQUENCE STRATIGRAPHIC FRAMEWORK

The recent renaissance in sequence stratigraphy has built upon the original insights of workers such as Sloss (1963) to provide a conceptual framework from which to interpret depositional systems. The arrangement of systems tracts (Posamentier et al., 1988) provides a model that is allostratigraphically based and recognizes the genetic significance of depositional units. Although the model was originally constructed from seismic data principally obtained from passive margins, our work suggests such models are applicable to these Acadian foreland basin clastic deposits.

Figure 10 is our proposed sequence stratigraphic framework for the Mahantango Formation. It is schematic and does not reflect true thicknesses, lateral distances, or cyclicity. It does provide, however, a conceptual model that accurately depicts the depositional patterns inferred for the Mahantango Formation. Below the datum (Figure 9A), the basal facies occurrence in any superjacent cycle typically indicates a paleobathymetrically shallower environment relative to the basal facies occurrence in the subjacent cycle. This pattern helps define large-scale cycles whose superimposed small-scale cycles similarly mimic this pattern. Thus, the large-scale cycles are prograding parasequences which define highstand systems tracts (Posamentier et al., 1988). The facies present everywhere immediately above the datum (Figure 9A) indicate a sudden transition into a much deeper, more distal environment. The depositional cycles (both large-and small-scale) above the datum, with the exception of the Rockville exposure, do not typically shoal into environments as shallow as those common at the tops of cycles below the datum. Additionally, the superimposition of the large-scale cycles above the datum reflects an overall retrogradational trend rather than a progradational one. These depositional patterns define a transgressive systems tract (Posamentier et al., 1988). The superimposition of a transgressive systems tract on a highstand systems tract (Figure 10) defines a Type II sequence, and the datum thus becomes a Type II boundary. Nowhere have we recognized fluvial incision associated with sediment bypassing as required for a Type I boundary.

Given this framework, it is evident that the Mahantango Formation, as formally defined lithostratigraphically, straddles two genetically distinct depositional systems. The Turkey Ridge through Montebello Members are part of the highstand systems tract, whereas the strata of the Sherman Ridge Member are the initial deposits of the overlying transgressive systems tract. Within this sequence stratigraphic framework, we tentatively suggest that: 1) the underlying Marcellus black shale and the overlying Burkett black shale (and possibly the Tully carbonate interval) are condensed sections; and 2) the transgressive lag developed along the surface defining the Type II boundary represents the initiation of the Taghanic onlap (described in Dennison and Head, 1975, and Dennison, 1985).

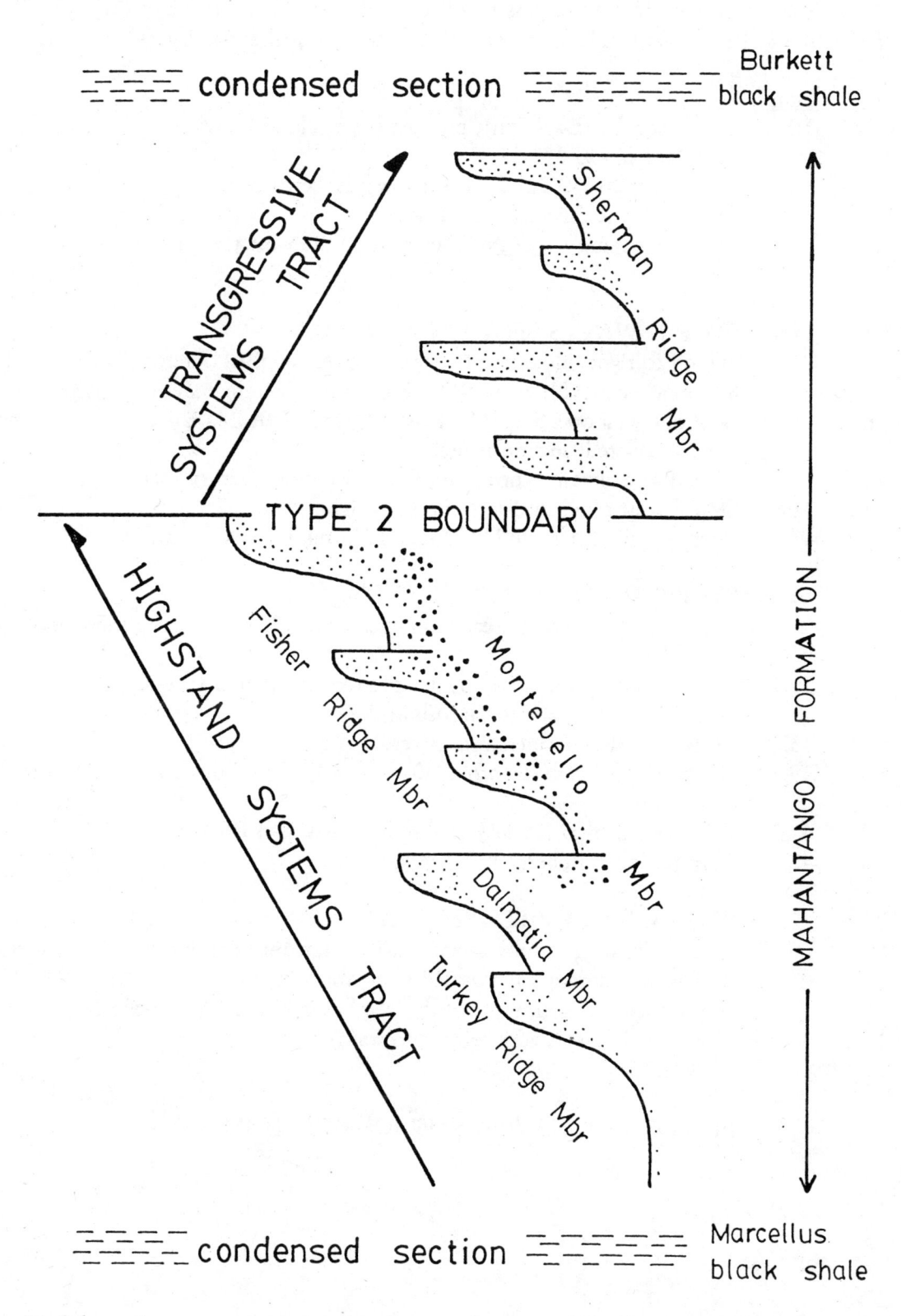

Figure 10. Schematic sequence stratigraphic framework proposed for the Mahantango Formation.

ROAD LOG

Road log begins at the overpass with I-81 and Front Street (next to Days Inn 3 miles north of Harrisburg, Pennsylvania, along the east side of the Susquehanna River).

Dist.	Cum.	
0.0	0.0	I-81–Front St. (Days Inn); proceed north along Front St.
3.0	3.0	Bear right to US 322-22,
0.3	3.3	Turn left onto US 322-22 (west towards Lewistown).
8.5	11.8	Cross over the Susquehanna River.
3.7	15.5	Watts exit ramp—**DO NOT** exit; continue west for 1.1 miles.
1.1	16.6	**STOP 1: WATTS SECTION** Pull off to right side of road near base (west end) of outcrop. Top of the Montebello Mbr in anticline; Sherman Ridge Mbr forms the remainder of roadcut; top is covered.
1.6	18.2	Continue west on US 22-322. Exit at Midway Exit. Mahantango along ramp; base and top are covered; section is faulted.
0.3	18.5	At end of exit ramp turn right, follow winding country road.
4.3	22.8	Stop sign at small intersection. Turn right.
0.3	23.1	Intersection with route US 11-15. Turn right (southward).
1.8	24.9	**STOP 2: GIRTY'S NOTCH SECTION** Turn left into parking area. Montebello and Sherman Ridge Mbrs; base and top are covered.
6.4	31.3	Continue south on US 11-15 to intersection with US 22-322. Take US 22-322 east to Harrisburg.
1.9	33.2	Cross over the Susquehanna River.
6.3	39.5	Hardee's for lunch (approx. 30 minutes). Continue east on US 22-322 towards Harrisburg.
1.8	41.3	Exit to right onto PA 443 to Fishing Creek and Rockville.
1.4	42.7	Turn left onto Roberts Valley Road.
0.2	42.9	**STOP 3: ROCKVILLE SECTION** Cross railroad, go under overpass. Turn left into parking area next to the electrical station. Mahantango is exposed along roadcut on the east berm of US 22-322 (base and top are covered; most of this section is in the Montebello Mbr).
1.6	44.5	Return to PA 443, turn left (southward) and return to I-81 intersection (Days Inn on left).

END OF TRIP. HAVE A SAFE JOURNEY HOME!

REFERENCES CITED

Dennison, J. M., 1985, Devonian eustatic fluctuations in Euroamerica: Discussion: Geological Society of America Bulletin, v. 96, p. 1595–1597.

——, and Hasson, K. O., 1976, Stratigraphic cross-section of Hamilton Group (Devonian) and adjacent strata along south border of Pennsylvania: American Association of Petroleum Geologists Bulletin, v. 60, p. 278–298.

——, and Head, J. W., 1975, Sea level variations interpreted from the Appalachian Basin Silurian and Devonian: American Journal of Science, v. 275, p. 1089–1120.

Duke, W. L., 1985, Hummocky cross-stratification, tropical hurricanes, and intense winter storms: Sedimentology, v. 32, p. 167–194.

——, 1990, Geostrophic circulation or shallow marine turbidity currents? The dilemma of paleoflow patterns in storm-influenced prograding shoreline systems: Journal of Sedimentary Petrology, in press.

——, and Prave, A. R., 1989, Lateral variation within "mixed-influenced" prograding shoreline sequences in the Middle Devonian Mahantango Formation in central Pennsylvania: Geological Society of America, Abstracts with Programs, v. 21, n. 6, p. 37.

Ellison, R. L., 1965, Stratigraphy and paleontology of the Mahantango Formation in south-central Pennsylvania: Pennsylvania Geological Survey, 4th series, General Geology Report 48, 298 p.

Faill, R. T., 1985, The Acadian Orogeny and the Catskill Delta, in Woodrow, D. L., and Sevon, W. D., eds., The Catskill Delta: Geological Society of America Special Paper 201, p. 31–50.

——, Hoskins, D. M., and Wells, R. B., 1978, Middle Devonian stratigraphy in central Pennsylvania—A revision: Pennsylvania Geological Survey, 4th series, General Geology Report 70, 28 p.

——, and Wells, R. B., 1974, Geology and mineral resources of the Millerstown quadrangle, Perry, Juniata, and Snyder Counties, Pennsylvania: Pennsylvania Geological Survey, 4th series, Atlas 136, 276 p.

Goldring, R., and Bridges, P., 1973, Sublittoral sheet sandstones: Journal of Sedimentary Petrology, v. 43, p. 736–747.

Hoskins, D. M., 1976, Geology and mineral resources of the Millersburg 15-minute quadrangle, Dauphin, Juniata, Northumberland, Perry, and Snyder Counties, Pennsylvania: Pennsylvania Geological Survey, 4th series, Atlas 146, 38 p.

Johnson, J. G., Klapper, G., and Sandberg, C. A., 1985, Devonian eustatic fluctuations in Euroamerica: Geological Society of America Bulletin, v. 96, p. 567–587.

Kaiser, W. R., 1972, Delta cycles in the Middle Devonian of central Pennsylvania: unpublished Ph.D. dissertation, Johns Hopkins University, Baltimore, MD, 183 p.

Posamentier, H. W., Jervey, M. T., and Vail, P. R., 1988, Eustatic controls onclastic deposition I—Conceptual framework: Sea level changes—An integrated approach: Society of Economic Paleontologists and Mineralogists Special Publication 42, p. 109–124.

Prave, A. R., and Duke, W. L., 1989, Prograding shoreline deposits in the Middle Devonian Mahantango Formation (central Pennsylvania, U.S.A.) reveal a vertical transition from storm-dominated sedimentation to tide-dominated sedimentation within individual shoreface sandstones: Proceedings of the 2nd International Research Symposium on Clastic Tidal Deposits, Calgary, Alberta, Canada, p. 77.

——, and ——, 1990, Sequence-capping oolitic ironstones in the Mahantango Formation (Middle Devonian) of Pennsylvania: Responses to basin dynamics: Geological Society of America, Abstracts with Programs, v. 22, n. 2, p. 63.

Quinlan, G. M., and Beaumont, C., 1984, Appalachian thrusting, lithospheric flexure, and the Paleozoic stratigraphy of the eastern interior of North America: Canadian Journal of Earth Sciences, v. 21, p. 973–996.

Reineck, H. E., and Singh, I., 1980, Depositional Sedimentary Environments, 2nd edition: Springer-Verlag, New York, 549 p.

Sarwar, G., 1984, Depositional model for the Middle Devonian Mahantango Formation of south-central Pennsylvania: unpublished M.S. thesis, State University of New York at Stony Brook, 251 p.

——, and Smoot, J. P., 1983, Depositional model for the Middle Devonian Mahantango Formation of south-central Pennsylvania: Geological Society of America, Abstracts with Programs, v. 15, n. 3, p. 127.

Scotese, C. R., Van der Voo, R., and Barret, S. F., 1985, Silurian and Devonian base maps: Philosophical Transactions of the Royal Society of London, B 309, p. 57–77.

Sloss, L. L., 1963, Sequences in the cratonic interior of North America: Geological Society of America Bulletin, v. 74, p. 93–114.

Vail, P. R., Mitchum, R. M., Jr., and Thompson, S., 1977, Seismic stratigraphy and global changes of sea level, part 4: Global cycles of relative changes of sea level, in Payton, C. E., ed., Seismic stratigraphy—Applications to hydrocarbon exploration: American Association of Petroleum Geologists Memoir 26, p. 83–97.

Willard, B., 1935, Hamilton Group of central Pennsylvania: Geological Society of America Bulletin, v. 46, p. 195–224.

9
GEOLOGY OF THE ROBERTSON RIVER IGNEOUS SUITE, BLUE RIDGE PROVINCE, VIRGINIA

Richard P. Tollo
George Washington University, Washington, DC 20052

Tamara K. Lowe
Stanford University, Stanford, CA 94305

Sara Arav
U.S. Environmental Protection Agency, New York, NY 10278

Karen J. Gray
U.S. Geological Survey, Reston, VA 22092

INTRODUCTION

This field trip examines the geology of the Robertson River Igneous Suite (RRS) over a distance of nearly 70 miles, extending from northern to central Virginia. This guide is based on the results of detailed field mapping, petrographic studies, and geochemical analyses that have been directed toward investigating the nature and significance of the lithologic diversity of the RRS (Tollo, 1986; Wallace and Tollo, 1986; Tollo and Arav, 1987; in press). Stops in nine of the ten lithologic units within the RRS demonstrate the compositional complexity that characterizes the suite as well as the type of field evidence and integrated studies that are necessary to define the geologic history of this area.

GEOLOGY OF THE BLUE RIDGE ANTICLINORIUM IN VIRGINIA

Introduction

The Blue Ridge anticlinorium is a prominent uplift extending from southern Pennsylvania southwest across Maryland and Virginia (Figure 1). The northwest limb is overturned and the entire structure plunges gently toward the northeast, where it is overlain by Lower Paleozoic sedimentary strata (Espenshade, 1970). The Blue Ridge physiographic province coincides with the anticlinorium in Pennsylvania and Maryland, but in Virginia is restricted to the western limb and western portions of the core.

Stratigraphy

The anticlinorium is comprised of a core zone that includes a variety of Middle Proterozoic gneisses overlain on the east and west limbs by a series of Upper Proterozoic metavolcanic and metasedimentary rocks and lower Paleozoic sedimentary strata. A number of units have been proposed within the multiply deformed gneissic terrane by various workers (Bloomer and Werner, 1955; Allen, 1963; Clarke, 1984; Lukert and Halladay, 1980; Bartholomew and Lewis, 1984; Sinha

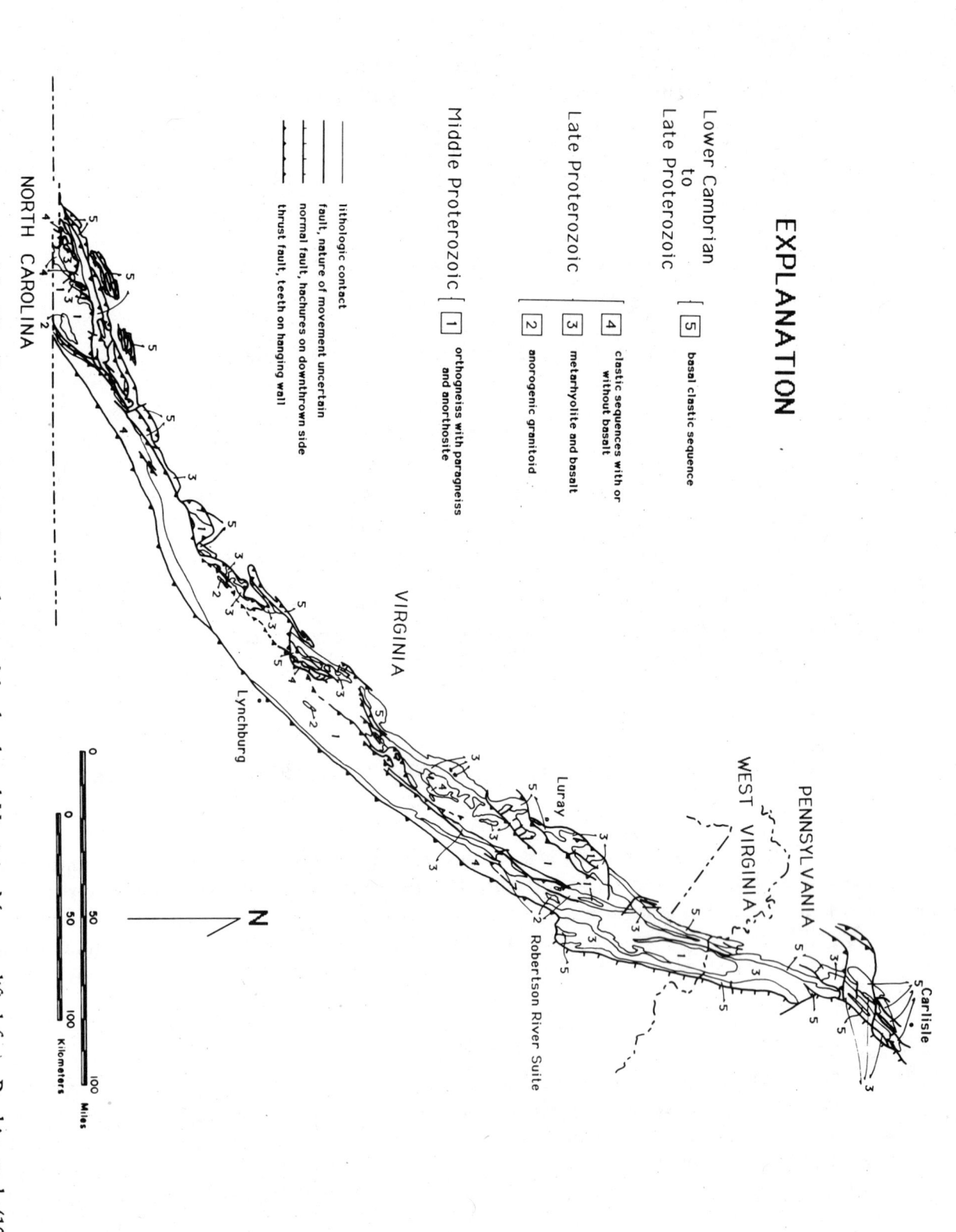

Figure 1. Generalized geologic map of the Blue Ridge province in Pennsylvania, Maryland and Virginia. Map modified from Rankin et al. (1990).

and Bartholomew, 1984), but the extent and mutual age relationships of these deformed rocks remain poorly constrained due to a lack of detailed mapping and petrologic studies. Nevertheless, a considerable range of rock types has been recognized. Throughout the main area of exposure, extending from the vicinity of Roanoke northward to Pennsylvania (Shenandoah massif of Rankin et al., 1989), orthogneiss is by far the most abundant lithologic type. Of these orthogneisses, felsic to intermediate, non-layered gneisses with or without orthopyroxene are the most common (Rankin et al., 1989). However, layered gneisses of non-plutonic affinity also occur. These gneisses are generally interpreted to be volcaniclastic or sedimentary in origin (Herz and Force, 1984) and may locally represent the country rock into which the protoliths of the orthogneisses intruded (Lukert and Halladay, 1980; Bartholomew and Lewis, 1984).

Granitoids, syenitoids, and felsites of the Robertson River Igneous Suite (RRS) intrude gneisses of the Blue Ridge core along a broad zone in Virginia extending from near Upperville nearly 70 miles southwest to the northern suburbs of Charlottesville (Lukert and Halladay, 1980; Mitra and Lukert, 1982; Conley, 1989; Tollo et al., in preparation). These distinctive rocks were originally named the Robertson River Formation by Allen (1963) for exposures in central Virginia. However, Tollo and Arav (in press) proposed the designation "Robertson River Igneous Suite" in order to emphasize the lithologic diversity that characterizes this belt. These rocks range in composition from alkali syenite to alkali feldspar granite to granite. Fine-grained varieties include probable rhyolite exposed within the Battle Mountain Complex near Massies Corner (Wallace and Tollo, 1986; Hawkins, unpublished data). This suite is part of a compositionally distinct group of intrusives of Late Proterozoic age that occurs within Grenville-age basement terranes throughout the central and southern Appalachians from New York to northwestern North Carolina (Rankin and Tollo, 1987). Within the Blue Ridge province of Virginia, the Robertson River Igneous Suite is by far the largest of a group of intrusives that includes, among others, the Rockfish River, Suck Mountain, Stewartsville, Mobley Mountain, and Irish Creek plutonic bodies (Bartholomew and Lewis, 1984; Herz and Force, 1984).

The Mechum River Formation includes a thick sequence of phyllite, meta-arkose, and metaconglomerate exposed within a narrow belt extending for approximately 60 miles southwest from Ben Venue, Virginia (Gooch, 1954; 1958). This formation is inferred to nonconformably overlie the Middle Proterozoic basement, although the contact separating these units along the western edge of the outcrop belt is typically obscured by subsequent Paleozoic faulting (Gooch, 1958). Schwab (1974) suggested that clasts of granite and syenite within the Mechum River Formation were of RRS affinity, indicating that Mechum River sedimentation post-dated the emplacement and exhumation of that plutonic suite. On the basis of relative structural position, the Mechum River Formation may be, at least in part, correlative with the Swift Run and Lynchburg Formations (Schwab, 1974).

Metavolcanic and metasedimentary rocks unconformably overlie basement gneiss along the east and west limbs of the Blue Ridge anticlinorium in northern Virginia (Espenshade, 1970). The Catoctin Formation, a thick sequence of greenstone and intercalated slate and phyllite, occurs along the western limb. This sequence comprises a thick succession of tholeiitic basalt lava flows and terrestrial clastic deposits (Reed, 1969) throughout Virginia, but in Pennsylvania and Maryland includes significant volumes of rhyolite (Freedman, 1967; Fauth, 1968; 1978; Rankin, 1975). Gathright (1976) mapped a thin, discontinuous series of metasedimentary strata below the Catoctin Formation in Virginia as the Swift Run Formation. However, Rankin et al. (1989) considered these rocks as part of the Catoctin Formation. On the southeast limb, a thick series of metasedimentary rocks of non-marine origin (Fauquier Formation) occurs between the basement and greenstone southwest of Warrenton. These rocks interfinger toward the southwest with a thicker series of submarine turbidites (Lynchburg Formation) including abundant graywacke (Wehr, 1983). Wehr and Glover (1985) interpreted the transition from the non-marine rocks of the Fauquier Formation to the marine rocks of the Lynchburg Formation as indicative of an oblique cross-section through the Late Proterozoic rifted margin of Laurentia. Quartz sandstone and coarse clastics forming the basal sequence of the Chilhowee Group conformably overlie greenstone of the Catoctin Formation on both anticlinorium limbs in Virginia, Maryland, and Pennsylvania. These strata represent the base of the Late Proterozoic–early Paleozoic marine transgression sequence of the Appalachian basin (Rodgers, 1970).

Structure and Metamorphism

The overall structure of the Blue Ridge province in northern Virginia is a northeast-plunging anticlinorium with the northwest limb overturned (Gathright, 1976). Thrust faults associated with the structure generally strike northeast-southwest, dip toward the southeast, and show southeast-over-northwest motion (Gathright, 1976; Mitra and Lukert, 1982). Folding within the anticlinorium repeats the stratigraphic sequence. However, such folds are rarely visible in the basement units. Nevertheless, a variety of structural features present throughout the anticlinorium defines the multiple-event deformational history characteristic of the Blue Ridge province.

Rocks of the cover sequence and those within the core that are not dominated modally by quartz and feldspar (such as greenstone dikes) typically show a pervasive southeast-dipping cleavage (the South Mountain cleavage of Mitra and Elliott, 1980). In contrast, the coarse-grained, quartz- and feldspar-rich gneisses and granites of the core are characterized by planar zones of mylonite, termed ductile deformation zones (DDZ's) by Mitra (1978). The DDZ's vary in scale from microscopic to tens of meters (Mitra, 1978) and are locally abundant on an outcrop scale. They have been interpreted as zones of high strain that account for a significant portion of the total shortening within the Blue Ridge basement (Mitra, 1979). In most cases, the DDZ's offset the prominent foliation of the basement gneisses (Mitra and Lukert, 1982). Mitra and Elliott (1980) and Mitra (1978; 1979) outlined evidence suggesting that the development of the DDZ's and regional cleavage was synchronous. They further proposed that the deformation was Late Proterozoic in age by noting the asymptotic orientation of the cleavage relative to the basal thrust faults of the Blue Ridge.

The Mechum River Formation, exposed in the central area of core gneisses, has been interpreted to be an infolded syncline (Batesville syncline of Gooch, 1958) or down-faulted block (Schwab, 1974). Mitra and Lukert (1982) noted that faults border the Mechum River outcrop belt along much of its length and proposed that these bounding structures were reactivated as northwest-verging thrust faults during late Paleozoic orogenesis. Correlation of the Mechum River Formation with Late Proterozoic clastic rocks of the anticlinorium limbs is the basis of the infolded/down-faulted structural model for the inlier. However, this model is based largely on relative stratigraphic position and is unconstrained by specific faunal evidence.

Many of the basement units within the Blue Ridge anticlinorium preserve evidence of granulite facies metamorphism during Grenville orogenesis (Espenshade, 1970). The orthopyroxene-bearing charnockite mineral assemblages and the pervasive foliation of the core gneisses were formed during this widespread metamorphic episode (Mitra and Lukert, 1982). Evidence for a subsequent metamorphic event of probable Paleozoic age includes greenschist facies mineral assemblages in the Late Proterozoic units of the Blue Ridge province and retrograde reaction assemblages developed in the Middle Proterozoic gneisses and charnockites (Mitra and Lukert, 1982).

Bartholomew et al. (1981) suggested that the central Blue Ridge terrane is comprised of two massifs separated by the Rockfish Valley fault zone. The massifs display differences in inferred peak metamorphic conditions, as well as in the rock type and field/textural relations of charnockite plutons. Using relict mineral assemblages, Bartholomew et al. (1981) and Sinha and Bartholomew (1984) proposed that the shallower (lower pressure, upper amphibolite to lower granulite facies conditions) Lovingston massif (located toward the east) was juxtaposed against the deeper (higher pressure, granulite facies conditions) Pedlar massif (located toward the west) by Paleozoic thrust movement along the Rockfish Valley fault zone. In this model, the massifs represent a pair of crustal blocks that expose rocks metamorphosed at two distinct depths during Grenvillian orogenesis. The Rockfish Valley fault zone in this model is a significant intra-basement thrust resulting from probable late Paleozoic compression. However, the significance of the observed mineral assemblages and sense of movement along the Rockfish Valley fault zone are topics of considerable debate. Evans (1987) suggested that the difference in mineral assemblages resulted from a widespread post–Late Proterozoic hydration event which affected only rocks of the Lovingston massif. Alternatively, Bailey (1990) suggested that the Lovingston massif experienced a pervasive, possibly Late Proterozoic, amphibolite facies event that obliterated most of the evidence for the earlier Grenvillian metamorphism and may have been related to emplacement of the Late Proterozoic felsic intrusives observed throughout the massif. Simpson and Kalaghan (1989) used data from a detailed study in southern Virginia to suggest that the entire Rockfish Valley and associated Fries fault zones formed as a result of Late Precambrian extension and were reactivated only locally as thrusts during Paleozoic time. Bailey (1990) also noted that some mylonites within the Lovingston massif are characterized by overall geometries and sense-of-movement indicators that are consistent with Late Proterozoic extension. Further recognition and characterization of lithologic evidence for Late Proterozoic extension versus one or more Paleozoic compressional events have important implications for models of the tectonic evolution of the Blue Ridge province; thus, additional detailed structural and petrologic studies are needed.

Age Relations

Isotopic analyses of gneisses comprising the Blue Ridge basement in Virginia indicate ages within the range 1000–1150 Ga (Tilton et al., 1960; Davis et al., 1962; Rankin et al., 1983; Herz and Force, 1984; Pettingill et al., 1984; Sinha and Bartholomew, 1984). Although precise age determinations are locally in accord with observed field relations (Sinha and Bartholomew, 1984), in most cases it is not possible to delineate significant differences between lithologic units. This problem is a result of difficulties in reconciling disparate ages obtained by different laboratories and techniques as well as the general lack of detailed field mapping and petrologic studies throughout the area. Nevertheless, the data are sufficient to identify one or more widespread high-grade metamorphic/plutonic events and to establish a general correlation in timing between the Virginia Blue Ridge and other well-known areas of Grenville-age metamorphism such as the Adirondack massif in New York and the Grenville terrane of southeastern Canada. Evidence for pre-Grenville crust is preserved by detrital zircons with a U/Pb upper intercept age of 1870 ± 200 Ma from the Stage Road layered gneiss in central Virginia (Sinha and Bartholomew, 1984).

Determination of the age and tectonic significance of the cover sequence and the relationship of some of these rocks to the Late Proterozoic intrusives within the Blue Ridge core has been the subject of numerous studies. Rankin et al. (1969) obtained an age of 810 Ma from a suite of five zircon samples collected from extrusive units within the Catoctin (Pennsylvania), Mount Rogers (Virginia), and Grandfather Mountain (North Carolina) Formations. However, this "date" is inconsistent with numerous ages obtained by isotopic analysis of possibly consanguineous felsic intrusives from the Crossnore area, North Carolina (680–710 Ma, Odom and Fullagar, 1984) and lithologically similar units from the Robertson River Igneous Suite in northern and central Virginia (640–730 Ma, Tollo et al., in press). The 810 Ma age also implies a major temporal hiatus that is not supported by field relations between the rift-related volcanic rocks of Catoctin affinity and the passive margin transgressive sequence represented by the Chilhowee Group.

More recently, Badger and Sinha (1988) obtained an Rb/Sr isochron age of 570 ± 36 Ma for greenstone samples collected from the Catoctin Formation in central Virginia. These data are consistent with the occurrence of Rusophycus fossils of Early Cambrian age identified by Simpson and Sundberg (1987) near the base of the Chilhowee sequence (Unicoi and Hampton Formations), which overlies the Catoctin Formation without major unconformity. In addition, U/Pb analyses of zircons collected from rhyolite units occurring near the base of the Catoctin sequence in the northern Blue Ridge indicate an age of approximately 600 Ma for the onset of volcanism in this area (J. Aleinikoff, personal communication, 1990).

Collectively, these data indicate that widespread plutonism and associated volcanism spanned an extended time interval from 730 to 570 Ma at the end of the Proterozoic. As noted by Badger and Sinha (1988), it is presently unclear whether the magmatism occurred in two discrete stages or continuously throughout an extended period of time. Nevertheless, the association of abundant metaluminous to peralkaline plutonism (Robertson River–Crossnore suites) with voluminous tholeiitic basalt and rhyolitic volcanism (Catoctin–Mount Rogers–Grandfather Mountain Formations), as well as local terrigenous sedimentation (Mechum River–Fauquier–Grandfather Mountain Formations), represents important regional evidence of profound continental rifting of Laurentian crust during the Late Proterozoic (Rankin 1975; 1976).

ROBERTSON RIVER IGNEOUS SUITE

Introduction

The Robertson River Igneous Suite (RRS) includes a variety of metaluminous to peralkaline granitoids, syenitoids, and felsites exposed in the core of the Blue Ridge anticlinorium (BRA) within a narrow belt extending from near Upperville in northern Virginia 70 miles (110 km) southwest to the area of Charlottesville (Figure 2). A small gap of less than one mile (0.6 km) occurs south of Castleton, Virginia, where phyllites, volcanogenic sedimentary rocks, and several basement units are exposed. Broad-scale reconnaissance mapping by previous workers (Allen, 1963; Lukert and Nuckols, 1976; Johnson and Gathright, 1978; Lukert and Halladay, 1980; Lukert and Banks, 1984; Clarke, 1984; Conley, 1989) has established that rocks of the RRS are intrusive into Grenville-age (1000–1150 Ma) gneisses of the anticlinorium core and are intruded by numerous greenstone dikes that may have served as feeders to the voluminous Catoctin Formation of the BRA cover sequence. The age of the RRS is thus broadly constrained by field evidence as younger than Grenvillian metamorphism and older than at least most of the late Precambrian Catoctin Formation. Published isotopic age determinations (Rankin, 1975; Clarke, 1981; Lukert and Banks, 1984; Mose and Nagle, 1984; Mose and Kline, 1986; Tollo et al., in press) span a considerable range (570–730 Ma), but are generally consistent with the observed field relations.

Field Relations

The sinuous boundaries of the RRS belt represent both intrusive and fault contacts with various basement gneisses and Late Proterozoic metasedimentary units of the BRA cover sequence (Mitra and Lukert, 1982; Lukert and Banks, 1984). The elongate dimension of the RRS outcrop belt is oriented obliquely to the trend of the anticlinorium in Virginia (Figure 1). Bartholomew (in press) suggested that the N25E trend of the RRS belt is indicative of emplacement within a regional fracture pattern resulting from Late Proterozoic extension oriented along a N65W azimuth. Rocks of the RRS characteristically lack the pervasive foliation typical of the Grenville-age gneisses, but locally show extensive development of ductile deformation zones of various scales, as noted by Mitra (1979). The younger age of the RRS as compared to the Grenvillian age gneisses is demonstrated by: 1) undeformed granitic dikes crosscutting the gneissic country rocks; 2) gneissic xenoliths included within RRS rocks; 3) inliers (representing possible screens and roof pendants) of gneiss mapped

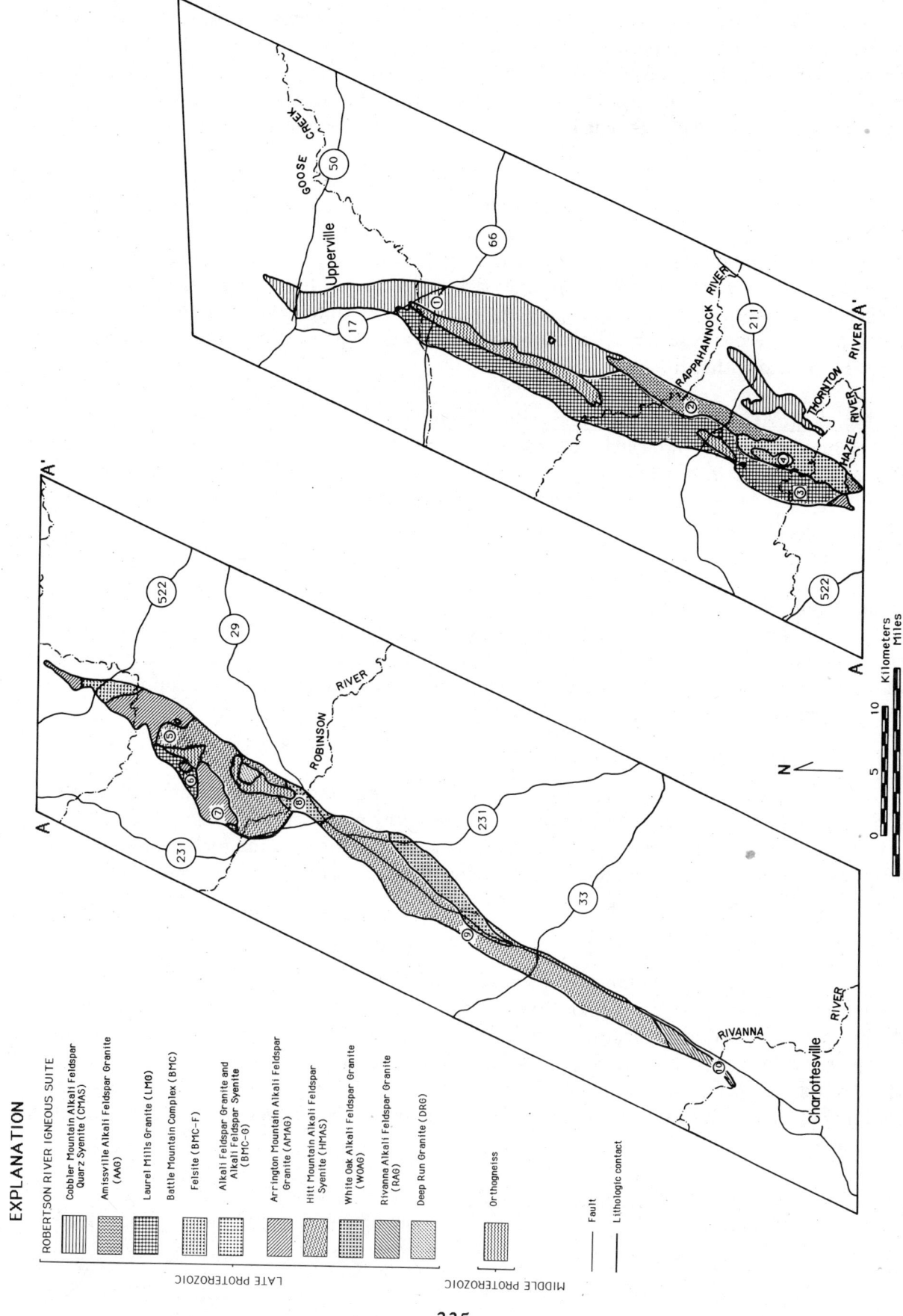

Figure 2. Geologic map showing the lithologic units of the Robertson River Igneous Suite.

within the principal outcrop belt of the RRS; and 4) shatter zones comprised of granitoid intruding dismembered gneiss (Lukert and Nuckols, 1976; Lukert and Halladay, 1980; Lukert and Banks, 1984; Clarke, 1984; Arav, 1989; Tollo et al., in preparation). Although the general nature of much of the RRS outcrop belt has been described by various authors (Allen, 1963; Lukert and Nuckols, 1976; Johnson and Gathright, 1978; Lukert and Halladay, 1980; Lukert and Banks, 1984; Clarke, 1984; Conley, 1989), until recently the detailed field relationships of the constituent intrusive rocks and the nature of the contacts with adjacent lithologic units were poorly constrained.

Lithologic Units

The RRS includes at least nine mappable lithologic units that can be distinguished on the basis of field, petrographic, and geochemical characteristics. The rock types range from granite to alkali feldspar granite to alkali feldspar syenite (Table 1). Most units contain amphibole with fluorite and allanite as accessories. In addition to the lack of pervasive foliation in the RRS, these mineralogic characteristics assist in distinguishing rocks of this suite from most lithologic units comprising the Blue Ridge basement. The following is a brief description of the lithologic units that have been recognized within the RRS:

Cobbler Mountain Alkali Feldspar Quartz Syenite (CMAS):

This unit is exposed throughout a broad area in the northern segment of the RRS outcrop belt (Figure 2) where it is in contact with several types of Grenville-age basement rocks and the Laurel Mills Granite, which it intrudes. The CMAS is composed largely of quartz-poor syenitoid typically bearing calcic amphibole, prominent mesoperthite phenocrysts, and accessory fluorite (Arav, 1989).

Amissville Alkali Feldspar Granite (AAG):

Rocks of the AAG are exposed discontinuously along the eastern edge of the northern RRS segment (Figure 2). This lithologic unit, which typically contains prominent quartz phenocrysts, is characterized by the distinctive ferromagnesian mineral assemblage of riebeckite + aegirine and, like the CMAS, contains abundant fluorite (Arav, 1989).

Laurel Mills Granite (LMG):

Rocks assigned to the LMG are locally exposed along the western edge of both segments of the RRS outcrop belt (Figure 2). The LMG is characterized by prominent calcic amphibole that is locally replaced by the assemblage biotite + stilpnomelane. This coarse-grained rock type is the only widespread lithologic unit within the RRS that contains subequal amounts of primary alkali feldspar and plagioclase (Tollo and Arav, in press).

Battle Mountain Complex (BMC):

The Battle Mountain Complex (Wallace and Tollo, 1986) includes a variety of medium-grained alkali feldspar granites and syenites closely associated with abundant, fine-grained to aphanitic felsite occurring within the central area of the RRS outcrop belt (Figure 2). Detailed field mapping indicates that the felsites occur as numerous dikes within the coarse-grained rocks of the complex (Hawkins, unpublished data). The mineralogic flow banding, miarolitic cavities, and possible lithophysae that locally characterize the felsites are indicative of the subvolcanic to volcanic nature of the complex.

Arrington Mountain Alkali Feldspar Granite (AMAG):

The AMAG occurs in both segments of the RRS outcrop belt (Figure 2) and includes medi-

Table 1. Field and petrographic characteristics of the lithologic units of the Robertson River Igneous Suite.

Lithologic Unit	Rock Type	Texture Grain Size	Kspar:plag ratio	Ferromagnesian Phase(s)	Distinguishing Characteristics
Cobbler Mountain	alkali feldspar quartz syenite	porphyritic medium	8:1 to 10:1	amphibole rare pyroxene	mesoperthite phenocrysts low quartz content
Amissville	alkali feldspar granite	porphyritic medium	9:1 to 20:1	riebeckite aegirine	ferromagnesian assemblage quartz phenocrysts
Laurel Mills	granite	inequigranular coarse	1:1 to 4:1	amphibole	coarse grain size inequigranular texture
Battle Mountain Complex	alkali feldspar granite	inequigranular medium	10:1 to >15:1	aegirine	low color index abundance of fluorite
	felsite	allotriomorphic fine	10:1 to >15(?):1	aegirine	fine grain size local flow banding
Arrington Mountain	alkali feldspar granite	equigranular medium	4:1 to 40:1	amphibole	eu- to subhedral mesoperthite equigranular texture
Hitt Mountain	alkali feldspar syenite	inequigranular medium to coarse	6:1 to >100:1	amphibole	low quartz content prominent amphiboles
White Oak	alkali feldspar granite	inequigranular coarse and fine	8:1 to >100:1	amphibole	associated coarse- and fine-grain types abundant prismatic amphibole
Rivanna	alkali feldspar granite	equigranular medium	4:1 to 10:1	biotite (muscovite)	low color index locally abundant fluorite
Deep Run	granite	equigranular medium	3:1 to 4:1	biotite (muscovite)	prominent clusters of biotite

um-grained, equigranular alkali feldspar granite with biotite ± amphibole. The textural and mineralogic characteristics of this unit are remarkably consistent throughout the outcrop area. This unit can be distinguished from the LMG, with which it is locally in contact, by its characteristic equigranular texture and high alkali feldspar:plagioclase ratio (Lowe, 1990).

Hitt Mountain Alkali Feldspar Syenite (HMAS):

This poorly exposed unit has been mapped across a broad area of the southern segment of the RRS outcrop belt (Figure 2) and includes medium- to coarse-grained, amphibole-bearing alkali feldspar quartz syenite and alkali feldspar syenite (Lowe, 1990). Although the unit is characterized by variable grain size and includes both fine-grained dikes and coarse-grained pegmatoids, the ubiquitous quartz-poor nature of the mineralogic assemblage is distinctive.

White Oak Alkali Feldspar Granite (WOAG):

The WOAG is comprised of coarse-grained, amphibole-bearing alkali feldspar granite locally associated with a fine-grained variety of identical mineralogic assemblage (Lowe, 1990). Both the coarse- and fine-grained varieties are typically present on the scale of individual outcrops. This unit occurs exclusively in the southern segment of the RRS outcrop belt (Figure 2).

Rivanna Alkali Feldspar Granite (RAG):

This lithologic unit is exposed only in the southernmost part of the RRS outcrop belt (Figure 2) and includes equigranular, biotite- (± muscovite) bearing alkali feldspar granite. Low color index is a distinguishing characteristic of this unit; fluorite is locally abundant.

Deep Run Granite (DGR):

The DRG includes medium-grained, equigranular biotite- (± muscovite) bearing granite exposed within the southern segment of the RRS outcrop belt (Figure 2). Exposures of this unit are typically poor and are limited to a restricted area. This unit is tentatively included within the RRS because of its association with other RRS units and its characteristic relatively undeformed fabric.

Petrology and Geochemistry

Most of the lithologic units of the RRS are characterized by high alkali feldspar:plagioclase ratios and, as a result, plot in the alkali feldspar granite, alkali feldspar quartz syenite, and alkali feldspar syenite fields of modal quartz-alkali feldspar-plagioclase diagrams (Figure 3a-i). Textural evidence further indicates that most of these units crystallized under hypersolvus conditions with plagioclase feldspar becoming a liquidus phase late in the crystallization sequence (Arav, 1989). The Laurel Mills Granite, with subequal amounts of alkali feldspar and plagioclase, is the most widespread exception, plotting within the granite field separate from all other RRS lithologic units (Figure 3c). Limited data available for the Deep Run Granite (Figure 3i) indicate a similar, low alkali feldspar:plagioclase ratio for this unit as well.

Amphibole is a characteristic ferromagnesian phase in most units and typically shows petrographic evidence indicative of a primary origin. In most cases, the nature of the amphibole is a reflection of the original bulk composition of the unit. Calcic amphiboles (nomenclature of Leake, 1978) are most common in the rocks analyzed to date and range from hastingsite to ferro-edenitic hornblende in composition (Arav, 1989; Gray, 1990). Fe-rich riebeckite occurs only in the AAG where it may be both late primary and subsolidus in origin (Tollo and Arav, in press). Aegirine and aegirine-augite are present in the AAG, some of the BMC units, and rarely in the CMAS where the presence of primary pyroxene is indicative of the low pH_2O conditions of initial crystallization (Tollo

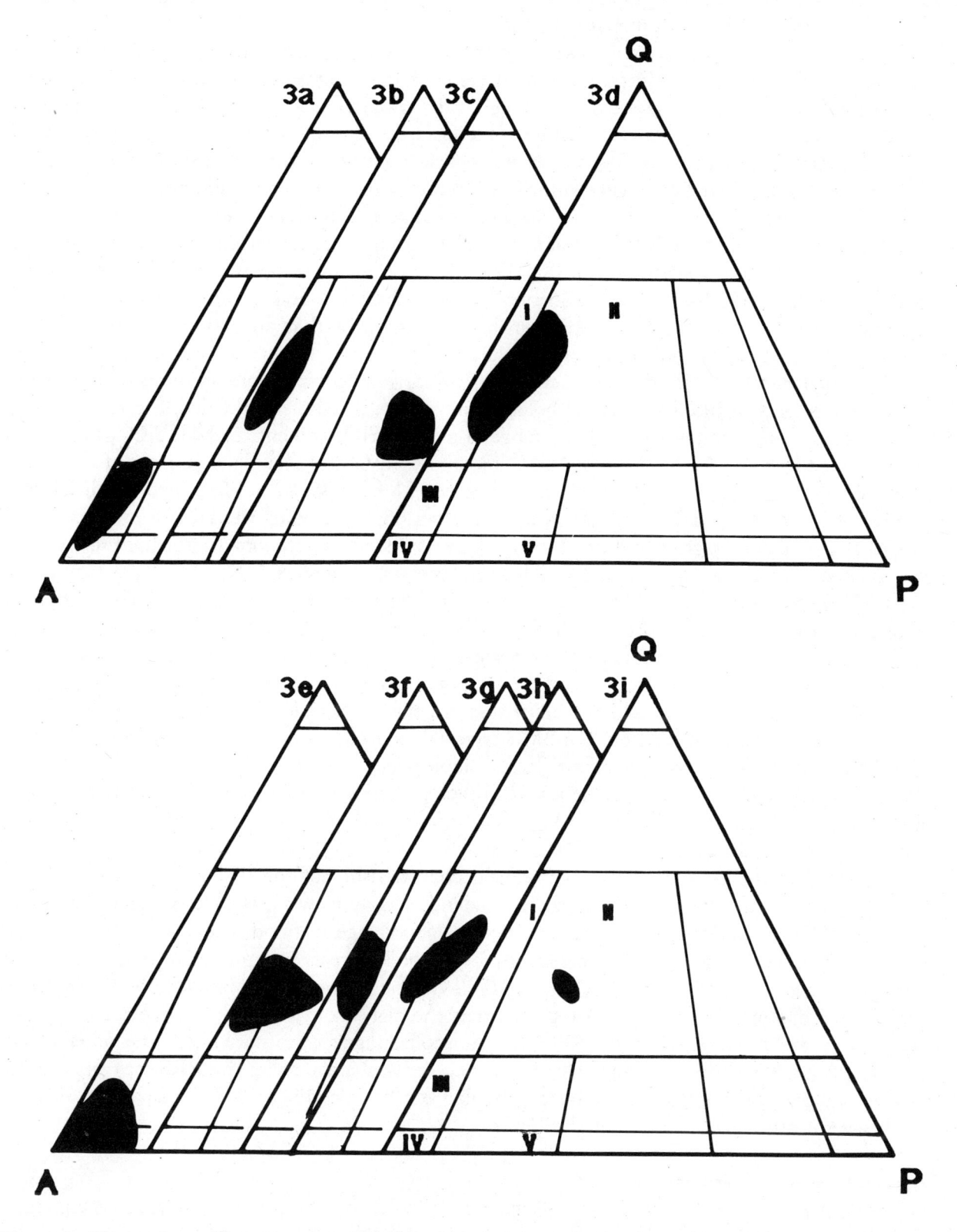

Figure 3. Plots of modal quartz(Q)-alkali feldspar(A)-plagioclase(P) for the lithologic units of the Robertson River Igneous Suite. Plots include data from CMAS (3a), AAG (3b), LMG (3c), BMC-G (3d), HMAS (3e), AMAG (3f), WOAG (3g), RAG (3h), and DRG (3i). Compositional field labels include I: alkali feldspar granite, II: granite, III: alkali feldspar quartz syenite, IV: alkali feldspar syenite, V: syenite (after Streckeisen, 1973).

and Arav, in press). Biotite is common in units throughout the suite (Table 1) and in some units displays textural evidence of a primary origin.

The effects of post-crystallization alteration are locally abundant in various RRS units. Evidence of late-stage deuteric alteration is most common in the more compositionally alkaline units. Veins of riebeckite and intergrowths consisting of riebeckite rims around grains of primary sodic pyroxene occur in the AAG and probably result from deuteric fluid migration. Similarly, Clarke (1984) interpreted intergrowths of quartz + magnetite (probably in the CMAS) as psuedomorphic after original pyroxene and as having formed as a result of deuteric recrystallization.

Diverse petrographic evidence of Paleozoic metamorphism is abundant throughout the suite and includes development of discrete grains of epidote group minerals within plagioclase, formation of albite-rich rims on mesoperthite grains, and creation of complex compositional zoning patterns in amphiboles. Gray (1990) discussed evidence for the subsolidus breakdown of calcic amphibole resulting in biotite + amphibole + stilpnomelane intergrowths that are most likely the result of relatively low-grade metamorphic recrystallization.

Geochemical data for the RRS define a broad range of compositions in terms of both major and trace elements (Figure 4a-f). The lithologic units of the RRS can be divided into the following broad compositional groups based on the range of average SiO_2 content: 1) high SiO_2 group (>74 wt. %) including the AAG, AMAG, RAG, DGR, and both felsites and alkali granites of the BMC; 2) moderate SiO_2 group (69–73 wt. %) including the LMG, WOAG, CMAS, and alkali feldspar syenites of the BMC; and 3) low SiO_2 group (less than 66%) including the HMAS and associated autoliths. The high SiO_2 group is characterized by the lowest concentrations (normalized to constant SiO_2 content to facilitate comparison) of Al_2O_3 (Figure 4a), total alkalis ($Na_2O + K_2O$, Figure 4b), and CaO (Figure 4c). The HMAS, which locally displays textural evidence of cumulate processes involving feldspar, is characterized by the highest values of Al_2O_3, total alkalis, and CaO.

In general, the concentration of trace elements does not reflect the variation in SiO_2 content (Table 2; Figure 4d-f). Concentrations of the high field strength cations Nb (Figure 4d), Zr (Figure 4e), and Ce (Figure 4f) show considerable variation between (and locally within) individual units. Samples collected from units within the BMC typically exhibit the highest concentrations of Zr and Nb (Table 2), and dikes associated with this complex locally show extreme enrichment in these elements (Tollo and Lowe, 1990). Fine-grained lithologic types from the HMAS (see Stops 5 and 6) also show high concentrations (> 1000 ppm) of Zr but, interestingly, have markedly lower concentrations of Nb relative to the BMC.

Many of the petrologic and geochemical characteristics of the RRS are typical of A-type granitoids worldwide. These characteristics include: 1) largely hypersolvus feldspar assemblages; 2) presence of calcic and/or sodic amphibole; 3) local abundance of fluorite; 4) low CaO and MgO contents; 5) high Fe/Mg ratio; 6) agpaitic ratio ~1.0; and 7) marked enrichment in selected trace elements including Zr, Nb, Zn, Y, Ga, and Ce (Loiselle and Wones, 1979; Pitcher, 1983; Whalen et al., 1987). Tollo and Arav (in press) reported that the high Ga values and, in particular, the high Ga/Al ratios are characteristic of the RRS and serve to distinguish this suite from most other similar occurrences in the world. The trace element enrichments characterizing the RRS may be a result, at least in part, of a petrogenetic history similar to that described for a comparable suite in southeastern Australia by Collins et al. (1982).

Age and Tectonic Significance

The mineralogic and geochemical characteristics of the RRS are typical of A-type granites (Whalen et al., 1987); these characteristics, coupled with field and textural data, imply that the RRS was emplaced in an anorogenic tectonic setting (Loiselle and Wones, 1979; Pitcher, 1983). Rankin (1975; 1976) proposed that a similar granitic suite (Crossnore) in the vicinity of Mount Rogers,

Table 2. Geochemical data for samples from the field trip stops.

Analysis #	1	2	3	4a	4b	5a	5b	6a	6b	7	8	9	10
Stop #	1	2	3	4	4	5	5	6	6	7	8	9	10
Map Unit	CMAS	AAG	LMG	BMC-G	BMC-F	HMAS	HMAS	LMG	HMAS	AMAG	WOAG	HMAS	RAG
SiO2	72.14	73.65	71.93	69.94	76.18	59.36	64.37	72.38	61.06	76.68	72.51	63.79	77.85
TiO2	0.27	0.08	0.29	0.35	0.18	0.78	0.47	0.36	0.30	0.18	0.35	0.17	0.09
Al2O3	13.53	12.17	13.82	13.47	10.70	15.22	15.71	13.43	18.11	11.96	12.96	17.75	12.26
Fe2O3*	3.28	4.10	3.03	4.93	3.75	11.55	6.29	3.60	6.42	2.28	3.69	4.90	0.93
MnO	0.05	0.04	0.04	0.11	0.05	0.26	0.14	0.05	0.11	0.04	0.09	0.09	0.01
MgO	0.16	0.14	0.28	0.19	0.39	0.12	0.12	0.24	0.16	0.05	0.12	0.08	0.16
CaO	0.35	0.07	0.76	1.22	0.25	2.66	1.41	1.17	2.12	0.22	1.19	1.34	0.83
Na2O	4.99	4.90	3.99	5.03	3.20	5.04	5.34	3.56	6.37	3.39	3.82	6.28	3.81
K2O	4.84	4.63	5.39	4.63	4.59	4.86	5.73	5.51	4.85	5.47	5.42	5.80	4.51
P2O5	0.01	0.00	0.06	0.05	n.d.	0.13	0.06	0.08	0.05	0.01	0.05	0.05	0.01
Total	99.62	99.78	99.59	99.92	99.29	99.96	99.64	100.36	99.55	100.28	100.20	100.24	100.45
analyses:	2	2	4	2	2	2	2	2	2	2	2	2	2
Nb	35	139	42	269	313	71	73	44	191	116	109	55	84
Zr	213	676	353	1440	2224	825	1260	574	1993	534	572	340	132
Zn	160	205	189	398	241	252	133	141	184	178	291	105	127
Ni	4	9	6	12	7	13	9	7	11	12	12	8	10
Cr	12	13	21	14	4	4	9	13	1	15	11	6	11
V	2	n.d.	5	4	2	3	3	4	1	2	1	1	1
Ce	198	91	244	367	208	229	492	229	269	227	355	493	152
Ba	64	n.d.	415	122	n.d.	381	148	524	282	158	257	73	206
A.I.	0.99	1.07	0.90	0.99	0.96	0.89	0.95	0.88	0.87	0.96	0.94	0.94	0.91

*total iron expressed as Fe2O3   n.d.: not detected   A.I.: agpaitic index = (Na+K)/Al

Sample descriptions:
1 gray, medium-grained, inequigranular alkali feldspar quartz syenite (sample RRSA-74)
2 light gray, medium-grained, inequigranular, aegirine + riebeckite-bearing alkali feldspar granite (sample RRS90-7)
3. gray, coarse-grained, inequigranular, amphibole-bearing granite (sample RR85-24)
4a light gray, medium-grained, inequigranular, aegirine-bearing alkali feldspar granite (sample RRDH-5)
4b dark gray, fine-grained, flow-banded felsite (sample RRDH-26B)
5a dark gray, fine-grained, inequigranular, amphibole-bearing alkali feldspar syenite (sample RR89-16A)
5b light gray, coarse-grained, inequigranular, amphibole-bearing alkali feldspar syenite (sample RR90-60)
6a gray, coarse-grained, inequigranular, amphibole-bearing granite (sample RR89-29)
6b dark gray, fine-grained, equigranular to porphyritic, amphibole-bearing syenitoid (sample RR89-30)
7. light gray, medium-grained, equigranular, amphibole + biotite-bearing alkali feldspar granite (sample RR90-79)
8 gray, coarse-grained, inequigranular, amphibole-bearing alkali feldspar granite (sample RR90-53)
9 light gray, medium- to coarse-grained, inequigranular alkali feldspar syenite (sample RR90-95)
10 white to light gray, medium-grained, biotite-bearing alkali feldspar granite (sample RR90-101)

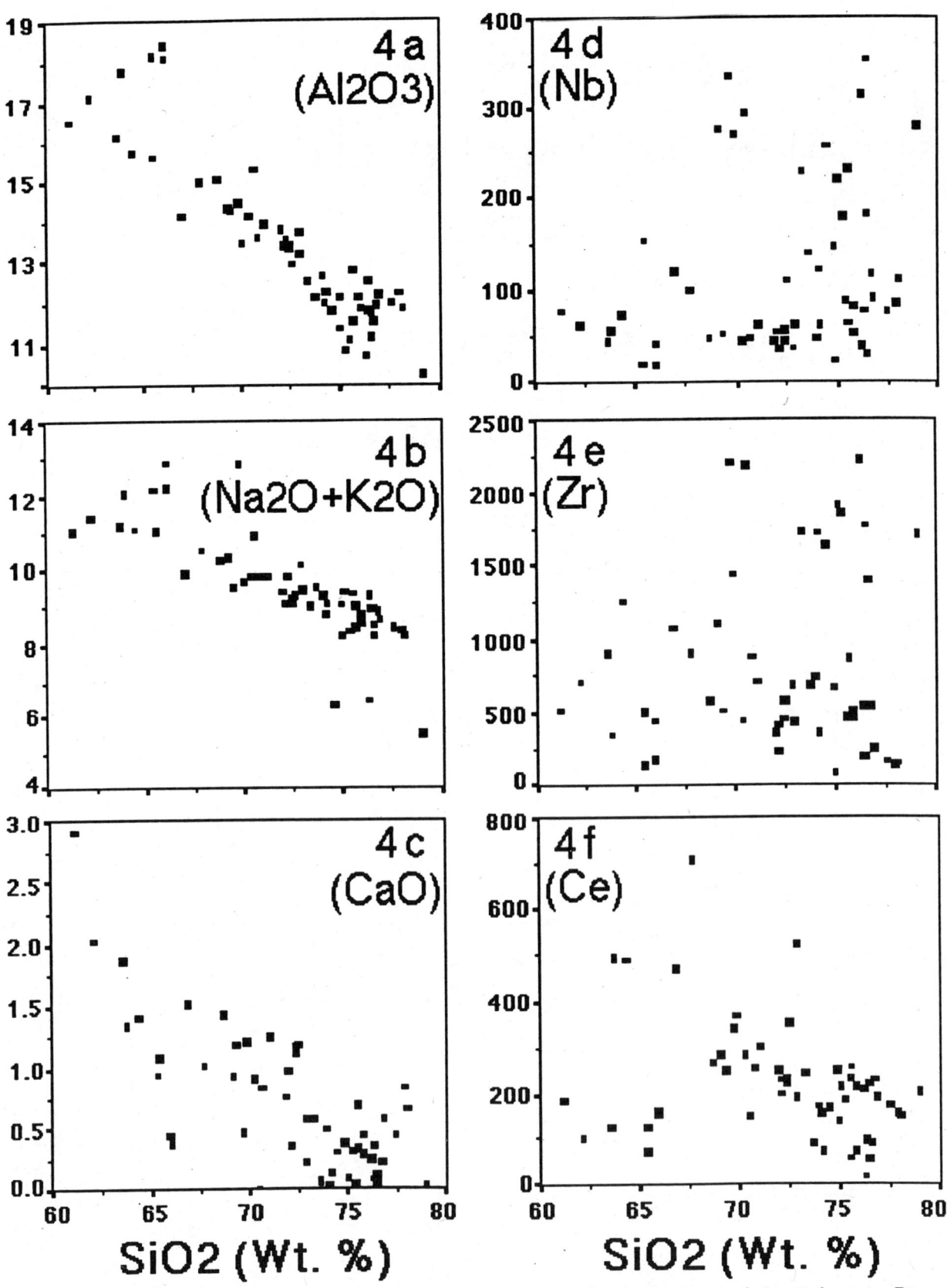

Figure 4. Major and trace element variation diagrams for the lithologic units of the Robertson River Igneous Suite. Plots of Al_2O_3 (4a), Na_2O+K_2O (4b), CaO (4c), Nb (4d), Zr (4e), and Ce (4f) versus SiO_2. Major elements are expressed in weight %; trace elements in ppm.

Virginia, and the widespread, bimodal metavolcanic sequence of the Blue Ridge province preserve a record of magmatic activity associated with the early stages of the Late Proterozoic rifting of Laurentia. The petrochemical characteristics of the RRS are consistent with this model, and the suite represents the largest occurrence of such Late Proterozoic rocks in the southern and central Appalachian Blue Ridge. The age and time span of RRS plutonism is particularly important to models of the Late Proterozoic tectonic evolution of the Appalachian orogen because of the diversity of magma compositions (Table 2), direct link to associated volcanic and subvolcanic products (Battle Mountain Complex), and inferred stratigraphic relationships with metasedimentary strata of probable rift-related origin (Mechum River Formation).

New U/Pb zircon isotopic data for units from the northern segment of the RRS indicate that magmatism spanned a considerable interval beginning with intrusion of the LMG at 729 Ma and continuing through emplacement of the AAG at approximately 640 Ma (Tollo et al., in press). Plutons of the Crossnore Suite in North Carolina and Virginia define a similar time span (710 to possibly 650 Ma, Odom and Fullagar, 1984). The limited data available for the RRS indicate that magmatism began with metaluminous magmas (LMG) and culminated with peralkaline types (AAG), corresponding to the sequence documented in other petrologically similar localities worldwide (Barker et al., 1975). The time span of RRS magmatism is comparable to that of other anorogenic suites including the White Mountain magma series of New England (USA) (Foland and Faul, 1977). The present database for Late Proterozoic intrusive rocks of the Blue Ridge province is, however, inadequate to define regional compositional trends and to determine whether magmatism associated with Laurentian rifting progressed throughout the interval or was restricted to two or more discrete episodes. Nevertheless, the data do indicate that anorogenic magmatism of Late Proterozoic age was considerably more varied in composition and more widespread in occurrence than envisioned by Rankin (1976). Further definition of the extent, temporal relations, and compositional diversity of the Late Proterozoic magmatic suite is critical to determining the nature of thermal anomalies, mechanisms of crustal melting, and style of lithospheric dynamics preceding and associated with rifting.

ROAD LOG

The field trip begins at the intersection of Routes 17 and 29/211 in Warrenton, Virginia. Cumulative mileage measurements start at this intersection. Incremental mileage is given in parentheses. The field trip stops are located on a generalized map of the Robertson River Igneous Suite in Figure 2.

Cumulative
mileage   Description

0.0   Intersection of Routes 17 and 29/211 in Warrenton, VA. Proceed north on Route 17.
10.7 (10.7) Intersection of Routes 17 and I-66. Turn left to continue west on Route 17/I-66.
15.7 (5.0) Exit 5 from Route I-66. Bear right onto Route 55/17 and proceed north toward Delaplane.
16.7 (1.0) Intersection of Routes 55 and 17 south of Delaplane. Turn left onto Route 55.
17.3 (0.6) Intersection of Routes 731 and 55. Turn right onto Route 55 and proceed west.
17.5 (0.2) Stop 1: Pull onto right shoulder. Beware of oncoming traffic.

STOP 1: Cobbler Mountain Alkali Feldspar Quartz Syenite

An unusually fresh exposure of the Cobbler Mountain Alkali Feldspar Quartz Syenite (CMAS), designated as the petrochemical type locality of this unit, is exposed here. This stop is located in the northern portion of the CMAS outcrop area approximately 0.5 miles (0.8 km) east of the contact with a large gneissic inlier that separates the CMAS from the Laurel Mills Granite (LMG) to the west (Figure 2). The CMAS is a gray, medium-grained, inequigranular alkali feldspar quartz syenite (ranging to alkali feldspar granite and syenite, see Figure 3a) composed of abundant mesoperthite, anhedral quartz, rare plagioclase, and amphibole that is locally replaced by quartz and Fe-Ti oxide phases (Arav, 1989). Fluorite is a common accessory mineral. Pyrite and other sulfide phases are present throughout the outcrop, both as disseminated grains and along veins. In hand specimen, prominent eu- to subhedral mesoperthitic alkali feldspar is characteristic and is most easily observed on weathered surfaces. Field evidence from this area clearly indicates that the CMAS is intrusive into the LMG and the strongly foliated gneiss comprising the country rock to the east. At Big Cobbler Mountain, located 4.5 miles (7.2 km) south-southwest of this stop, the CMAS is cut by two sets of dikes, one of which is a fine-grained variety of the CMAS and the other an aegirine + riebeckite-bearing granitoid that is mineralogically and geochemically identical to the Amissville Alkali Feldspar Granite (AAG). At this roadcut, the CMAS is cut by three NE-striking greenstone dikes. Such dikes are particularly abundant in the northernmost area of the RRS outcrop belt.

Petrographic examination of samples collected from this outcrop indicates that the rock is composed of: 1) abundant, medium-grained, eu- to subhedral mesoperthite; 2) fine- to medium-grained, anhedral quartz exhibiting recrystallization textures; 3) rare, predominantly fine-grained plagioclase; and 4) intergrowths of quartz + green biotite + Fe-Ti oxides ± stilpnomelane formed after primary amphibole. Accessory phases include fluorite, zircon, and allanite commonly rimmed by epidote. An alkali feldspar:plagioclase ratio of 25:1 is typical of samples from this exposure (Table 1; Figure 3a). Microprobe analyses indicate that the amphibole (X=pale yellow-green, Y=yellow-green, Z=dark green) in the CMAS is predominantly hastingsite to hastingsitic hornblende (terminology of Leake, 1978) in composition (Arav, 1989). Textural evidence suggests that the amphibole is primary in origin (Figure 5), although in many samples this mineral is replaced by an intergrowth of secondary phases, as described previously. Aegirine-augite, also of primary origin, has been described from one location on Big Cobbler Mountain.

The CMAS is characterized by a broad range in geochemical composition (67–75 wt. % SiO_2). Data for a sample collected from this roadcut (analysis 1, Table 2) closely approximate the overall average for this unit (Tollo and Arav, in press) with the exception of the relatively low concentrations of Nb and Zr. The variation in major element composition characterizing the CMAS encompasses the entire range shown by the moderate silica group of RRS units. Field and geochemical evidence indicates that the CMAS is comprised of multiple, as yet unidentified, intrusive units (Arav, 1989). The range in observed compositions is probably a manifestation of this lithologic complexity.

Zircons from this locality define two morphological populations including: 1) elongate grains with dominant prismatic faces and 2) stubby grains with dominant {301} dipyramidal faces (Tollo et al., in press). The zircons with prominent dipyramidal faces occur locally as overgrowths on cores with prominent prismatic faces. The elongate grains are interpreted to be xenocrysts, and a concordant age of 750 Ma is a minimum for their source. Preliminary isotopic data for the stubby grains suggest an age of approximately 723 Ma for the CMAS. These data confirm field relationships

(Tollo and Arav, in press) indicating that the CMAS is resolvably younger than the LMG.

Proceed west on Route 55 to bottom of the hill.

17.8	(0.3)	Reverse direction and proceed east on Route 55.
18.3	(0.5)	Intersection of Routes 731 and 55. Turn left onto Route 55.
18.8	(0.5)	Intersection of Routes 55 and 17. Turn right onto Route 55/17.
19.5	(0.7)	Bear left to continue on Route I-66 (east).
23.4	(3.9)	Exit 6 off Route I-66. Exit right from highway.
23.6	(0.2)	Turn right at end of offramp and immediately right again onto Route 647. Proceed south.
27.5	(3.9)	Intersection of Routes 724 and 647 at Ada. Continue southwest on Route 647. Road is located on Middle Proterozoic gneiss. Prominent hills to the west comprise the Big and Little Cobbler Mountains, which are underlain mostly by alkali feldspar quartz syenite of the CMAS.
33.2	(5.7)	Intersection of Routes 688 and 647 near Jerry's Shop. Continue southwest on Route 647.
36.9	(3.7)	Intersection of Routes 637 and 647. Turn left onto Route 637 and proceed south.
38.4	(1.5)	Prominent exposures on side of road of coarse-grained amphibole-bearing granite of the Laurel Mills type.
40.2	(1.8)	Stop 2: Turn left into private driveway.

STOP 2: Amissville Alkali Feldspar Granite

The summit of Poes Mountain, located near the eastern border of the RRS, is underlain by abundant outcrops of unusually fresh Amissville Alkali Feldspar Granite (AAG). This unit has been traced continuously 5 miles (8 km) south-southwest from this locality to the area of Battle Mountain, where it is in contact with rocks of the Battle Mountain Complex (Figure 2). The AAG is typically composed of medium- to coarse-grained, eu- to subhedral quartz phenocrysts; medium-grained, subhedral alkali feldspar; rare, fine-grained, subhedral plagioclase; and fine- to medium-grained, subhedral, green aegirine locally associated with riebeckite. Fluorite is a common accessory phase in many samples. Samples collected from the vicinity of Poes Mountain differ texturally from most of the typically porphyritic AAG in that the rocks of this area are characterized by a medium-grained, non-porphyritic, inequigranular texture.

Samples collected from this area are petrographically identical to most of the AAG in that they are composed of: 1) medium-grained, subhedral microcline-mesoperthite; 2) fine- to medium-grained, anhedral quartz; 3) fine-grained, subhedral plagioclase; 4) fine- to medium-grained, subhedral aegirine (X=pale green, Y=yellow-green, Z=pale yellow-green); 5) fine-grained, subhedral riebeckite (X=dark blue, Y=gray-blue, Z=light brown); and 6) fine-grained, subhedral, brown biotite (secondary). Textural evidence indicates that aegirine is probably a primary magmatic phase (Figure 6). The riebeckite, which typically occurs as either medium-grained, anhedral patches or as fine, acicular grains, is either late magmatic or subsolidus in origin. Zircon is present in trace quantities; fluorite is abundant. The average alkali feldspar:plagioclase ratio of samples from this area is 10:1, which is typical of the AAG unit (Table 1; Figure 3b).

The AAG and some lithologies of the Battle Mountain Complex (BMC) are characterized by the only true peralkaline compositions within the RRS. The AAG is part of the high-SiO_2 (average 75.6 wt. %) group of the RRS and exhibits enrichments in Nb and Zr that are comparable

Figure 5. Photomicrograph of calcic amphibole (dark-colored, high relief) occurring within the Cobbler Mountain Alkali Feldspar Quartz Syenite. Scale at bottom equals 1 mm.

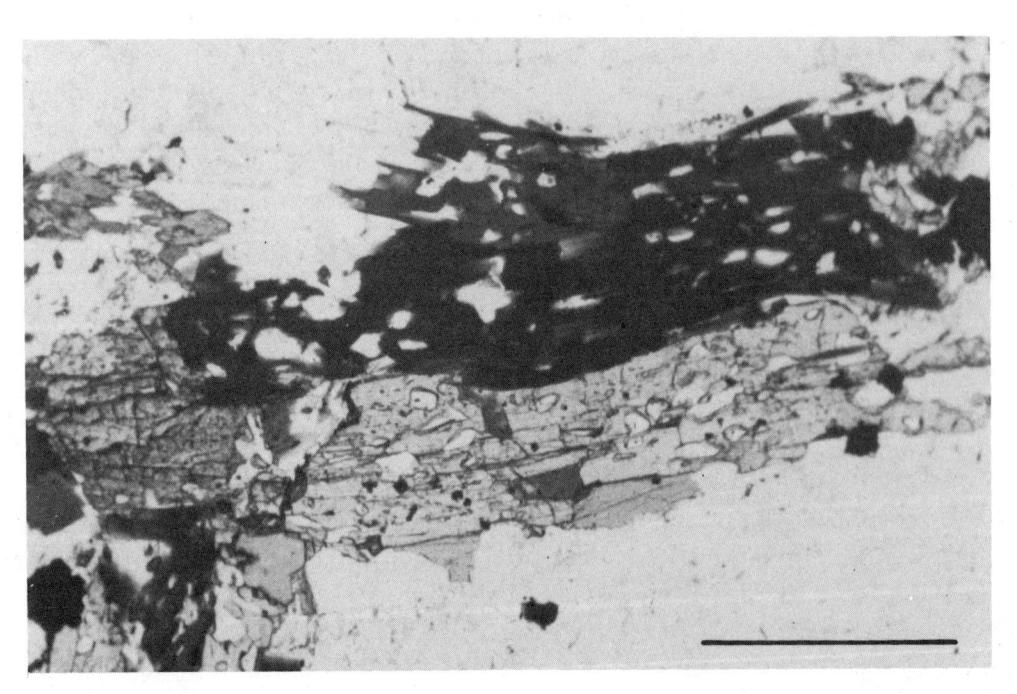

Figure 6. Photomicrograph of aegirine (light-colored, high relief) occurring in association with riebeckite (dark) in the Amissville Alkali Feldspar Granite. Scale at bottom equals 1 mm.

to most of the other units. The AAG is also characterized by low CaO, high Fe/Mg, and marked enrichment in light rare earth elements (Tollo and Arav, in press), similar to peralkaline anorogenic granites from other localities worldwide (Loiselle and Wones, 1979; Whalen et al., 1987). The chemical similarity and physical association of the AAG to rocks of the BMC strongly suggest that these lithologic units are petrologically related and together may represent the shallow plutonic and subvolcanic to possibly volcanic (see Stop 4) remnants of a silicic magmatic center.

Zircons from a sample collected at the AAG type locality, 2 miles (3.2 km) south of Poes Mountain, define a single population that is typically anhedral in shape. Preliminary data indicate an age of 640 Ma for the AAG (Tollo et al., in press), which, when considered within the context of new ages recently determined for other units of the RRS (see Stops 1 and 3), suggests that RRS magmatism spanned a considerable time interval. The limited data obtained thus far for the RRS indicate that the AAG is the youngest intrusive unit of the (northern) RRS and that suite magmatism may have culminated with the development of a peralkaline volcanic complex (see Stop 4).

Proceed back to Route 637. Turn left onto Route 637.

42.3	(2.1)	Intersection of Routes 637 and 645. Bear right to continue south.
42.7	(0.4)	Intersection of Routes 645 and 211. Turn right onto Route 211 and proceed northwest.
44.0	(1.3)	Intersection of Routes 640 and 211. Roadcut adjacent to westbound lanes of Route 211 has provided fresh samples of Amissville Alkali Feldspar Granite. Tollo et al. (in press) obtained a nearly concordant U/Pb age of 640 Ma for a sample from here. Rankin (1975, citing T. W. Stern) reported a Pb^{207}/Pb^{206} age of 650 Ma.
50.5	(6.5)	Intersection of Routes 729 and 211 at Ben Venue. Turn left onto Route 729 and proceed south.
51.8	(1.3)	Saprolite exposure of gneiss adjacent to road.
54.0	(2.2)	The two peaks to the east comprise Little Battle and Battle Mountains. Both hills are underlain by various felsites and granitoid units of the Battle Mountain Complex.
55.8	(1.8)	Intersection of Routes 729 and 618. Turn right onto Route 618 and proceed west.
55.9	(0.1)	Exposure adjacent to road on right is comprised of Laurel Mills Granite. A dike of amphibole-bearing felsite cuts the granite at the eastern end of the outcrop.
56.5	(0.6)	Bridge on Route 618 over Thornton River at Laurel Mills village.
57.3	(0.8)	Intersection of Routes 619 and 618. Bear left to remain on Route 618.
57.5	(0.2)	Stop 3: Pull onto right shoulder.

STOP 3: Laurel Mills Granite

This roadcut, located 0.6 miles (1.0 km) from the western border of the RRS (Figure 2), is an exceptionally fresh exposure of the Laurel Mills Granite (LMG) and is designated as the petrochemical type locality of this unit. The LMG in this area is in contact with the Arrington Mountain Alkali Feldspar Granite (AMAG) to the south and with the Amissville Alkali Feldspar Granite (AAG) and Battle Mountain Complex (BMC) to the east (Figure 2). At this locality, the LMG is composed of quartz, alkali feldspar, plagioclase, and amphibole that is partially to wholly replaced by biotite + stilpnomelane. As is typical of the LMG, abundant planar structural features including narrow ductile deformation zones and brittle faults and joints are present. Pyrite is locally abundant both along these features and disseminated throughout the granite.

Petrographic examination of samples from this outcrop illustrates many of the features characteristic of this unit. The LMG is composed of: 1) coarse-grained, anhedral quartz exhibiting pronounced recrystallization textures; 2) coarse-grained, sub- to euhedral microcline-microperthite; 3) medium-grained, saussuritized plagioclase containing abundant inclusions of epidote group minerals; and 4) remnants of coarse-grained amphibole (X=yellow-tan, Y=pale green, Z=dark green) which have been partially or entirely replaced by intergrowths of brown biotite and orange-brown stilpnomelane. The presence of biotite + stilpnomelane has been interpreted to be a product of retrograde metamorphic breakdown of the original amphibole (Gray, 1990). Accessory phases include zircon, allanite, apatite, ilmenite, and magnetite. The alkali feldspar:plagioclase ratio for samples from this roadcut is 2:1, which is typical of the LMG lithologic unit (Table 1; Figure 3c). The LMG is the only widespread unit within the RRS with subequal amounts of primary alkali feldspar and plagioclase.

Electron microprobe analyses of amphiboles from this roadcut and from other samples of the LMG unit (Gray, 1990) indicate that this phase has largely retained an igneous composition ranging from hastingsite to ferro-edenitic hornblende (nomenclature of Leake, 1978). Chemical zoning patterns are typically complex and appear to result from a combination of igneous and retrograde metamorphic reactions.

Geochemical data for samples from this roadcut are typical of the LMG (Table 2). This unit is characterized by moderate SiO_2 content (69–73 wt. %), CaO contents of approximately 1 weight percent, $Na_2O > K_2O$, and agpaitic ratios indicative of metaluminous compositions. Concentrations of the trace elements Nb, Zr, Zn, and Ce are typically moderate, relative to the range exhibited by the other RRS units. Analyses of multiple samples indicate that the composition of the LMG is remarkably consistent (compare, for example, analyses 3 and 6a, Table 2) throughout the area of exposure.

New U/Pb isotopic data for zircons collected from this roadcut define a nearly concordant age of 729 Ma (Tollo et al., in press). This age indicates the LMG to be the oldest lithologic unit present within the northern segment of the RRS. This age is consistent with the observed field relationships.

Continue south on Route 618.

57.8	(0.3)	Saprolite on left exposes weathered Laurel Mills Granite and a greenstone dike characterized by prominent cleavage.
58.0	(0.2)	Intersection of Routes 649 and 618. Reverse direction and proceed north on Route 618 toward Laurel Mills village.
60.3	(2.3)	Intersection of Routes 618 and 729. Continue straight (east) on Route 729.
61.2	(0.9)	Intersection of Routes 729 and 640. Bear right to continue south on Route 729/640.
61.3	(0.1)	Intersection of Routes 729 and 640. Turn left onto Route 640 and immediately turn left into lot of old school house. Park and assemble for hike to summit of Battle Mountain (Stop 4).

STOP 4: Battle Mountain Complex

The area between Little Battle and Battle Mountains is underlain by the main body of the Battle Mountain Complex (Wallace and Tollo, 1986), which is comprised of closely associated hypersolvus alkali feldspar granites (Figure 3d), subvolcanic felsites, and rhyolite. These rocks represent the highest level of RRS magmatism and preserve the only remaining evidence of probable volcanic products. This stop consists of a traverse to the summit area of Battle Mountain and involves detailed examination of four outcrops exhibiting the lithologic diversity characteristic of this complex.

The distinct nature of the rocks in the Battle Mountain area was first recognized by Lukert and Halliday (1980), who mapped the area as a separate part of the Robertson River Formation. Subsequent detailed field mapping indicates that rocks of the Battle Mountain Complex (BMC) intrude the Laurel Mills Granite (LMG) and that the complex truncates the contact between the LMG and the Amissville Alkali Feldspar Granite (AAG), suggesting that the BMC represents the youngest phase of magmatism within the northern segment of the RRS (Wallace and Tollo, 1986; Arav, 1989; Tollo and Arav, in press; Hawkins, unpublished data). In addition, numerous fine-grained dikes with petrochemical characteristics similar to the felsites of the BMC cut many of the RRS lithologic units throughout both segments of the outcrop belt (Tollo et al., in preparation). The BMC and AAG together comprise the only major rock units characterized by porphyritic textures and peralkaline composition within the RRS.

Stop 4a:

This large exposure is composed of medium-grained, inequigranular alkali feldspar granite, one of the more widespread lithologic units in this portion of the BMC. The granite is closely associated with abundant felsite in this area, and detailed field mapping demonstrates that the felsites are younger than the granites throughout the complex (Hawkins, unpublished data). Petrographic examination indicates that the alkali granite at this stop is composed of: 1) abundant, eu- to subhedral mesoperthite; 2) anhedral quartz; 3) relict aegirine (locally replaced by intergrowths of quartz + magnetite + zircon); 4) fine-grained, green biotite (secondary); 5) zircon; and 6) fluorite. Albite is present throughout the rock as overgrowths on primary mesoperthite.

Stop 4b:

This exposure is composed of light- and dark-colored, sparsely porphyritic felsite. Medium-grained granite occurs toward both the east and west and is closely intermixed with the felsite throughout the area. The felsite is a fine-grained, allotriomorphic intergrowth of alkali feldspar, quartz, rare plagioclase, green biotite, acicular aegirine, magnetite, and zircon. Rare, medium-grained microcline-mesoperthite is the only phenocrystic phase. Sparse xenoliths of medium-grained, equigranular granite and small inclusions characterized by a distinctive trachytic texture and composed solely of albite occur within the felsite. The light- and dark-colored patches that occur locally throughout the rock coincide with areas of distinct petrographic characteristics. The light-colored areas are dominated by feldspar and are characterized by irregularly shaped areas of felty texture. Fluorite is locally more abundant, and sparse Fe-Ti oxides are very fine-grained in these areas. The dark-colored patches are characterized by a smaller percentage of modal feldspar, sharp grain boundaries, and relatively coarser-grained Fe-Ti oxides. The mineralogical differences suggest that these patches result from the affects of fluids active under subsolidus conditions.

Stop 4c:

This outcrop is composed of medium-grained, inequigranular alkali feldspar granite similar to that exposed at Stop 4a. This alkali granite differs from that at Stop 4 solely in preserving more of the primary pyroxene. Geochemical data for a sample collected from this outcrop (analysis 4a, Table 2) indicate that this alkali granite has a relatively low silica content and a borderline metaluminous/peralkaline composition. The trace element concentrations, indicating enrichment in Zr, Nb, and Ce relative to the other units of the RRS, are typical of the granitoid portions of the BMC.

Stop 4d:

The large boulder and underlying bedrock in this area are composed of dark gray, fine-grained to sparsely porphyritic felsite/rhyolite. The rock is composed of an allotriomorphic intergrowth of alkali feldspar, quartz, very fine-grained allanite, zircon, green biotite (secondary), and magnetite (secondary). The pervasive light- and dark-colored banding is defined by differences in petrographic characteristics. The dark-colored bands are richer in feldspar, which typically exhibits brown staining, whereas the light-colored bands are richer in quartz and contain mostly unstained feldspar. The modal differences are interpreted to reflect igneous flow segregation occurring in the near surface or surface environment. The staining is probably post-crystallization in origin, possibly enhancing slight differences in composition between the two populations of feldspar. Numerous, very fine-grained accessory phases (including zircon) occur in bands that crosscut the large-scale flow banding.

Geochemical data for a sample collected from this area (analysis 4b, Table 2) demonstrate the high SiO_2 content and marked enrichment in the trace elements Zr and Nb that are typical of most of the BMC felsite. The average concentration of Zr in the BMC is more than twice the concentration in most other units of the RRS. Dike rocks believed to be associated with the complex locally exhibit further enrichments of nearly an order of magnitude and probably represent the product of a variety of processes including volatile complexing coupled with extreme fractionation.

The fine-grain size and flow banding in the felsite of this area suggest that these rocks were part of a rhyolitic volcanic complex. The close spatial association and similarity in peralkaline composition of the AAG and alkali granitoids of the BMC further indicate that together these units may comprise the hypabyssal portions of a siliceous volcanic center that can presently be traced over an area of at least 20 square miles (51 km^2).

Turn right onto Route 640 and proceed to intersection. Turn left onto Route 729 and proceed south.

62.7	(1.4)	Bridge over Thornton River. Continue south on Route 729.
63.9	(1.2)	Intersection of Routes 729 and 615. Bear right onto Route 615 and proceed southwest.
64.4	(0.5)	The peak located to the right is Castleton Mountain, which is underlain mostly by alkali feldspar granite of the Battle Mountain Complex.
66.1	(2.2)	Intersection of Routes 615 and 617. Turn left to continue on Route 615 and proceed south.
67.2	(1.1)	Outcrops on both sides of road expose phyllites and arkosic conglomerates of the Mechum River Formation.
68.7	(1.5)	Turn left onto bridge across Blackwater Creek to continue on Route 615. Fine-grained, volcanogenic rocks within the Mechum River Formation are exposed in the creek. Geochemical data indicate a close compositional correspondence between these rocks and felsites of the Battle Mountain Complex (Hutson, 1990).
70.5	(1.8)	Intersection of Routes 615 and 650. Bear left onto Route 650 and proceed south.
71.3	(0.8)	View toward west is of the western limb of Blue Ridge anticlinorium. The high, jagged peak toward the southwest is Old Rag Mountain, which is underlain by the Old Rag Granite, a coarse-grained granitoid with an age of 1115 Ma (Lukert, 1982).
71.6	(0.3)	Intersection of Routes 650 and 522. Continue straight and proceed south on 650.
72.3	(0.7)	Intersection of Routes 650 and 707. Turn right onto Route 707; proceed southwest.
74.1	(1.8)	Road is located close to geologic contact. Outcrops on right expose biotite phyllites of the Mechum River Formation; outcrops on left expose granitoids of the Robertson River Suite.

74.5   (0.4)   Intersection of Routes 707 and 644/607. Turn left onto Route 644/607 and proceed south.

76.2   (1.7)   Intersection of Routes 644 and 607. Bear right to remain on Route 607 and proceed south.

77.3   (1.1)   Stop 5: Turn right into private drive. Proceed to house for hike to outcrops.

STOP 5: Hitt Mountain Alkali Feldspar Syenite

The numerous exposures on the northwestern slope of Hitt Mountain, located less than 0.25 miles (0.4 km) from the Laurel Mills Granite (LMG)–Hitt Mountain Alkali Feldspar Syenite (HMAS) contact (Figure 2), provide an opportunity to examine an area of abundant autoliths present along the margin of the HMAS unit. The outcrops in this area contain numerous oval-shaped, fine-grained, dark gray autoliths (Figure 7) composed of alkali feldspar + amphibole and ranging in length from four to fourteen inches (10–25 cm). These autoliths, which are surrounded by coarse-grained, light-gray, alkali feldspar syenite containing perthitic feldspar, minor quartz, and amphibole, appear to be limited to the vicinity of Hitt Mountain. The similarity in mineralogy between the fine-grained autoliths and the surrounding coarse-grained matrix, the lack of a chilled margin in the coarse-grained HMAS adjacent to the autoliths, and the rounded edges of these enclaves are characteristics suggesting that this area is part of the chilled margin of a magma chamber in which the (completely crystallized, but still warm) border zone was intruded by a petrochemically related melt.

Petrographic and geochemical analyses of representative samples of the autoliths and coarse-grained matrix indicate the consanguineous nature of the autoliths and surrounding syenitoid and support their grouping within the HMAS unit. The coarse-grained alkali feldspar syenite from this exposure contains: 1) rare, medium- to coarse-grained, anhedral quartz displaying recrystallization textures; 2) rare, fine-grained, subhedral plagioclase; 3) medium- to coarse-grained, subhedral microcline-microperthite; and 4) medium- to coarse-grained, anhedral amphibole (X=pale yellow-green; Y=dark-green; Z=green-blue) locally replaced by biotite ± stilpnomelane. Accessory phases include zoned allanite, stilpnomelane, and zircon with minor inclusions. The alkali feldspar:plagioclase modal ratio exceeds 100:1 for the coarse-grained matrix (Table 1; Figure 3e). The autoliths are characterized by 1) rare, fine-grained, anhedral quartz; 2) fine-grained, subhedral plagioclase; 3) fine-grained, subhedral microcline-microperthite containing inclusions of amphibole ± stilpnomelane; and 4) fine-grained, subhedral amphibole (X=pale yellow-green; Y=dark-green; Z=green-blue). Accessory phases include fine-grained, subhedral allanite rimmed by epidote, rare stilpnomelane, and rare zircon. Coarser-grained perthitic feldspars occur locally as phenocrysts.

Geochemical analyses of the autoliths and the associated coarse-grained matrix confirm their correlation with the HMAS and provide further evidence for the chilled margin model. It should be noted that, while the analysis for the autoliths was obtained from the exposures on the northwestern slope of Hitt Mountain, the analysis for the coarse-grained alkali feldspar syenite of this area could only be obtained from the southeastern slope. However, the southeastern exposures are petrographically identical to those from the northwest except that stilpnomelane and biotite are not present in the southeast. The analyses from this area (analyses 5a and 5b, Table 2) demonstrate the geochemical characteristics indicative of the HMAS unit. Relative to the other RRS lithologic types, both the autoliths and the matrix syenitoid have: 1) very low silica contents; 2) elevated Fe_2O_3* (total iron expressed as Fe_2O_3), Al_2O_3, and CaO contents; and 3) relatively high Zr concentrations. Relative to the enclosing syenitoid matrix, the autoliths are lower in SiO_2, total alkalis, and the trace elements Zr and Ce, indicating that these fine-grained rocks are less geochemically evolved than the associated coarser-grained variety. These data, and the lower concentrations of total Fe (Fe$_2$O$_3$*, Table 2) and CaO in the coarse-grained matrix, further suggest that fractionation of calcic amphibole was an important factor in developing the matrix composition from an initial magma similar in composition to the autoliths.

Retrace route along driveway to road. Turn right onto Route 607 and proceed southwest.

78.2	(0.9)	Intersection of Routes 607 and 606. Turn right onto Route 606/607 and proceed west.
78.5	(0.3)	Intersection of Routes 607 and 606. Continue west on Route 606.
78.8	(0.3)	Intersection of Routes 606 and 605 at Novum. Continue straight on Route 606.
79.4	(0.6)	Road passes through gneissic inlier within the Arrington Mountain Alkali Feldspar Granite. Shatter zone comprised of granite intruding gneiss is exposed in field on right.
79.7	(0.3)	Stop 6: Turn left into private drive and proceed to house for walk to outcrop.

STOP 6: Laurel Mills Granite Intruded by Hitt Mountain Alkali Feldspar Syenite Dike

The large pavement outcrop near the pond, located within 0.25 miles (0.4 km) of the Laurel Mills Granite (LMG)–Arrington Mountain Alkali Feldspar Granite (AMAG) contact (Figure 2), provides a critical age relationship for two of the RRS lithologic units. The dominant lithology of the outcrop is a gray, coarse-grained, inequigranular granite composed of blue-gray quartz, pale green saussuritized plagioclase, perthitic alkali feldspar, and amphibole typically replaced by intergrowths of biotite ± stilpnomelane ± magnetite. The petrographic and geochemical characteristics of this rock are nearly identical to those observed at the LMG type locality (Stop 3), although at this location the granite contains numerous quartz- and feldspar-filled vugs suggesting a shallow level of emplacement. Near the southern edge of the outcrop the LMG is cut by two dikes. The older dike strikes approximate N-S and is composed of medium-grained, white to light gray, amphibole-bearing alkali feldspar syenite of probable Hitt Mountain Alkali Feldspar Syenite (HMAS) affinity. This dike is, in turn, cut by a second, younger dike that is approximately 3 feet (1 m) in width and strikes N72W. This second dike, a finer-grained variety of the first, is also correlated with the HMAS unit and, at the western edge of the outcrop, is offset and dismembered by a small fault/ductile deformation zone. Numerous xenoliths of the surrounding LMG are included within the second dike (Figure 8). These fragments, which are concentrated along the inner edges of the dike, have rounded edges, suggesting that the liquidus temperatures for the two rock types may have been similar and that assimilative reactions involving the xenoliths and surrounding melt were arrested by quenching. Mafic phases, dominantly biotite, concentrated along the perimeter of this second dike are possibly the product of cooling or, more likely, are a result of ductile deformation along the dike-country rock contact.

The granite of this outcrop is characterized in thin section by: 1) coarse-grained, anhedral quartz showing undulatory extinction and recrystallization textures; 2) coarse-grained, sub- to euhedral plagioclase locally replaced by discrete crystals of epidote and zoisite; 3) coarse-grained, sub- to euhedral microcline-microperthite; 4) medium-grained, anhedral amphibole (X=pale yellow-green; Y=dark-green; Z=green-blue) locally replaced by biotite ± stilpnomelane ± magnetite ± sphene; and 5) fine-grained, subhedral biotite (X=pale yellow-brown; Y=Z=brown-green) probably formed after amphibole. Accessory phases include zoned or metamict allanite, stilpnomelane, and zircon. The alkali feldspar:plagioclase ratio for this exposure is 3:1, a value which is typical of this unit (Table 1; Figure 3c).

Petrographic correlation of the second syenitic dike with other samples from the HMAS is

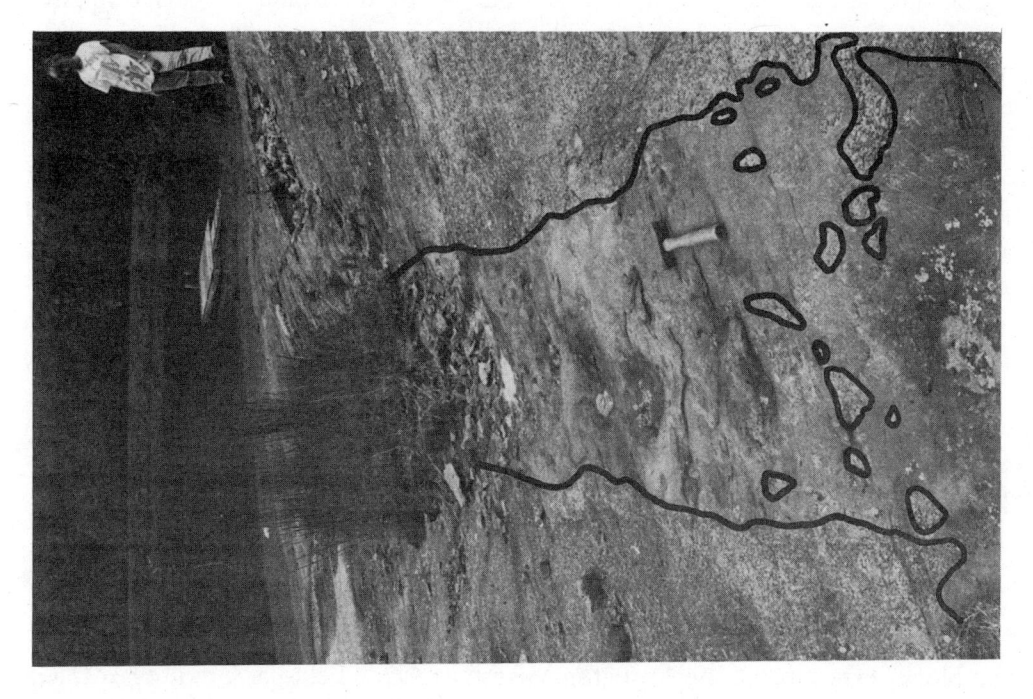

Figure 8. Photograph of dike of fine-grained Hitt Mountain Alkali Feldspar Syenite intruding coarse-grained Laurel Mills Granite at Stop 6. Numerous, irregularly-shaped, coarsegrained xenoliths occur within and at the margins of the dike in the foreground. Hammer is 14 inches (36 cm) in length.

Figure 7. Photograph of fine-grained syenitoid autoliths within coarse-grained alkali feldspar syenite exposed on northwestern slope of Hitt Mountain at Stop 5. Bar scale at bottom denotes 4 inches (10 cm).

made difficult by the fine grain size and by the presence of untwinned feldspar. However, this rock is mineralogically similar to others from the HMAS unit as it is characterized by: 1) rare, anhedral quartz; 2) subhedral plagioclase containing inclusions of epidote and biotite; 3) subhedral microcline; and 4) eu- to subhedral amphibole locally replaced by biotite ± stilpnomelane. Accessory phases include fine, euhedral allanite rimmed by epidote and subhedral zircon. Coarser-grained microperthite occurs locally as phenocrysts.

The granite of this exposure is geochemically indistinguishable from other samples of the LMG unit (compare analyses 3 and 6a, Table 2). Geochemical analysis of the second dike provides an additional basis for correlating this rock with the HMAS unit. In accord with other analyses of this unit (compare analyses 6b and 5b, Table 2), this dike is characterized by an extremely low silica content, elevated Al_2O_3, Fe_2O_3, and CaO values (manifested by the abundance of calcic amphibole), and pronounced enrichment in Nb and Zr relative to most of the suite.

Retrace route along driveway to road. Turn left onto Route 606.

80.3	(0.6)	Intersection of Routes 606 and 604. Bear left on Route 606/604.
81.0	(0.7)	Intersection of Routes 606 and 604. Turn right onto Route 606 and proceed west.
83.2	(2.2)	Stop 7: Park on right side of road. Assemble for hike to outcrop in woods.

STOP 7: Arrington Mountain Alkali Feldspar Granite

The numerous blocky exposures of the Arrington Mountain Alkali Feldspar Granite (AMAG) in the woods along the eastern slope of Arrington Mountain are located less than 1 mile (0.6 km) from the western contact of the RRS (Figure 2). This unit has been traced discontinuously for 11 miles (18 km) toward the north and is the petrologic link between the northern and southern segments of the RRS outcrop belt. The AMAG is a light gray, medium-grained, equigranular alkali feldspar granite (Figure 3f) composed of quartz, rare, saussuritized plagioclase, perthitic alkali feldspar, and amphibole that is usually completely replaced by biotite. The AMAG and the closely associated Laurel Mills Granite (LMG) locally appear quite similar in the field, and care must be taken to distinguish the two units. In practice, the finer grain size, equigranular texture, scarcity of plagioclase, and local occurrence of fluorite are characteristics of the AMAG that serve to distinguish this unit from the LMG. As is typical throughout the AMAG, the exposures in this area are cut by numerous fine-grained granitic dikes containing euhedral amphibole and fluorite. The majority of these dikes are presumed to be of AMAG affinity; however, geochemical data indicate that some are similar in composition to felsites of the Battle Mountain Complex (Lowe, 1990).

Petrographically, the samples from these exposures are similar to others of the AMAG. In thin section, this unit is characterized by: 1) medium-grained, anhedral quartz exhibiting recrystallization textures; 2) rare, fine-grained, subhedral plagioclase locally replaced by epidote and zoisite; 3) medium-grained, subhedral microcline-microperthite; 4) fine-grained, anhedral amphibole (X=pale yellow-green, Y=dark green, Z=green-blue) occurring in clusters with stilpnomelane; and 5) fine-grained, anhedral, brown biotite. Accessory phases include stilpnomelane intergrown with biotite, sub- to euhedral zoned allanite, discrete crystals of epidote clustered with ferromagnesian phases, and fine-grained, subhedral zircon with rare prismatic inclusions. Fluorite, which is present in 50 percent of the AMAG samples, is not present in samples from this particular outcrop. The alkali feldspar:plagioclase ratio for this exposure is 38:1, which is the maximum value for this unit (Table 1; Figure 3f) (Lowe, 1990).

Geochemical data for a sample obtained from this exposure are similar to analyses of other AMAG samples (Table 2). The AMAG unit is characterized by a high silica content, relatively low Fe_2O_3 and CaO contents (the latter of which is demonstrated by the near absence of plagioclase), an agpaitic ratio indicative of a borderline metaluminous/peralkaline composition, and concentrations of Nb, Zr, Zn, and Ce that are comparable to the rest of the suite.

Continue south on Route 606.

83.6	(0.4)	Intersection of Routes 606 and 609. Turn right onto Route 609 and proceed west.
84.3	(0.7)	Intersection of Routes 609 and 603 at Haywood. Turn left onto Route 603 and proceed southeast.
85.6	(1.3)	Intersection of Routes 603 and 604 at Decapolis. Bear right to remain on Route 603.
87.2	(1.6)	Intersection of Routes 603 and 638. Turn right onto Route 638 and proceed southwest.
87.9	(0.7)	Stop 8: Park on right shoulder of road near bridge.

STOP 8: White Oak Alkali Feldspar Granite

The large roadcut on the eastern bank of the Robinson River, located approximately 1.25 miles (2 km) from the eastern border of the RRS (Figure 2), is the petrochemical type locality of the White Oak Alkali Feldspar Granite (WOAG), a unit that has been traced discontinuously for 13 miles (21 km) to the south. As is typical of the WOAG, both coarse- and fine-grained, mineralogically identical varieties are closely intermixed within this outcrop. The coarser variety, which is the dominant lithology in this and most WOAG exposures, is a gray, coarse-grained, inequigranular alkali feldspar granite (Figure 3g) composed of quartz, alkali feldspar, and fine-grained, euhedral amphibole. Also typical of other WOAG exposures, this outcrop shows abundant evidence of deformation in the form of localized foliation and at least two sets of ductile deformation zones. Additionally, at the southern end of the roadcut, a pegmatite dike approximately 4 inches (10 cm) in width and composed of quartz + alkali feldspar can be observed cutting the coarse-grained WOAG.

In thin section, this unit contains: 1) medium- to coarse-grained, anhedral quartz; 2) fine-grained, an- to subhedral plagioclase; 3) medium-grained, anhedral microcline-microperthite; and 4) medium-grained, an- to subhedral amphibole (X=pale yellow green, Y= dark green, Z=green-blue). Accessory minerals include metamict allanite with epidote rims, fluorite, calcite, stilpnomelane (occurring along fractures in the alkali feldspar and in clusters within other ferromagnesian silicates), and sub- to euhedral zircon. The alkali feldspar:plagioclase ratio for the coarse-grained variety is in excess of 15:1 (Table 1; Figure 3g).

Geochemical data provide further support for the field- and petrographic-based distinctions. Although extreme intermixing of the fine-grained variety with the coarse has precluded sampling of the former for geochemical analysis, data were obtained for the coarse-grained variety. The WOAG of this exposure is compositionally similar to others of the unit in that it is characterized by a relatively low silica content, elevated CaO concentration (indicated by the abundance of calcic amphibole), an agpaitic ratio indicative of borderline metaluminous/peralkaline composition, and average to high concentrations of Nb, Zr, Zn, and Ce relative to the rest of the suite. It may be noted that the WOAG and LMG have similar major element contents, such as SiO_2, Al_2O_3, Fe_2O_3*, and total alkalis (compare analyses 3, 6a, and 8, Table 2). However, in terms of trace element concentrations, the WOAG is quite distinct from the LMG, especially in terms of Nb, Zr, and Ce.

Continue west on Route 638.

88.6	(0.7)	Intersection of Routes 638 and 635. Bear left to remain on Route 638. Continue south.
89.4	(0.8)	Intersection of Routes 638 and 231. Turn left on Route 231 and proceed south.
90.4	(1.0)	Intersection of Routes 231 and Alternate 29 in Madison. Turn right on Route 29/231 and proceed south.
91.7	(1.3)	Intersection of Routes Alternate 29 and 29. Turn right onto Route 29 and continue southwest.
100.2	(8.5)	Intersection of Routes 29 and 609. Turn right onto Route 609 and proceed west.
100.9	(0.7)	Stop 9: Turn right into private drive and continue to assembly point for hike to outcrop.

STOP 9: Hitt Mountain Alkali Feldspar Syenite

Located approximately 0.25 miles (0.4 km) from the western border of the RRS (Figure 2), this large outcrop is an excellent exposure of the Hitt Mountain Alkali Feldspar Syenite (HMAS), with the considerable variability in grain size and degree of deformation that is typical of this unit. The HMAS, which has been traced 9 miles (14 km) north and 12 miles (19 km) south from this area, is a light gray, fine- to coarse-grained, inequigranular alkali feldspar syenite (Figure 3e) characterized by rare interstitial quartz, perthitic alkali feldspar, and sub-to euhedral amphibole. Grain size ranges from fine to ultra-coarse to locally pegmatitic. Similarly, the degree of deformation also varies considerably within the exposure, ranging from localized, amphibole-defined lineation to protomylonite developed along ductile deformation zones (see, for example, the southwestern end of the exposure).

In thin section, the syenitoid is characterized by: 1) rare, fine-grained, anhedral quartz; 2) rare, fine-grained, an- to subhedral plagioclase; 3) fine- to coarse-grained, anhedral microcline-microperthite containing inclusions of epidote, allanite, and amphibole; and 4) coarse-grained, subhedral amphibole (X=pale yellow-brown Y=dark green, Z=green-blue) closely associated with stilpnomelane and Fe-Ti oxides. The alkali feldspar:plagioclase ratio is 9:1 for undeformed rock from this locality (Table 1; Figure 3e). Accessory phases include metamict allanite rimmed by epidote, stilpnomelane (occurring intergrown with amphibole and as a fracture filling), and zircon.

The syenitoid from this exposure is geochemically similar to other samples analyzed from the HMAS unit (see analyses 5b and 9, Table 2) in that it is low in silica, exhibits relatively high concentrations of Al_2O_3, Na_2O, K_2O, and CaO, and shows relatively low Nb content. The Zr content of the sample from this exposure is unusually low compared to other samples of the HMAS (compare analyses 5a, 5b, and 6b, Table 2).

Retrace route to road and turn left onto Route 609.

101.6	(0.7)	Intersection of Routes 609 and 29. Turn right onto Route 29 and proceed south.
104.0	(2.4)	Luck Stone Quarry on right exposes locally deformed Hitt Mountain Alkali Feldspar Syenite.
109.5	(5.5)	The prominent hill located to the right is Piney Mountain, which includes abundant exposures of Hitt Mountain Alkali Feldspar Syenite.
115.5	(6.0)	Outcrop in median of roadway and roadcut adjacent to northbound lanes of Route 29 expose finely laminated metamorphosed siltstone and shale of the Lynchburg

		Formation.
117.2	(1.7)	Intersection of Routes 29 and 631. Turn right onto Route 631 and proceed northwest.
117.7	(0.5)	Intersection of Routes 631 and 659. Turn right onto Route 659 and proceed north.
119.1	(1.4)	Stop 10: Entrance to South Rivanna Water Treatment Plant. Turn left and proceed to assembly point for drive to outcrop.

STOP 10: Rivanna Alkali Feldspar Granite

Located approximately 1.5 miles (2.5 km) from the southern termination of the RRS (Figure 2), the large roadcut along the upper level of the plant service road is a relatively fresh and undeformed exposure of the Rivanna Alkali Feldspar Granite (RAG) and is designated as the petrochemical type locality. The RAG is typically a white to light gray, medium-grained, equigranular, biotite- and fluorite-bearing alkali feldspar granite (ranging to quartz-rich granite, Figure 3h) that can be traced southward from this locality to the southern end of the RRS outcrop belt. The rock comprising this roadcut is typical of the RAG unit except for a greater than normal abundance of fluorite.

In thin section, this unit is characterized by: 1) fine- to medium-grained, anhedral quartz exhibiting recrystallization textures; 2) fine- to medium-grained, relatively rare plagioclase feldspar; 3) medium- to coarse-grained, an- to subhedral microcline-microperthite; 4) fine-grained, subhedral biotite (X=pale yellow-brown, Y=Z=light green-brown) occurring in clusters with muscovite ± stilpnomelane and containing inclusions of zircon; and 5) fine-grained, subhedral muscovite in clusters with biotite. Accessory phases include abundant fine- to medium-grained, anhedral fluorite, fibrous, orange-brown stilpnomelane, and rare, metamict allanite. The alkali feldspar:plagioclase ratio for samples from this exposure is 4:1, which is the lowest value observed in this unit (Table1; Figure 3h).

The high silica content, low Fe_2O_3* and CaO (manifested in the apparent absence of both plagioclase and calcic amphibole), and low Zr and Ce concentrations are all characteristic of this unit. In terms of A-type granite characteristics, the RAG exhibits the typically low CaO and MgO contents and elevated trace element concentrations relative to average I-, S-, and M-type compositions (Whalen et al., 1987). Although the Zr value of the RAG from this outcrop is significantly lower than those from other units within the suite and is, in fact, similar to the average Zr concentrations of I-, S-, and M-type granitic rocks, the elevated Nb, Zn, and Ce values for this unit all support its A-type classification as well as its inclusion in the RRS.

Retrace route to road and follow Routes 659 and 631 to Route 29. Route 29 provides convenient access to points north and south. End of field trip.

ACKNOWLEDGMENTS

This research was supported by National Science Foundation grant EAR89-16048, a contract from the Department of Mines, Minerals, and Energy of the Commonwealth of Virginia, and grants from the University Committee on Research of the George Washington University to RPT. The research of TKL, SA, and KJG was partially supported by awards from the Sigma Xi Society.

REFERENCES CITED

Allen, R. M., Jr., 1963, Geology and Mineral Resources of Greene and Madison Counties: Virginia Division of Mineral Resources Bulletin No. 78, 98 p.

Arav, S., 1989, Geology, geochemistry, and relative age of two plutons from the Robertson River Formation in northern Virginia: unpublished M.S. thesis, George Washington University, Washington, D.C., 125 p., plates.

Badger, R. L., and Sinha, A. K., 1988, Age and Sr isotopic signature of the Catoctin volcanic province: Implications for subcrustal mantle evolution: Geology, v. 16, p. 692–695.

Bailey, C. M., 1990, Proterozoic metamorphism in the Blue Ridge anticlinorium in central and northern Virginia: Geological Society of America Abstracts with Programs, v. 19, no. 4, p. 2.

Barker, F., Wones, D. R., Sharp, W. N., and Desborough, G. A., 1975, The Pikes Peak Batholith, Colorado Front Range, and a model for the origin of the gabbro-anorthosite-syenite-potassic granite suite: Precambrian Research, v. 2, p. 97–160.

Bartholomew, M. J., in press, Structural characteristics of the late Proterozoic (post-Grenville) continental margin of the ancestral North American craton, in Bartholomew, M. J., Hyndman, D. W., Mogk, D. W., and Mason, R., eds., Characterization and Comparison of Ancient (Precambrian-Mesozoic) Continental Margins—Proceedings of the 8th International Conference on Basement Tectonics, Butte, Montana: Kluwer Academic Publishers, Dordrecht, Holland.

——, Gathright, T. M., II, and Henika, W. S., 1981, A tectonic model for the Blue Ridge in central Virginia: American Journal of Science, v. 281, p. 1164–1183.

——, and Lewis, S. E., 1984, Evolution of Grenville massifs in the Blue Ridge geologic province, southern and central Appalachians, in Bartholomew, M. J., Force, E. R., Sinha, A. K., and Herz, N., eds., The Grenville Event in the Appalachians and Related Topics: Geological Society of America Special Paper 194, p. 229–254.

Bloomer, R. O., and Werner, H. J., 1955, Geology of the Blue Ridge region in central Virginia: Geological Society of America Bulletin, v. 66, p. 579–606.

Clarke, J. W., 1981, Billion-year-old rocks of the Blue Ridge anticlinorium of northern Virginia: A review: Virginia Journal of Science, v. 32, p. 127.

——, 1984, The core of the Blue Ridge anticlinorium in northern Virginia, in Bartholomew, M. J., Force, E. R., Sinha, A. K., and Herz, N., eds., The Grenville Event in the Appalachians and Related Topics: Geological Society of America Special Paper 194, p. 153–160.

Collins, W. J., Beams, S. D., White, A. J. R., and Chappell, B. W., 1982, Nature and origin of A-type granites with particular reference to southeastern Australia: Contributions to Mineralogy and Petrology, v. 80, p. 189–200.

Conley, J. F., 1989, Stratigraphy and structure across the Blue Ridge and Inner Piedmont in central Virginia: Field Trip Guidebook T207, International Geologic Conference, American Geophysical Union, Washington, D.C., 23 p.

Davis, G. L., Tilton, G. R., and Wetherill, G. W., 1962, Mineral ages from the Appalachian province in North Carolina and Tennessee: Journal of Geophysical Research, v. 67, p. 1987–1996.

Espenshade, G. H., 1970, Geology of the northern part of the Blue Ridge anticlinorium, in Fisher, G. W., Pettijohn, F. J., Reed, J. C., Jr., and Weaver, K. N., eds., Studies in Appalachian Geology—Central and Southern: Wiley Interscience, New York, p. 199–211.

Evans, N. H., 1987, Post-Grenville metamorphism in the central Virginia Blue Ridge: Origin of the contrasting Lovingston and Pedlar terranes: Geological Society of America Abstracts with Programs, v. 19, no. 2, p. 83.

Fauth, J. C., 1968, Geology of the Caledonia Park quadrangle area, South Mountain, Pennsylvania: Pennsylvania Department of Internal Affairs Atlas 129a, 133 p., map.

——, 1978, Geology and mineral resources of the Iron Springs area, Adams and Franklin Counties, Pennsylvania: Pennsylvania Department of Environmental Resources Atlas 129c, 72 p., map.

Foland, K. A., and Faul, H., 1977, Ages of the White Mountain intrusives—New Hampshire, Vermont, and Maine, USA: American Journal of Science, v. 277, p. 888–904.

Freedman, J., 1967, Geology of a portion of the Mount Holly Springs quadrangle, Adams and Cumberland Counties, Pennsylvania: Pennsylvania Department of Environmental Resources Progress Report 169, 66 p., map, plates.

Gathright, T. M., II, 1976, Geology of the Shenandoah National Park, Virginia: Virginia Division of Mineral Resources Bulletin 86, 53 p., maps.

Gooch, E. O., 1954, Infolded metasediments near the axial zone of the Catoctin Mountain–Blue Ridge anticlinorium: unpublished Ph.D. dissertation, University of North Carolina, Chapel Hill, 29 p.

——, 1958, Infolded metasedimentary rocks near the axial zone of the Catoctin Mountain–Blue Ridge anticlinorium in Virginia: Geological Society of America Bulletin, v. 69, p. 569–574.

Gray, K. J., 1990, Mineral chemistry of amphiboles from the Laurel Mills Granite of the Late Proterozoic Robertson River Formation, central Appalachians, northern Virginia: Geological Society of America Abstracts with Programs, v. 22, no. 71, p. A345.

Herz, N., and Force, E. R., 1984, Rock suites in Grenvillian terrane of the Roseland district, Virginia, in Bartholomew, M. J., Force, E. R., Sinha, A. K., and Herz, N., eds., The Grenville Event in the Appalachians and Related Topics: Geological Society of America Special Paper 194, p. 187–214.

Hutson, F. E., 1990, Evidence for volcanogenic sedimentation in an outlier of the Mechum River Formation in the axial zone of the Blue Ridge anticlinorium, central Virginia: Geological Society of America Abstracts with Programs, v. 22, no. 71, p. A173.

Johnson, S. S. and Gathright, T. M., III, 1978, Geophysical characteristics of the Blue Ridge anticlinorium in central and northern Virginia, in Contributions to Virginia Geology — III: Virginia Division of Mineral Resources Publication 7, p. 23–36.

Leake, B. E., 1978, Nomenclature of amphiboles: American Mineralogist, v. 63, p. 1023–1052.

Loiselle, M. C., and Wones, D. R., 1979, Characteristics and origin of anorogenic granites: Geological Society of America Abstracts with Programs, v. 11, p. 468.

Lowe, T. K., 1990, Petrology and geochemistry of the Robertson River Suite, Castleton, Woodville, Brightwood and Madison 7.5' quadrangles, Virginia: unpublished senior honors thesis, George Washington University, Washington, D.C., 38 p.

Lukert, M. T., 1982, Uranium-lead isotopic age of the Old Rag Granite, northern Virginia: American Journal of Science, v. 282, p. 391-398.

——, and Banks, P. O., 1984, Geology and age of the Robertson River pluton, in Bartholomew, M. J., Force, E. R., Sinha, A. K., and Herz, N., eds., The Grenville Event in the Appalachians and Related Topics: Geological Society of America Special Paper 194, p. 161-166.

——, and Halladay, C. R., 1980, Geology of the Massies Corner quadrangle, Virginia: Virginia Division of Mineral Resources Publication 17, text and 1:24,000 scale map.

——, and Nuckols, E. B., III, 1976, Geology of the Linden and Flint Hill quadrangles, Virginia: Virginia Division of Mineral Resources Report of Investigations no. 44, 83 p., plates.

Mitra, G., 1978, Ductile deformation zones and mylonites: The mechanical processes involved in the deformation of crystalline basement rocks: American Journal of Science, v. 278, p. 1057-1084.

——, 1979, Ductile deformation zones in Blue Ridge basement rocks and estimation of finite strains: Geological Society of America Bulletin, Part I, v. 90, p. 935-951.

——, and Elliott, D., 1980, Deformation of basement in the Blue Ridge and the development of the South Mountain cleavage, in Wones, D. R., ed., The Caledonides in the USA: Virginia Polytechnic Institute and State University Memoir no. 2, p. 307-311.

——, and Lukert, M. T., 1982, Geology of the Catoctin-Blue Ridge anticlinorium in northern Virginia, in Lyttle, P. T., ed., Central Appalachian Geology, NE-SE GSA 1982 Field Trip Guidebooks: American Geological Institute, Falls Church, Virginia, p. 83-108.

Mose, D. G., and Kline, S. W., 1986, Resetting of Rb-Sr isochrons by metamorphism and ductile deformation—Blue Ridge province, VA: Geological Society of America Abstracts with Programs, v. 18, p. 257.

——, and Nagle, S., 1984, Rb-Sr age for the Robertson River pluton in Virginia and its implication on the age of the Catoctin Formation, in Bartholomew, M. J., Force, E. R., Sinha, A. K., and Herz, N., eds., The Grenville Event in the Appalachians and Related Topics: Geological Society of America Special Paper 194, p. 167-173.

Odom, A. L., and Fullagar, P. D., 1984, Rb-Sr whole rock and inherited zircon ages of the plutonic suite of the Crossnore Complex, southern Appalachians, and their implications regarding the time of opening of the Iapetus Ocean, in Bartholomew, M. J., Force, E. R., Sinha, A. K., and Herz, N., eds., The Grenville Event in the Appalachians and Related Topics: Geological Society of America Special Paper 194, p. 255-261.

Pettingill, H. S., Sinha, A. K., and Tatsumoto, M., 1984, Age and origin of anorthosites, charnockites, and granulites in the Central Virginia Blue Ridge: Nd and Sr isotopic evidence: Contributions to Mineralogy and Petrology, v. 85, p. 279–291.

Pitcher, W. S., 1983, Granite type and tectonic environment, in Hsu, K., ed., Mountain Building Processes: Academic Press, London, p. 19–40.

Rankin, D. W., 1975, The continental margin of eastern North America in the southern Appalachians: The opening and closing of the proto-Atlantic Ocean: American Journal of Science, v. 275-A, p. 298–336.

——, 1976, Appalachian salients and recesses: Late Precambrian continental breakup and the opening of the Iapetus Ocean: Journal of Geophysical Research, v. 81, p. 5605–5619.

——, Drake, A. A., Jr., Glover, L., III, Goldsmith, R., Hall, L. M., Murray, D. P., Ratcliffe, N. M., Read, J. F., Secor, D. T., and Stanley, R. S., 1989, Pre-orogenic terranes, in Hatcher, R. D., Jr., Thomas. W. A., and Viele, G. W., eds., The Appalachian-Ouachita Orogen in the United States: Volume F-2, The Geology of North America: Geological Society of America, Boulder, Colorado, p. 7–100.

——, ——, and Ratcliffe, N. M., 1990, Geologic map of the U.S. Appalachians showing the Laurentian margin and the Taconic orogen: Plate 2, The Appalachian-Ouachita Orogen in the United States: Volume F-2, The Geology of North America, Geological Society of America, Boulder, Colorado.

——, Stern, T. W., McLelland, J., Zartman, R. E., and Odom, A. L., 1983, Correlation chart for Precambrian rocks of the eastern United States, in Harrison, J. E., and Peterman, Z. E., eds., Correlation of Precambrian rocks of the United States and Mexico: U.S. Geological Survey Professional Paper 1241, p. E1–E10.

——, ——, Reed, J. C., Jr., and Newell, M. F., 1969, Zircon ages of felsic volcanic rocks in the upper Precambrian of the Blue Ridge, Appalachian Mountains: Science, v. 166, p. 741–744.

——, and Tollo, R. P., 1987, Late Proterozoic anorogenic granitoids and Iapetan rifting of Laurentia as preserved in the Appalachian orogen: Geological Society of America Abstracts with Programs, v. 19, p. 813.

Reed, J. C., Jr., 1969, Ancient lavas in Shenandoah National Park near Luray, Virginia: U.S. Geological Survey Bulletin 1265, 43 p., map.

Rodgers, J., 1970, The Tectonics of the Appalachians: Wiley Interscience, New York, 271 p.

Schwab, F. L., 1974, Mechum River formation: Late Precambrian(?) alluvium in the Blue Ridge province of Virginia: Journal of Sedimentary Petrology, v. 44, p. 862–871.

Simpson, C., and Kalaghan, T., 1989, Late Precambrian crustal extension preserved in Fries fault zone mylonites, southern Appalachians: Geology, v. 17, p. 148–151.

Simpson, E. L., and Sundberg, F. A., 1987, Early Cambrian age for syn-rift deposits of the Chilhowee Group of southwestern Virginia: Geology, v. 15, p. 123–126.

Sinha, A. K., and Bartholomew, M. J., 1984, Evolution of the Grenville Terrane in the central Appalachians, in Bartholomew, M. J., Force, E. R., Sinha, A. K., and Herz, N., eds., The Grenville Event in the Appalachians and Related Topics: Geological Society of America Special Paper 194, p. 175–196.

Streckeisen, A. L., 1973, Plutonic rocks: Classification and nomenclature recommended by the IUGS Subcommission on the Systematics of Igneous Rocks: Geology, v. 18, no. 10, p. 26–30.

Tilton, G. R., Wetherill, G. W., Davis, G. C., and Bass, M. N., 1960, 1,000-million-year-old minerals from the eastern United States and Canada: Journal of Geophysical Research, v. 65, p. 4173–4179.

Tollo, R. P., 1986, Constituent granitoids of the Robertson River Formation in the northern Virginia Blue Ridge: Geological Society of America Abstracts with Programs, v. 18, p. 269.

——, Aleinikoff, J. N., and Gray, K. J., in press, New U/Pb zircon isotopic data from the Robertson River Igneous Suite: Implications for the duration of Late Proterozoic anorogenic magmatism: Geological Society of America Abstracts with Programs.

——, and Arav, S., 1987, Petrochemical comparison of two plutons from the Robertson River Formation in northern Virginia: Geological Society of America Abstracts with Programs, v. 19, p. 133.

——, and ——, in press, The Robertson River Igneous Suite: Late Proterozoic, anorogenic granitoids of unique petrochemical affinity, in Bartholomew, M. J., Hyndman, D. W., Mogk, D. W., Mason, R., eds., Characterization and Comparison of Ancient (Precambrian-Mesozoic) Continental Margins—Proceedings of the 8th International Conference on Basement Tectonics, Butte, Montana: Kluwer Academic Publishers, Dordrecht, Holland.

——, and Lowe, T. K., 1990, Compositional diversity in the Late Proterozoic Robertson River Suite, Blue Ridge province, Virginia: Geological Society of America Abstracts with Programs, v. 22, no. 7, p. A343.

——, ——, Arav, S., and Gray, K. J., in preparation, Lithologic and petrochemical map of the Robertson River Igneous Suite, northern Virginia: U.S. Geological Survey, scale 1:100,000.

Wallace, P. J., and Tollo, R. P., 1986, The Battle Mountain felsite: Evidence of an anorogenic (A-type) subvolcanic complex in the core of the Blue Ridge anticlinorium, northern Virginia: EOS, Transactions, American Geophysical Union (abstract), v. 67, p. 385.

Wehr, F. E., 1983, Geology of the Lynchburg Group in the Culpeper and Rockfish River areas, Virginia: Ph.D. dissertation, Virginia Polytechnic and State University, Blacksburg; University Microfilms International, Ann Arbor, Michigan, 245 p., plates.

——, and Glover, L., III, 1985, Stratigraphy and tectonics of the Virginia–North Carolina Blue Ridge: Evolution of a late Proterozoic–early Paleozoic hinge zone: Geological Society of America Bulletin, v. 96, p. 285–295.

Whalen, J. B., Currie, K. L., and Chappell, B. W., 1987, A-type granites: Geochemical characteristics, discrimination and petrogenesis: Contributions to Mineralogy and Petrology, v. 95, p. 407–419.

10
LATE PROTEROZOIC SEDIMENTATION AND TECTONICS IN NORTHERN VIRGINIA

Stephen W. Kline
George Mason University, Fairfax, VA 22030

Peter T. Lyttle
U.S. Geological Survey, Reston, VA 22092

J. Stephen Schindler
U.S. Geological Survey, Reston, VA 22092

INTRODUCTION

The Blue Ridge geologic province in northernmost Virginia comprises the area from Bull Run Mountain on the east to the Blue Ridge on the west. This field trip is confined to the eastern limb of the Blue Ridge anticlinorium and concentrates primarily on stratigraphic, sedimentologic, and structural problems relating to Late Proterozoic metasediments produced during rifting of the margin of ancestral North America. The provenance of these metasediments, as well as their contact relations with the underlying Middle Proterozoic granitic basement, is explored. We will visit a number of northwest-trending normal faults, which we interpret to be Late Proterozoic in age and related to rifting. We will also discuss the relative and absolute ages of important tectonic fabrics seen in outcrop and debate whether any of these structures are directly related to Late Proterozoic extensional tectonics. Although the information conveyed in this field guide is based mainly on recent mapping by the authors, our work benefitted greatly from the excellent earlier mapping by Gilbert Espenshade (1986) and Frederick Wehr (1983, 1985).

REGIONAL GEOLOGY

The Blue Ridge geologic province (Figure 1) in northern Virginia is traditionally defined as the rocks that constitute the Blue Ridge anticlinorium, including the plutonic complex in its core and the flanking low-grade metasediments and metavolcanic rocks. The basement consists mainly of Middle Proterozoic granitoids, presumably Grenville in age (Clarke, 1984), plus the Robertson River Igneous Suite, a series of somewhat alkaline plutonic rocks that intruded between 730 and 640 Ma ago (Rankin, 1975; Lukert and Halladay, 1980; Clarke, 1981; Tollo et al., 1991). The overlying metasedimentary and metavolcanic cover sequence is similar in gross aspect on either side of the anticlinorium. Arkosic metasediments lie nonconformably on the basement. On the east limb these metasediments are mainly the Fauquier Formation and equivalent rocks, while on the west limb a much thinner, discontinuous section consists of the Swift Run Formation (Gooch, 1958). Above the metasediments is the Catoctin Formation, which is principally metabasalt (greenstone and greenschist) with occasional interlayers of metasediments. On the western limb in northern Virginia and especially in Maryland, the Catoctin contains felsic metavolcanics.

Chilhowee Group clastic metasediments lie above the Catoctin Formation on both limbs of the anticlinorium. On the western limb, in spite of local omission by thrusting, a stratigraphic sequence can be traced from the earliest Cambrian Chilhowee Group into the unmetamorphosed

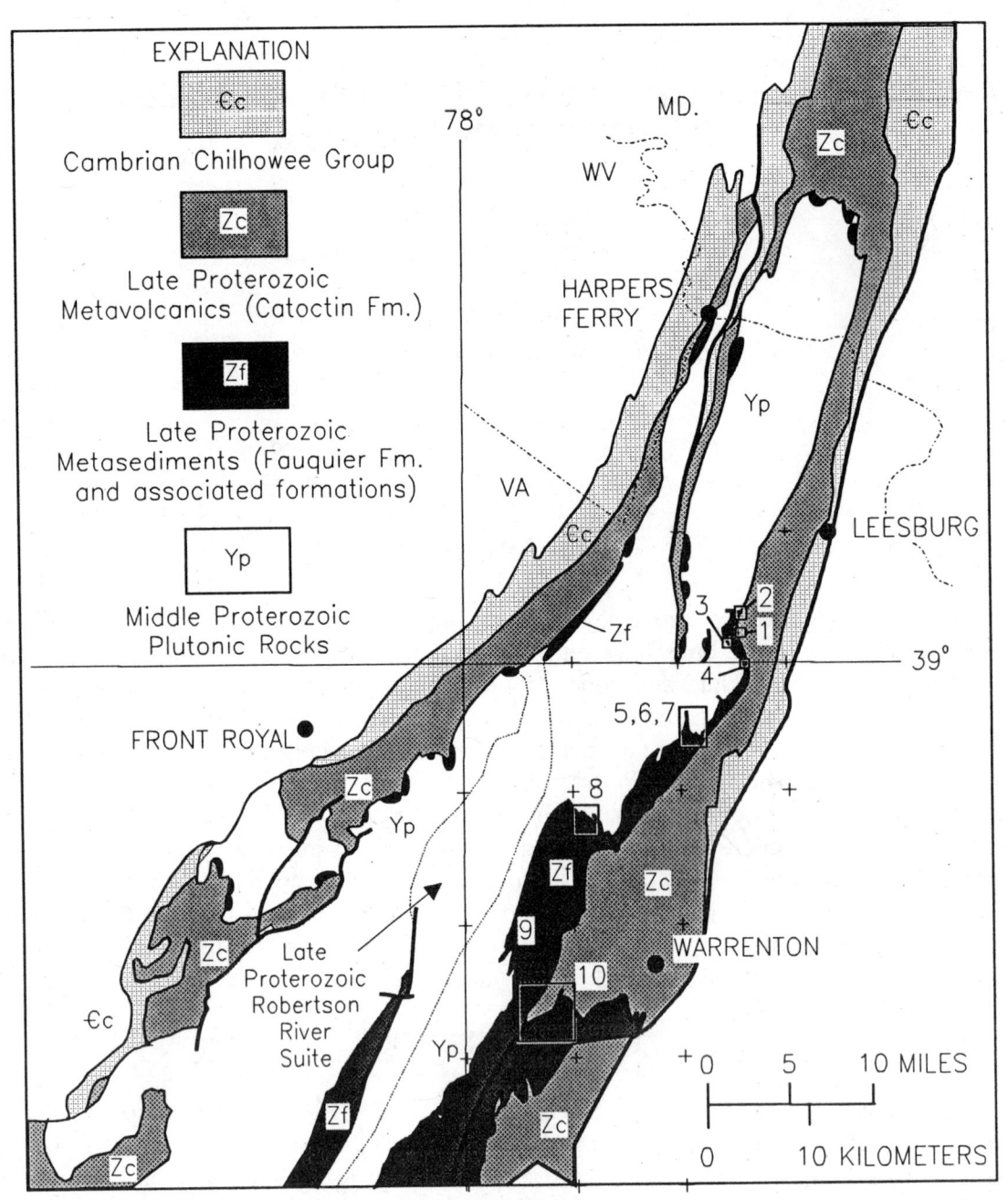

Figure 1. Simplified geologic map of Blue Ridge geologic province in northernmost Virginia. Modified from Espenshade (1986). Numbers 1–10 are the field trip stops.

A		B
		Charlottesville Formation
		——— thrust fault ———
		Ball Mountain Formation
		——— thrust fault ———
Fauquier Formation	metarhythmite and dolomitic units	Monumental Mills Formation
	meta-arkose and metasiltstone units	
	meta-arkose and metaconglomerate units	Bunker Hill Formation

Figure 2. Correlation chart of Late Proterozoic metasediments in the field trip area showing relations between units of A, Espenshade (1986) in the north and B, Wehr (1985) in the south.

Paleozoic continental shelf sedimentary rocks of the Valley and Ridge province (Simpson and Eriksson, 1989). On the east, however, the Chilhowee rocks are truncated by normal faults and sedimentary cover of the Mesozoic Culpeper basin, which obscures the relationship between the Blue Ridge province and the crystalline rocks of the Piedmont geologic province.

ROCKS OF THE FIELD TRIP

Basement Terrane

The plutonic rocks of the basement are overprinted by several tectonic fabrics to extremely varying degrees. However, rocks in thin sections commonly have relict plutonic fabrics. In outcrop, pegmatitic and aplitic dikes, dikelets, and irregular segregations are common. Veins and pods of bluish quartz also occur. The basement rocks in the vicinity of the first eight field trip stops are almost certainly correlative with the Middle Proterozoic Marshall Metagranite (Espenshade, 1986). The granite consists of a medium- to fine-grained phase and a less-abundant coarse-grained phase. Our observations (Kline et al., 1990b) support the conclusions of Espenshade (1986) that both the coarse and finer grained granites are from the same magma. The granites have a variable amount of mafic minerals (principally biotite), and some are very leucocratic.

The metagranites of the basement terrane are cut by numerous greenstone dikes, and in the field trip area most have a northeasterly trend. Many earlier workers have suspected these dikes were feeders to Catoctin volcanism (e.g., Reed and Morgan, 1971), but no one has ever directly traced a dike into a Catoctin flow. Geochemical data suggests that these dikes are co-magmatic with the Catoctin Formation (Kline and Gottfried, unpublished data).

Fauquier Formation

We use the name Fauquier Formation as defined by Espenshade and Clarke (1976) and Espenshade (1986). Parker (1968) referred to the same suite of rocks in the Lincoln quadrangle of Virginia as the Swift Run Formation. Wehr (1985), working to the south in the Culpeper, Virginia, area, mapped the Bunker Hill Formation and the Monumental Mills Formation and included them in his Lynchburg Group. His units correlate well with subunits of the Fauquier Formation as mapped by Espenshade (1986) in the Marshall quadrangle. Wehr's Bunker Hill Formation is equivalent to Espenshade's meta-arkose unit, and the Monumental Mills Formation is equivalent to Espenshade's meta-arkose and metasiltstone unit plus his metarhythmite and dolomitic units (Figure 2).

The Fauquier Formation rests nonconformably on Marshall Metagranite. The nonconformable relationship is clearly demonstrated by basal boulder and cobble conglomerates containing clasts of Marshall Metagranite and perhaps other nearby granites (see Stops 3 and 6). The basal conglomerate is not present everywhere.

The lower part of the Fauquier Formation in northern Virginia consists primarily of coarse, feldspathic metasandstone, with lesser amounts of finer metasandstone and phyllite, plus the boulder and cobble conglomerate described above. The coarse metasandstone commonly is conglomeratic with scattered granules and pebbles of quartz. Some of the most highly feldspathic metasandstone associated with the cobble conglomerate at the base has very indistinct bedding and is easily confused with the underlying Marshall Metagranite. Cross stratification is common, especially higher in the section (see Stop 7).

In the upper part of the Fauquier Formation, coarse meta-arkose becomes less abundant, and medium- to fine-grained metasandstone becomes the dominant rock type. The metasandstone is interlayered with even finer constituents (metasiltstone and metamudstone) which become increasingly abundant upwards. Cross bedding is not common in the upper part.

The uppermost Fauquier Formation consists almost entirely of very fine-grained clastic metasediment and metacarbonate. These rocks crop out poorly in most places and occupy a topographic low. In the Marshall quadrangle, Virginia, Espenshade (1986) called the uppermost unit of the Fauquier Formation metarhythmite. It consists of laminated to thinly bedded metasiltstone composed of alternating light and dark layers composed primarily of quartzose silt and sericite, respectively. In some of the better exposures, graded lamination and soft-sediment slump structures are well preserved (Stop 8). In the Middleburg quadrangle outcrops are extremely rare and the unit is detected in most places only by chips of phyllite in soil B-horizons. Where outcrops do occur, the phyllite is not as finely laminated as described by Espenshade (1986), but bedding can be discerned.

The carbonate at the top of the Fauquier Formation crops out in a number of places in the Marshall quadrangle and in the Lincoln quadrangle. It has been found in the Middleburg quadrangle only near the northern border in the deep valley of a tributary of Goose Creek. Espenshade (1986) mapped it as lenses only where it crops out, but considered that it is likely more continuous than shown on the map. Parker (1968) also mapped the carbonate as lenses. We consider the carbonates to be more extensive but poorly exposed.

Espenshade (1986) described the carbonate as being dolomitic. Parker (1968) called it limestone. Two of the field trip stops (Stops 1 and 4) demonstrate that in some places it is composed of calcite and in others of dolomite. The calcitic rocks at Stop 1 are structureless, but the dolomitic rocks at Stop 4, in places, appear to be oolitic, and in one place have sparry calcite "eyes" that may be pseudomorphs of gypsum crystals. Espenshade (1986) notes a locality where the dolomite has a texture that resembles an intraformational conglomerate.

The carbonate rocks are interlayered with the fine clastics in the upper part of the Fauquier Formation, and also with metabasalt near the base of the overlying Catoctin Formation (see Stops 1 and 4). Parker (1968) included some carbonate layers in the lower Catoctin Formation, but he considered their inclusion to be of a tectonic origin. Observations at Stop 1 do not support this. A possible genetic link between Catoctin volcanism and carbonate precipitation is discussed under Stop 1. The interlayering of sedimentary and volcanic components indicates a gradational change from sedimentation to volcanism and hence a conformable stratigraphic sequence. The regionally consistent stratigraphic succession of a thin unit of fine clastics and carbonates underlying the Catoctin Formation also diminishes the likelihood that any significant erosion could have occurred between Fauquier and Catoctin time (Espenshade, 1986).

Lynchburg Group

The Lynchburg Group of Wehr (1985) consists of five formations. Going upward in section, the depositional environments of the lower three formations show a transition from nonmarine alluvial plain to fairly deep-water submarine fan (Wehr, 1985). The lower two formations, the Bunker Hill and Monumental Mills Formations, correlate well with the Fauquier Formation, as we mapped it and as discussed above. Overlying the Monumental Mills Formation, Wehr mapped the Thorofare Mountain Formation, which appears to be absent in the field trip area. Overlying these three units are the Ball Mountain and Charlottesville Formations, which Wehr (1985) interprets as sediment gravity flows deposited in deep water. Both the Ball Mountain and Charlottesville Formations contain tabular concordant bodies of mafic and ultramafic rocks. Although never exposed, the basal contacts of both the Ball Mountain and Charlottesville Formations are interpreted to be thrust faults. In the Culpeper, Virginia, area, Wehr shows that the Ball Mountain Formation truncates underlying units and suggests either an unconformity or a thrust fault. However, in other areas Wehr notes that the transition from the Thorofare Mountain Formation to the Ball Mountain Formation appears conformable and gradational. At the southern end of the field trip area, the Ball Mountain Formation appears to be thrust over the Catoctin Formation and over itself.

Catoctin Formation

The Catoctin Formation consists of a thick section of greenstone and greenschist above the Fauquier Formation. These rocks are undoubtedly metabasalt as indicated by the common preservation of primary textures and structures, including vesicles and amygdules, porphyritic texture in places, and flow-top breccia. The Catoctin lavas are tholeiitic in composition (Badger and Sinha, 1988). Interior parts of flows are commonly more chlorite-rich, and are greenschist; at the tops of flows the rocks are more epidotized and thus are less foliated and have greater preservation of primary textures.

A prominent belt of metabasaltic breccia occurs near the base of the Catoctin, especially from the Marshall quadrangle southward through the Brandy Station quadrangle. Espenshade (1986) has documented two chemically different breccia zones, a lower one with low Ti and high Mg, and an upper one with higher Ti and lower Mg. The upper breccia is chemically similar to the upper part of the Catoctin Formation. The breccia was considered to be agglomerate by Furcron (1939) and agglutinate by Espenshade (1986). Kline et al. (1990a) reinterpreted the breccia as various types of "hyaloclastite" (brecciated basalt produced by rapid thermal contraction of subaqueously extruded lava), also termed "broken-pillow breccia" and "aquagene tuff" (Carlisle, 1963; Dimroth et al., 1978).

At several horizons in the high-Ti metabasalts of the Catoctin Formation are thin, discontinuous lenses of metasediment consisting of micaceous quartzite, quartz-muscovite schist, and phyllite.

STRUCTURAL GEOLOGY

Cover Sequence

In most of the field trip area, structure and outcrop pattern in the Fauquier Formation is very simple. Both bedding and cleavage have fairly constant attitudes, striking northeast and dipping moderately to the east. However, in several places the common northeast trend of the lower contact of the Fauquier Formation with the underlying Marshall Metagranite is interrupted by an abrupt change to a more northwest trend (Figure 1). This abrupt change in trend is not seen in the upper contact of the Fauquier Formation with the overlying Catoctin Formation, which maintains a constant northeast trend. In the Marshall quadrangle, the upper units of the Fauquier Formation and possibly the Fauquier/Catoctin contact are offset, but not nearly as much as the Fauquier/basement contact. The northwest-trending offsets are therefore accompanied by significant changes in thickness of the Fauquier Formation through truncation of the lower part.

These observations suggest that the northwest-trending offsets are synsedimentary normal faults related to rifting and formation of pull-apart basins in Late Proterozoic time. This interpretation was previously suggested by Espenshade (1986) to explain the Horner Run fault. Some faults (e.g., Stop 5) stopped moving before the end of Fauquier sedimentation because upper units in that formation are not offset, but the Horner Run fault (Stop 8) continued to move throughout the deposition of the Fauquier Formation and possibly into early Catoctin time. Some areas remained as structural highs throughout Fauquier time, so that the Catoctin Formation was deposited directly on Marshall Metagranite with no intervening Late Proterozoic sediments (Stop 2). Synsedimentary normal faulting is also a good explanation for an inlier of basement rocks within the Fauquier Formation near Halfway, Virginia (Stop 6). Tectonic breccia that is possibly associated with these faults is discussed below.

Basement Rocks

The oldest structure recognized in the Middle Proterozoic Marshall Metagranite is a discontinuous lineation of aggregates of ferromagnesian minerals, which can be quite strong in some places. This lineation clearly predates Fauquier Formation sedimentation (see Stop 3).

Most outcrops of granite possess a variably developed penetrative foliation that is either a single anastomosing feature or two foliations that intersect at a low angle. The overall trend of the foliation is northeast, but varies to north-northwest. Dip is moderate to steep to the east. The anastomosing foliation commonly produces an augen texture. Augen can either be polycrystalline granite in fine- to medium-grained rocks, or alkali feldspar megacrysts in coarse-grained granite. This foliation apparently developed under greenschist facies conditions, because it consists of abundant white mica, clinozoisite, and green biotite derived from retrograde alteration of the primary igneous minerals (see Stop 3). In many places, the earlier lineation is found in the plane of this foliation.

Previous workers have generally regarded greenschist facies deformation fabrics in the basement rocks as Paleozoic in age (for example, Bartholomew and Lewis, 1984). The discussion of Stop 3 describes evidence that the penetrative greenschist fabric of the basement granites in this area is largely, though not exclusively, pre-Paleozoic in age. There are also phyllonitic zones with greenschist assemblages in these granites that are not found in the Fauquier Formation. We interpret the formation of the phyllonites as predating Fauquier Formation deposition. This suggests that there was a greenschist facies tectonic event in the Proterozoic. It is not likely that greenschist facies conditions prevailed in the basement granites during the (earlier) Grenville orogeny, because the Marshall Metagranite intruded under deep-seated conditions. Evans (1984) thought that greenschist facies fabric in the Lovingston gneiss of central Virginia was formed during crustal extension accompanying Late Proterozoic continental rifting (Rankin, 1975). Erosion following the Grenville orogeny would have brought the Marshall Metagranite to a much shallower level by the onset of Late Proterozoic extension. Regional extension could impart a fabric under greenschist facies conditions. If much of that fabric formed prior to or during early stages of intrusion of the Robertson River Igneous Suite, that would explain the general lack of penetrative fabric in those rocks (Lukert and Banks, 1984).

Proponents of a Paleozoic age for the greenschist facies foliation in the Blue Ridge basement point out that the foliation is generally parallel to the regional cleavage in the cover rocks of Paleozoic age and in greenstone dikes within the basement. However, if a foliation formed in the basement prior to the Paleozoic, that fabric could have influenced subsequent strain patterns. Nearly coaxial Paleozoic strain may have enhanced the preexisting fabric or may be the origin of the commonly observed two cleavages at a low angle to one another in the basement granites.

Espenshade (1986) recognized two zones of phyllonite in the Marshall Metagranite, which he believed to be Paleozoic in age. Several northeast-trending zones of variably phyllonitized Marshall Metagranite occur in the Middleburg quadrangle, ranging from about a meter to 1.5 km wide, including a zone at Stop 5 of this trip (Kline et al., 1990b). Foliation in these zones strikes northeast and dips southeast. Kinematic indicators denoting sense of shear have only been found in one place. At the Glenwood Racetrack, north of Middleburg, phyllonitic foliation dips more steeply than an included lens of less-deformed phyllonite. Both have a down-dip lineation. This relationship indicates a normal-slip sense of motion (southeast side down). The age of the phyllonite zones must predate deposition of the lower part of the Fauquier Formation. At Stop 5 a phyllonite zone 2 km long is truncated by a Late Proterozoic northwest-trending fault separating Marshall Metagranite and the Fauquier Formation. The rocks of the Fauquier Formation, along strike with the phyllonite zone to the southwest, are unaffected.

Most basement outcrops have an assortment of narrow shear zones referred to as ductile deformation zones (DDZs) by Mitra (1979). The DDZs are normally separate from one another, but branching and anastomosing DDZs are common. DDZs are normally only 1-2 mm thick, but range up to several centimeters in thickness. In most outcrops the DDZs dip steeply, and a wide range of strikes is typical. This tectonic fabric was formed under greenschist facies conditions with abundant growth of micas at the expense of feldspars. The DDZs crosscut and offset all other structures in the Marshall Metagranite. Mitra (1979) interpreted the DDZs as having formed under compression during the Paleozoic Alleghanian orogeny. However, our work indicates that some of these features are more consistent with extensional tectonics. Lineations in the plane of the DDZs in the Middleburg quadrangle consistently plunge down the dip (Kline et al., 1990b). Accompanying kinematic indicators show normal-slip sense of movement down to the east. For example, an oriented sample of protomylonite from an 8 cm-wide DDZ in the Marshall Metagranite has both asymmetric augen and C-S fabric that indicate a normal-slip sense of movement down to the east (Figure 3).

Espenshade (1986) described a tectonic breccia in the Marshall Metagranite, which he thought might be related to movement along the Horner Run fault (Stop 8). He noted that this breccia was most commonly found near the fault. If the Horner Run fault zone formed during the deposition of the Fauquier Formation, the Marshall Metagranite would have been at a shallow enough depth to develop brittle structures. Many occurrences of similar breccia have also been observed in the Middleburg quadrangle (Kline et al., 1990b), most commonly near northwest-trending faults of Late Proterozoic age. At this time it is not possible to prove whether the breccia is related to the faults.

While it is not certain that the breccia is related to Late Proterozoic normal faulting, we do feel confident that breccia is related to some DDZs. It is common for DDZs to progressively broaden along strike and grade into a zone of breccia a few centimeters or decimeters wide. Some large breccia zones (e.g., at Stop 6) are interpreted to be the result of swarms of closely spaced DDZs. In some of these, the granite is more-or-less intact, having many narrow DDZs; in others, fine-grained material increases to over 50 percent and clasts are rotated. Some of the DDZs, therefore, exhibit brittle features in addition to ductile ones, making the use of the term "DDZ" questionable. If breccia is indeed related to the northwest-trending faults, then both the breccias and at least some DDZs formed during Late Proterozoic extensional deformation. Possible evidence to the contrary is that no boulders with DDZs have been found in basal conglomerate of the Fauquier Formation. A 1.5 km wide phyllonitic zone in the northwest corner of the Middleburg quadrangle has map-scale areas of nonphyllonitic granite. These areas have some of the highest concentrations of DDZs in the quadrangle, many of them subparallel to the trend of the phyllonite zone (Kline et al., 1990a). This spatial relation suggests, but does not prove, a genetic link between the phyllonites and the DDZs.

The rocks exposed at Stop 10, which we have tentatively correlated with the Ball Mountain Formation of the Lynchburg Group of Wehr (1985), have rare mesoscopic folds and minor outcrop-scale thrust faults. These rocks are interpreted to be part of a major thrust sheet or stack of thrust sheets which moved west over the Fauquier Formation. Evidence for the thrust fault is seen in the truncation of units beneath the lower contact of the Ball Mountain Formation (Wehr, 1985). To the west of Stop 10, the Ball Mountain Formation appears to be conformable upon the Monumental Mills Formation of Wehr (1985) (equivalent with the uppermost Fauquier Formation in this area), although we interpret this contact to be a thrust fault. In addition, to the north of Stop 10 the Ball Mountain Formation appears to be thrust over the Catoctin Formation. The amount of displacement on these thrust faults is not known. Rankin et al. (in press) suggest that the fault at the base of the Ball Mountain Formation represents a terrane suture of Taconic age; the western footwall is ancient North American basement and its nonconformable cover (Laurentia), and the eastern hanging wall is the Jefferson terrane. The Jefferson terrane consists of deep-water sediments with numerous tabular conformable bodies of mafic and ultramafic rocks conformably overlain by a fairly thick sequence of volcanics. Most previous workers (e.g., Nelson, 1962; Conley, 1978; Pavlides, 1990) have mapped these volcanic rocks as Catoctin Formation, and we agree with this assignment. Unpublished chemical analyses of some of these volcanics just south of the field trip area are compositionally similar to the high-Ti metabasalts of the Catoctin Formation. In the vicinity of Stop 10 there are

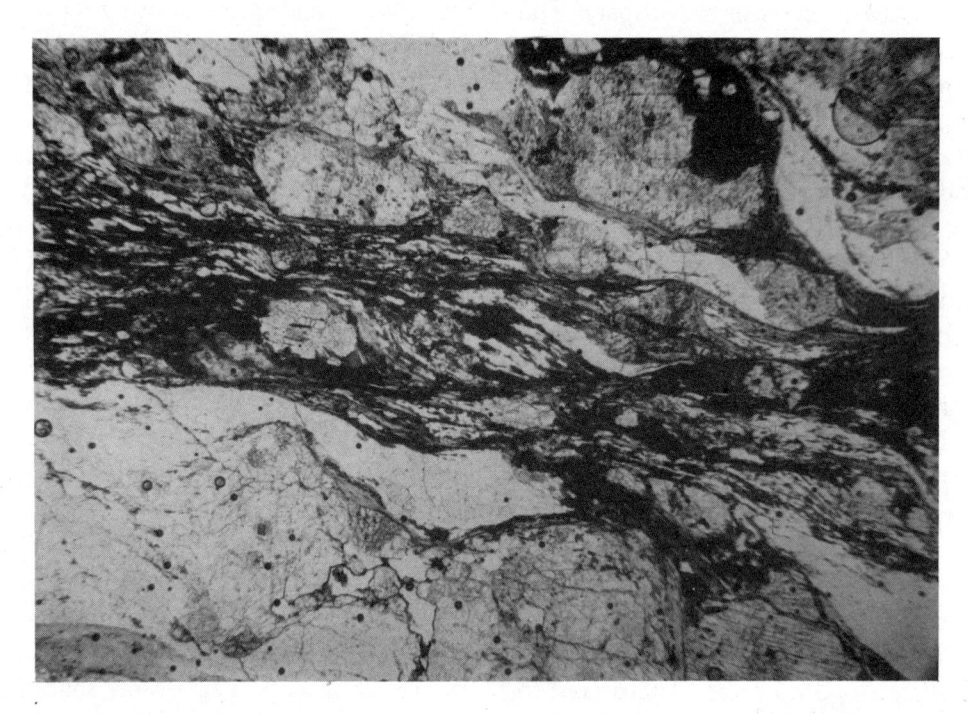

Figure 3. Photomicrograph (6 mm across, plane light) of oriented thin section, showing C-S fabric in ductile deformation zone in basement granite. Top of photograph is hanging wall, left is down dip or east. (Outcrop location is 3800 ft south of US 50, 2.1 miles ESE of US 50/VA 626 (north) intersection in Middleburg, on east bank of Little River).

several bodies of highly altered metagabbro (and possibly altered ultramafic rocks) that appear conformable with surrounding deep-water metasediments of presumed Late Proterozoic age. Chemical analyses of major and trace elements show that these rocks have compositions similar to the low-Ti metabasalts of the Catoctin Formation (Espenshade, 1986; Lyttle, unpublished data). Southeast of the field trip area in the Orange and Rapidan 7.5-minute quadrangles, Pavlides (1990) has mapped discontinuous lenses of siliciclastic rocks (quartz pebble conglomerate, quartzite, and siltstone) unconformably above the Catoctin Formation. These rocks have many similarities with rocks of the basal Chilhowee Group on both limbs of the Blue Ridge anticlinorium in northernmost Virginia. Based on the similar chemistries of the mafic volcanic rocks, our stratigraphic correlations, and the structures seen near Stop 10, we feel that the western boundary of the Jefferson terrane of Rankin et al. (in press) may well be a Taconic thrust fault of significant displacement. However, north of Culpeper, Virginia, a terrane distinct from Laurentia may be unwarranted.

TECTONIC HISTORY

The Marshall Metagranite is interpreted to have been syntectonically emplaced during the Grenville orogeny about 1010 Ma. It intruded a number of older granitoids and probably is one of the youngest units of Middle Proterozoic age in the Blue Ridge basement of northern Virginia (Clarke, 1984; Espenshade, 1986). During a period between 730 and 640 Ma ago, the Robertson River Igneous Suite intruded this basement terrane (Tollo et al., 1991). This magmatic event is believed to have accompanied crustal extension that led to the opening of the proto-Atlantic or Iapetus Ocean (Rankin, 1975).

This Late Proterozoic extension may be responsible for many of the structures present in the basement rocks of this area. In this interpretation, the penetrative foliation in the basement rocks formed early during crustal extension. The rocks were shallow enough for greenschist facies conditions. Strain occurred largely by flattening of original feldspar grains as they chemically altered to hydrous assemblages. Uplift and erosion accompanied this event, because clasts of the Robertson River Igneous Suite have been found in basal conglomerates of the Late Proterozoic cover sequence (Lukert and Banks, 1984). As the basement terrane rose to shallower levels, strain became more inhomogeneous and was concentrated in zones of phyllonite. Formation of phyllonites and possibly DDZs was accomplished through influx of water and retrograde degradation of feldspars. Rapid strain rates at a shallow depth produced breccia. Where normal-slip movement was greatest, subsiding basins formed.

As faults continued to move, considerable topographic relief was maintained. Coarse arkose and boulder and cobble conglomerate were deposited from uplands to form thick sequences in the basins. These sediments are believed to have been deposited in a braided alluvial environment (Wehr, 1985; Espenshade, 1986). Downdropping of basins ended during the closing stages of Fauquier Formation sedimentation, and some faults became inactive before others. Local topography became more subdued, and clastic sediments became finer and less arkosic. Some places remained as structural highs and either received a thinner sequence of Fauquier Formation sediments or none at all (see Stop 2).

Throughout northern Virginia, the uppermost Fauquier Formation is dominated by sediments deposited in low energy aqueous environments. Whether these deposits indicate the beginning of a marine transgression or a lacustrine environment is still a matter of debate. More rigorous sedimentalogical studies are needed to resolve the problem. The question is important in understanding the tectonic history of the Late Proterozoic rifting of the Laurentian margin. The faulting that produced the local basins seems to have ended near the time (approximately 600 Ma) of onset of eruption of tholeiitic basalts of the Catoctin Formation. We are inclined to agree with the interpretation of Wehr (1985) and that prior to Catoctin volcanism, the southern part of the area covered in this field trip had dropped significantly below sea level. The northern part of the field trip area experienced less subsidence, but also was below sea level.

In latest Late Proterozoic time (Badger and Sinha, 1988; Aleinikoff et al., 1991) Catoctin volcanism began. Some eruptions produced a significant proportion of hyaloclastite volcanic breccia and probably involved lava fountaining. In the northern part of the area, the build-up of lava flows may have created protected areas where carbonate precipitated. At first basalts were interlayered in places with fine clastics and carbonates, but soon volcanism dominated, creating a thick sequence of basalt flows with local coarse to fine clastic interlayers. This volcanism included the emplacement of sills of diabase and gabbro in the Fauquier Formation, possibly in the Catoctin lava flows, and many dikes in the Middle Proterozoic basement. In some basement rocks, the dikes make up 20 percent or more by volume of the rock (Espenshade, 1986; Schindler, 1990).

After Catoctin volcanism stopped, the region foundered to form an early Paleozoic passive margin. A transgressive succession of clastic sediments of the Lower Cambrian Chilhowee Group were deposited conformably on the Catoctin Formation.

In the field trip area, Paleozoic compression deformed and metamorphosed the cover sequence along with the basement. With westward transport of the Blue Ridge thrust sheet and the formation of the Blue Ridge anticlinorium, a number of thrust faults of fairly small displacement along the east limb of the anticlinorium formed. Regional greenschist facies foliation was produced in lithologies of appropriate composition. In most of the field trip area, we presume that the regional cleavage was pro-duced during the late Paleozoic Alleghanian orogeny. However, in the southern part of the field trip area, thrust faults of probable Taconic age produced significant tectonic shortening in the Late

Proterozoic metasediments. This juxtaposed the deep-water sediments of the Lynchburg Group with the shallower-water and fluvial sediments of the Fauquier Formation (Wehr, 1985; Rankin et al., in press; Lyttle, unpublished maps). We are presently employing $^{40}Ar/^{39}Ar$ isotopic techniques in an attempt to date muscovites formed in specific deformational events. Although significant faulting occurred nearby to the east forming the Mesozoic Culpeper basin, extension in Blue Ridge basement crust in the area of the field trip was relatively minor. This is indicated by a paucity of Mesozoic-age mafic dikes (Espenshade, 1986; Kline et al., 1990b).

ROAD LOG

It is necessary to obtain permission from landowners to visit all of the outcrops on this trip except for Stop 9, which is open to public access along a road. All of the landowners have been extremely cooperative, but misuse of their trust could jeopardize future visits.

This roadlog begins at the intersection of US 15 and Virginia State Route 733. This is 7.1 miles south of the intersection of VA 7 and US 15 (Leesburg, Virginia), and 4.6 miles north of the intersection of US 50 and 15 (Gilberts Corner, Virginia). In this log, all U.S. routes are abbreviated US, and all Virginia state routes are abbreviated VA.

0.0 Intersection of US 15 (James Madison Highway) and VA 733 (Limekiln Road). The western border fault of the Mesozoic Culpeper basin passes through this intersection and follows the east flank of Hogback Mountain. This fault juxtaposes Triassic conglomerate on the east and Cambrian quartzites of the Weverton Formation on the west.
Go west on VA 733.

3.8 Stop 1. Pull vehicles into small parking area on the right. Proceed up trail to north.

STOP 1. Carter's Quarry. Marble at the Catoctin/Fauquier Contact. Figure 4. Lincoln 7.5-minute quadrangle.

North of the main quarry pit is a smaller pit. At the north end of this smaller pit are good exposures near the contact between the Fauquier and Catoctin Formations. There is interlayering of phyllite, marble, and greenschist. Best exposure is during periods of dry weather when the pit is dry.

The contact between the marble and overlying greenschist is exposed here. Nearby in the Lincoln and Middleburg quadrangles, metadiabase and metagabbroic sills have been mapped by us and Parker (1968) near the contact between the Fauquier Formation and the Catoctin Formation. Examples of these sills are seen at Stop 4. The greenschist here is interpreted to be a metabasalt flow rather than a sill, because of the fine grained texture and the lack of metasomatic reaction at the sharp contact. Most, but not all, sills are coarser grained. In some areas, such as Stop 4, there are calc-silicate rocks produced by metasomatic reaction at the contact of a mafic sill and carbonate layer; such relations are not seen here.

Another occurrence of a probable metabasalt flow interlayered with phyllite and marble occurs 2.3 km south of here along Goose Creek. Such occurrences indicate that the transition from Fauquier Formation sedimentation to Catoctin volcanism was gradual, with continued sedimentation after the start of volcanism. Lenses of metasandstone (similar to metasandstone in the Fauquier Formation) and phyllite can also be found up section from here in the midst of the Catoctin Formation, and are a common occurrence in the Catoctin in the Marshall quadrangle (Espenshade, 1986).

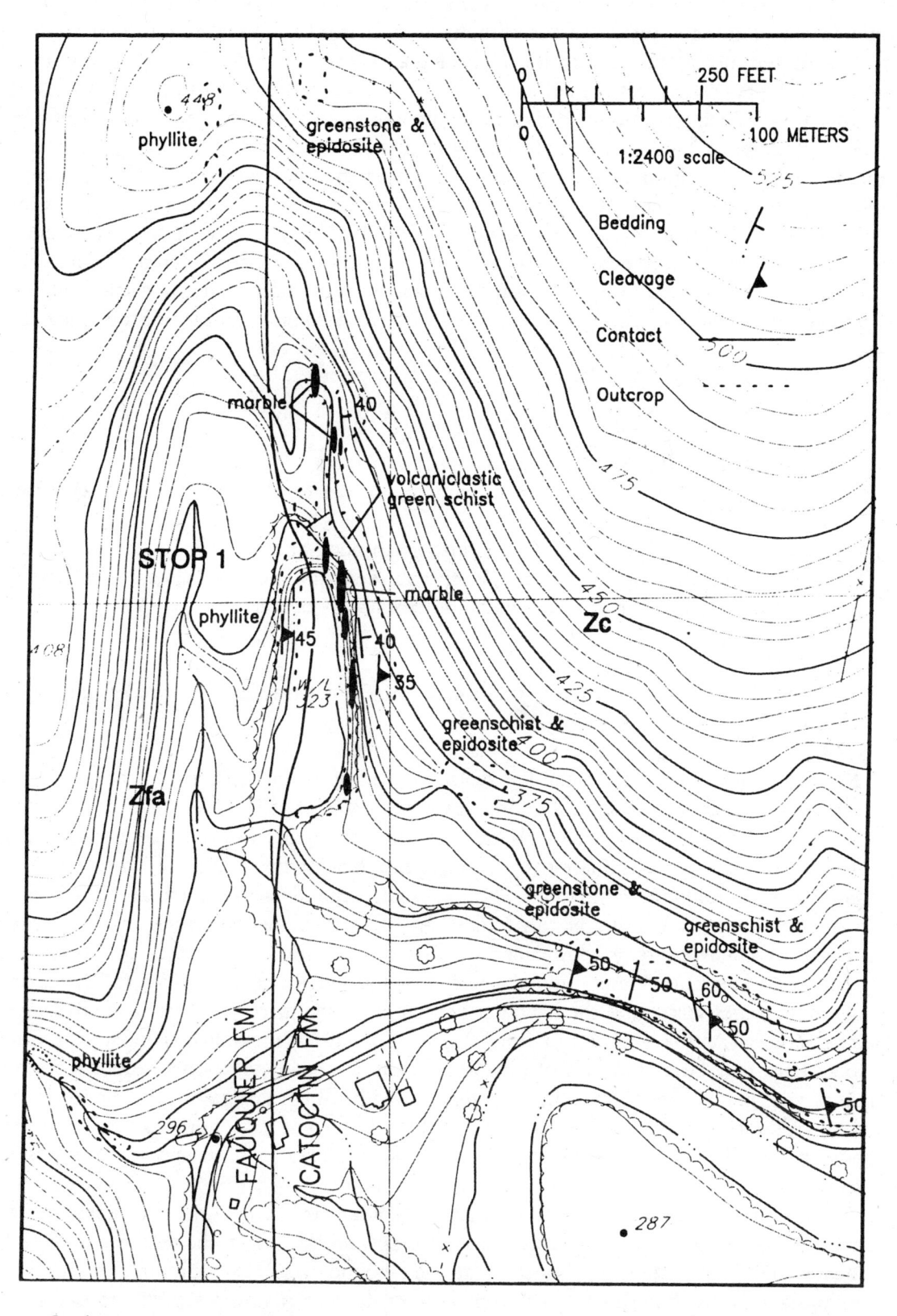

Figure 4. Geologic map in the vicinity of Stop 1. Carter's Quarry locality with contact between the Fauquier and Catoctin Formations.

The association of carbonates with the volcanics may be genetic. Restricted siliciclastic sedimentation is commonly required before carbonate deposition. A present-day analog of the environment that may have produced the carbonates near the Fauquier/Catoctin contact is found in the volcanic terrane of the Afar region in Ethiopia. There, volcanic flows block the influx of siliciclastic sediments and protected marine embayments occur. A stable substrate can then form on which carbonate-producing organisms thrive during periods of local volcanic quiescence (Barberi and Varet, 1970; Accademia Nazionale Dei Lincei, 1980). Parker (1968) found a number of occurrences of marble in the Catoctin Formation in the Lincoln quadrangle, and Espenshade (1986) mapped all marble lenses within the uppermost Fauquier Formation in the Marshall quadrangle. At this locality the lowest exposed Catoctin metabasalt is above the lowest marble. At present, it is uncertain whether the basalts played a role in formation of the carbonates, or whether carbonate precipitation was established independent of the volcanism. It is important to note that the calcite marble here has no primary sedimentary textures. This contrasts with the dolomitic marble at Stop 4, about 4 km to the south.

Continue west on VA 733.

4.8   Turn right on VA 763 (Steptoe Hill Road).
5.4   Highroad Program Center sign on left shows location of Vesper Hill (Stop 2). Continue straight on dirt road.
5.9   Go through gate, turn left immediately, and go through another gate.
6.3   Continue down hill past dining hall. When you see swimming pool ahead, veer right.
6.8   Stop 2. Park anywhere in the field with the cross in its center.

STOP 2. Vesper Hill. Basement/Cover Relations Along West-trending Normal Fault. Figure 5. Lincoln 7.5-minute quadrangle.

Proceed to outcrop A. This outcrop is typical of the Middle Proterozoic Marshall Metagranite, which crops out over an area of several kilometers west of Vesper Hill. This cliff exposure is a good vantage point to discuss basement/cover relations. About 300 ft south of the sharp bend in Beaverdam Creek is the northern termination of a long strike belt of the relatively undeformed clastic metasediments of the Fauquier Formation. These rocks consistently strike north and dip to the east, with no evident folds. The width of this strike belt of metasediments, which make up the hills southwest of this cliff exposure, is approximately 1070 m (3500 ft). To the west, basal conglomerate of the Fauquier Formation rests unconformably on basement. Up section to the east, the conglomerate becomes sparse and the Fauquier Formation is dominantly medium- to coarse-grained crossbedded arkose. At the very top of the Fauquier Formation, immediately below the Catoctin Formation, there may be a thin belt of phyllite. The total thickness of the Fauquier Formation, excluding several minor greenstone sills, is established with a high degree of certainty to be *greater* than 460 m (1500 ft). Yet along strike just south of this cliff the entire section of Fauquier Formation is abruptly truncated. This truncation is interpreted to be a fault whose approximate position is shown in Figure 5.

Outcrop A is a granite with a strong foliation generally trending north and dipping east. The foliation is defined by biotite and chlorite anastomosing around quartz and feldspar augen. These augen locally have strong downdip lineation at their boundaries. The foliation is parallel to the regional Appalachian trend in this area and is probably Paleozoic in age. In several places on the face of the cliff an older foliation, composed of sparse mafic segregations, strikes east and dips south. This

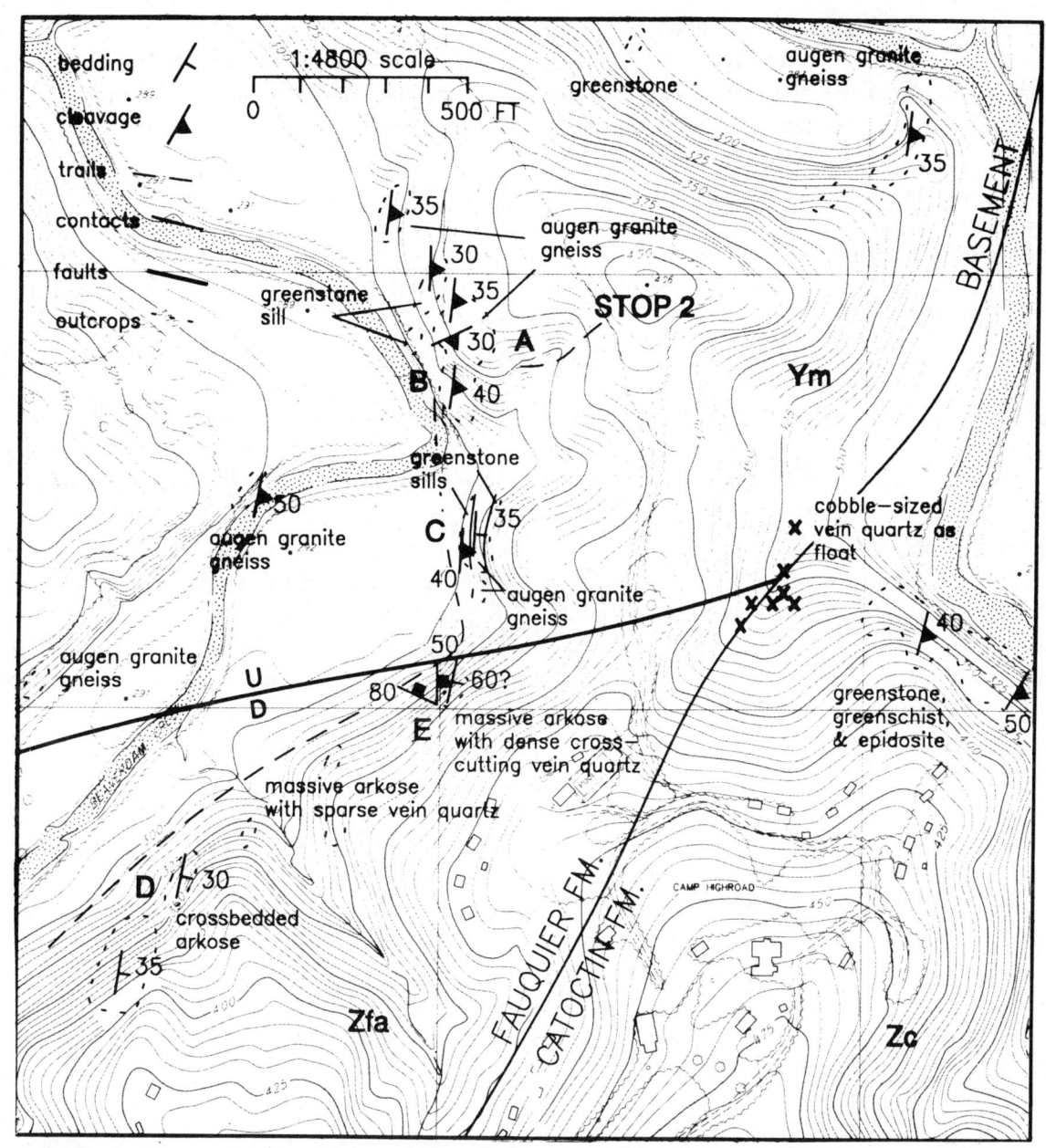

Figure 5. Geologic map in the vicinity of Stop 2. Vesper Hill locality with fault between the Fauquier Formation and the Marshall Metagranite.

foliation is cut by the more pervasive north-trending one and is thought to be Grenville in age. Relationships between similar fabrics will be discussed more fully at Stop 3.

At outcrop B a metadiabase dike intrudes the granite. The pervasive Paleozoic foliation parallels the contacts of this dike. This dike has a medium-grained interior and very fine grained margins.

Outcrop C shows two metadiabase dikes intruding the granite. Once again, the foliation in the granite is roughly parallel to the dike contacts.

Outcrop D is a large exposure of crossbedded meta-arkose typical of many exposures in the area. The meta-arkose is medium- to coarse-grained, unsorted, thin- to medium-bedded, with subangular quartz and feldspar grains exhibiting little deformation. Crossbeds are sparse and faint, but one example can be seen at the middle of the outcrop in a bed of medium thickness. Note the thin quartz veins at a high angle to bedding.

Outcrop E is a massive meta-arkose in which bedding is either faint or obliterated by deformation. Note the abundance of crosscutting quartz veins and proximity to the proposed fault.

Return via trodden path ascending hill above outcrop C.

Retrace route out of Highroad Program Center.

8.9 Turn right on VA 733 (Limekiln Road).
9.5 Turn left through white fence of Groveton Farm (telephone line follows farm road in field to the south). Stop 3.

STOP 3. Goose Creek Unconformity. Figure 6. Lincoln 7.5-minute quadrangle.

This stop consists primarily of two very large outcrops where important relationships between the Marshall Metagranite and the Fauquier Formation occur. The spectacular outcrop of the Fauquier Formation consists of a polymict, matrix-supported, boulder and cobble conglomerate (Figure 7a). Similar conglomerate is common at the base, but rarely so well exposed. Nearby is a large cliff exposure, on Goose Creek, of the Marshall Metagranite. Comparison of rocks in these outcrops yields information concerning the depositional environment at the beginning of Fauquier time, and provides clues to the age of some tectonic fabrics in both the Marshall Metagranite of basement complex and the overlying Fauquier Formation. Understanding of these relationships has been aided by an undergraduate research project by Dede Pitts at George Mason University.

Origin of Clasts

Examination of the outcrop and thin sections of core samples of representative boulders from the conglomerate shows that the majority, if not all, of the clasts are very locally derived. Clasts consist of four types: 1) rounded boulders and cobbles of leucocratic granite with scattered ferromagnesian minerals that define a lineation; 2) elongate boulders and cobbles with a gneissic texture; 3) cobbles and pebbles of blue quartz; and 4) round cobbles of pegmatite. Bedrock similar to all of these clasts can be found in the cliff exposure of granite on Goose Creek.

Thin sections of core samples of type 1 clasts are all, with one exception, mineralogically and texturally identical to granite exposed in the lower part of the nearby cliff. The chief constituents of these medium-grained granites are K-feldspar, having an exsolution texture of bleb-like albite lamellae and variably developed gridiron twinning, quartz, and heavily sausseritized plagioclase. K-feldspar constitutes about two-thirds of total feldspar content and commonly has inclusions of quartz and less commonly plagioclase. The fabric is distinctly allotriomorphic, with only hints of a rectangular shape to some feldspars. The original ferromagnesian minerals (less than 5 percent of the rock) appear to have been Fe-Ti oxides, reddish (titaniferous) biotite, and zircon. The oxides are mantled or totally replaced by sphene. Reddish biotite is variably degraded to mixtures of green biotite, sphene $\pm$ rutile, opaques, and quartz. One boulder is distinctly different from the others and from the adjacent basement exposure, being granitoid, but with plagioclase $>>$ K-feldspar, and a paucity of ferromagnesian minerals. This rock may correspond to scattered occurrences of leucocratic tonalite in the general area (W. Leo, personal communication).

Gneissic-textured clasts (type 2) (Figure 7B) are identical in hand sample and thin section to rocks near the top of the basement exposure. There a strongly developed penetrative foliation overprints the earlier fabric that is seen at the bottom of the exposure. This foliation is like the main foliation seen in the granite at Stop 2 (outcrop A). The rock of the lower part of the present outcrop

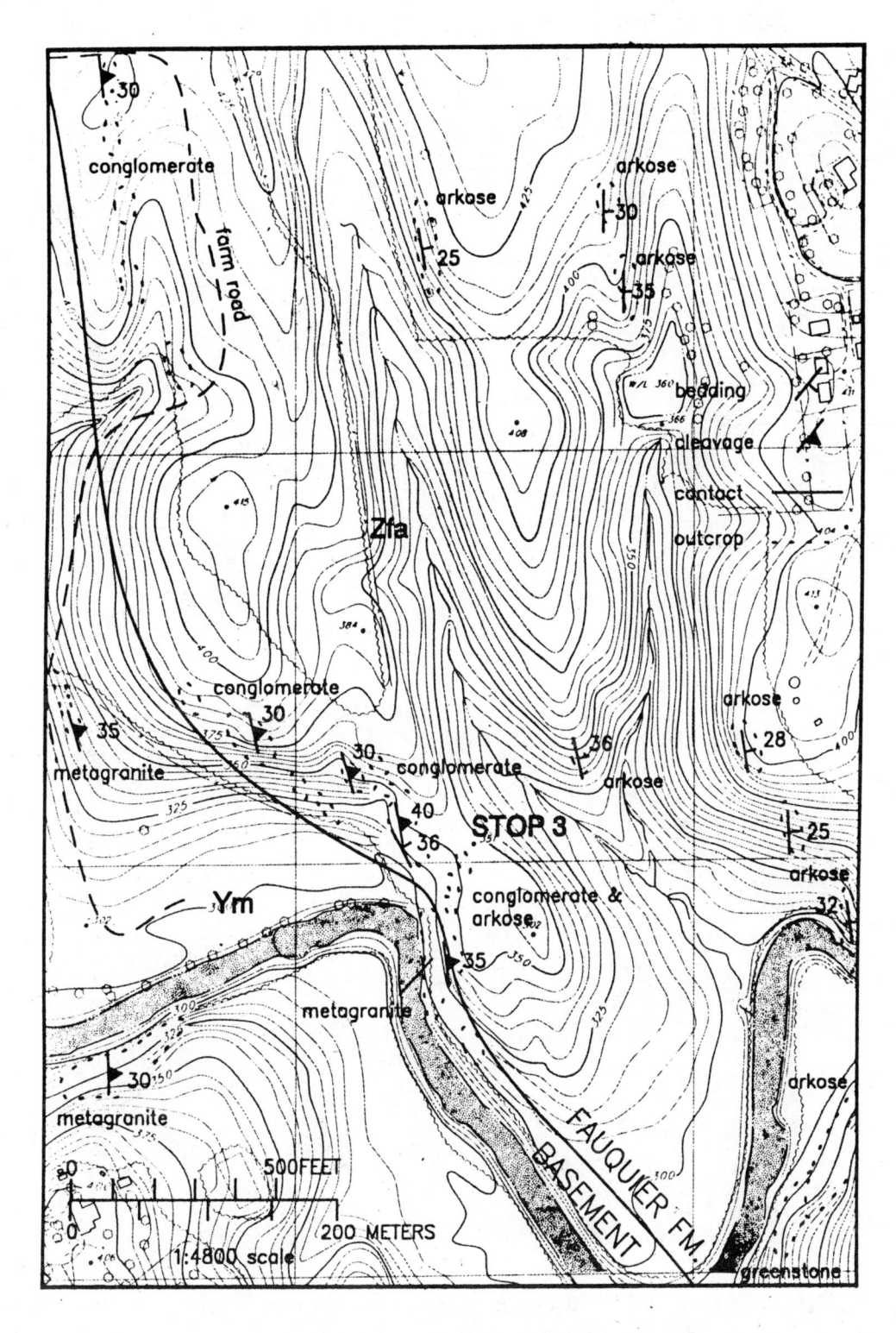

Figure 6. Geologic map in the vicinity of Stop 3. Goose Creek locality with the Fauquier Formation unconformably overlying the Marshall Metagranite.

Figure 7. Basal conglomerate of the Fauquier Formation conglomerate at Stop 3. A, Wide-angle view of the conglomerate. B, Close up of one clast from photo A.

is like the less-deformed lens at Stop 2. This rock contains variably spaced, subparallel shear planes, with nearly undeformed granite between them ("microlithons"). Near the shear planes, feldspars are more heavily altered and flattened. Quartz either remains as large strained grains amongst white mica and fine clinozoisite-epidote, or is broken down to subgrains. In the shear planes, a high concentration of strongly aligned white mica, green biotite, and clinozoisite-epidote occurs along with quartz, sphene, and sphene-mantled opaques. These are all derived from metamorphic alteration of primary constituents.

Type 3 clasts of blue quartz correspond to blue quartz veins and irregular segregations commonly found in the granites of this region. Examples can be seen at the very top of the adjacent basement outcrop. Type 4 pegmatite clasts look identical to small pegmatite veins in the local granite.

In addition to the clasts, the matrix of the conglomerate seems to be of very local origin. Very minor transport of detrital constituents is suggested because in many places the feldspar abundance in the matrix approaches that of the clasts. The boundary between clast and matrix in some places is hard to discern. The conglomerate is thought to be of fluvial origin.

Age of Tectonic Fabrics

Tectonic fabrics visible in rocks of this stop include: 1) a lineation in the rounded granite clasts of the conglomerate and in the lower part of the basement outcrop, defined by aggregates of ferromagnesian minerals; 2) a penetrative foliation in the upper part of the granite outcrop and in similar-looking clasts in the conglomerate; 3) discreet, more or less randomly spaced and variably oriented, narrow DDZs in basement granite; and 4) a foliation in the matrix of the conglomerate. The foliation in the matrix shows a lineation immediately adjacent to some clasts.

The cleavage in the matrix of the conglomerate is interpreted as Paleozoic in age. The rocks were deposited near the end of the Proterozoic, and the cleavage is parallel to the regional Paleozoic trend. The mineral lineation in the granite is clearly Proterozoic in age, presumably Grenvillian, because clasts of this granite in the conglomerate contain the lineation at an oblique angle to the foliation in the matrix. The DDZs in the granite are not found in the conglomerate clasts, suggesting that they could be Paleozoic in age (Mitra, 1979). However, a Proterozoic age for some of the DDZs is possible. The age of the penetrative fabric at the top of the granite outcrop has been more difficult to resolve. Geologists visiting this stop have interpreted that part of the outcrop as 1) basal Fauquier Formation nonconformably on basement; 2) a paleo-saprolite zone; 3) a detachment at the basement/Fauquier Formation contact; and 4) a Late Proterozoic deformation fabric. Combined evidence from both outcrops of this stop best supports the fourth interpretation. This foliation clearly postdates the less-distinct lineation of presumed Grenville age in the lower part of the outcrop. If one accepts that the foliation in type 2 clasts (Figure 7B) predates the foliation in the matrix of the conglomerate, and that it is the same as the foliation at the top of the basement outcrops, clearly it must predate Fauquier Formation deposition. We will spend time trying to convince you that these two claims are true.

The various clasts of the conglomerate indicate that some rocks in the local basement terrane were undeformed except for the early mineral lineation. Others were pervasively deformed with a fabric as seen in the Goose Creek outcrop. Basement rocks throughout this area show the same variation in degree of deformation. We suggest that the little-deformed granite at the Late Proterozoic surface underwent spheroidal weathering and was deposited with quartz and pegmatite clasts and highly feldspathic sand derived from the same general area. A few of the highly foliated granitic rocks weathered as flat slabs that were deposited in the same fluvial environment with a preferred orientation subparallel to bedding. Paleozoic metamorphism and deformation imparted a foliation to the matrix of the conglomerate subparallel to bedding and therefore subparallel to the clasts.

Retrace route to paved road. Continue west on VA 733.

9.9 Turn left on VA 734 (Snickersville Pike).
12.5 Turn left into Dresden Farm. Stop 4.

STOP 4. Marble Beds in Uppermost Fauquier Formation. (Optional Stop) Figure 8. Middleburg and Lincoln 7.5-minute quadrangles.

A string of outcrops along the east side of the valley on the Dresden Farm, east of Route 734, have additional features characteristic of the contact between the Fauquier and Catoctin Formations. Although the outcrops are small and scattered, we have mapped at least three layers of dolomitic marble in the upper part of the Fauquier Formation here. It is not known how extensive they are, but along strike between here and Goose Creek, 1.6 km to the north, a number of scattered outcrops occur. At Goose Creek three separate dolomite layers also crop out. It is likely that the dolomite layers are more-or-less continuous for at least 4 km in this area, and may not occur as pods as mapped by Parker (1968).

Interlayered between the dolomitic marbles are greenstone and phyllite. The greenstone crops out in several places, but the phyllite infrequently. Most of the greenstone has a relict gabbroic texture and is interpreted as sills of Catoctin mafic magmas. In the gully that follows the east-trending fence line just south of where the farm road crosses the intermittent stream, rocks of probable contact metasomatic origin occur. Please leave them in place—they are rare! They are coarse radiating aggregates of blade-like tremolite-actinolite crystals. Crystal size ranges up to 1.2 x 3.3 cm. These are thought to be calc-silicates produced by the reaction of silicate melt with carbonate country rocks.

The carbonate outcrops here are dolomitic, in contrast with the calcite of Stop 1. Oolitic texture is present in some places. The texture is also apparent in thin section, where oolite-size round structures make up the bulk of the rock, but characteristic concentric lamination is not present. In one outcrop, small specs of sparry calcite are scattered amongst the finer dolomite. These could be casts of gypsum crystals in the original dolomite. A lot more work needs to be done to work out details of these carbonates, but it appears that the depositional environment varied from place to place. The dolomitization and oolitic texture (and possible evaporite mineral casts) suggest that the carbonate here was deposited in an intertidal environment, while the carbonate at Stop 1 was subtidal.

Retrace route to VA 734 and continue southeastward.

12.8 Turn right on dirt road VA 627 (Macsville Road).
13.8 Take dogleg to the left.
15.3 Turn right (west) on US 50. Straight south in the distance is Bull Run Mountain, which consists of the resistant Weverton Quartzite.
15.6 Turn left on dirt road VA 627 (Parsons Road).
17.0 Turn left (southeast) on paved VA 776 (Landmark School Road).
18.6 Turn right (southwest) on dirt road VA 628.
20.6 Turn right into Meredyth Vineyards. Stop 5.

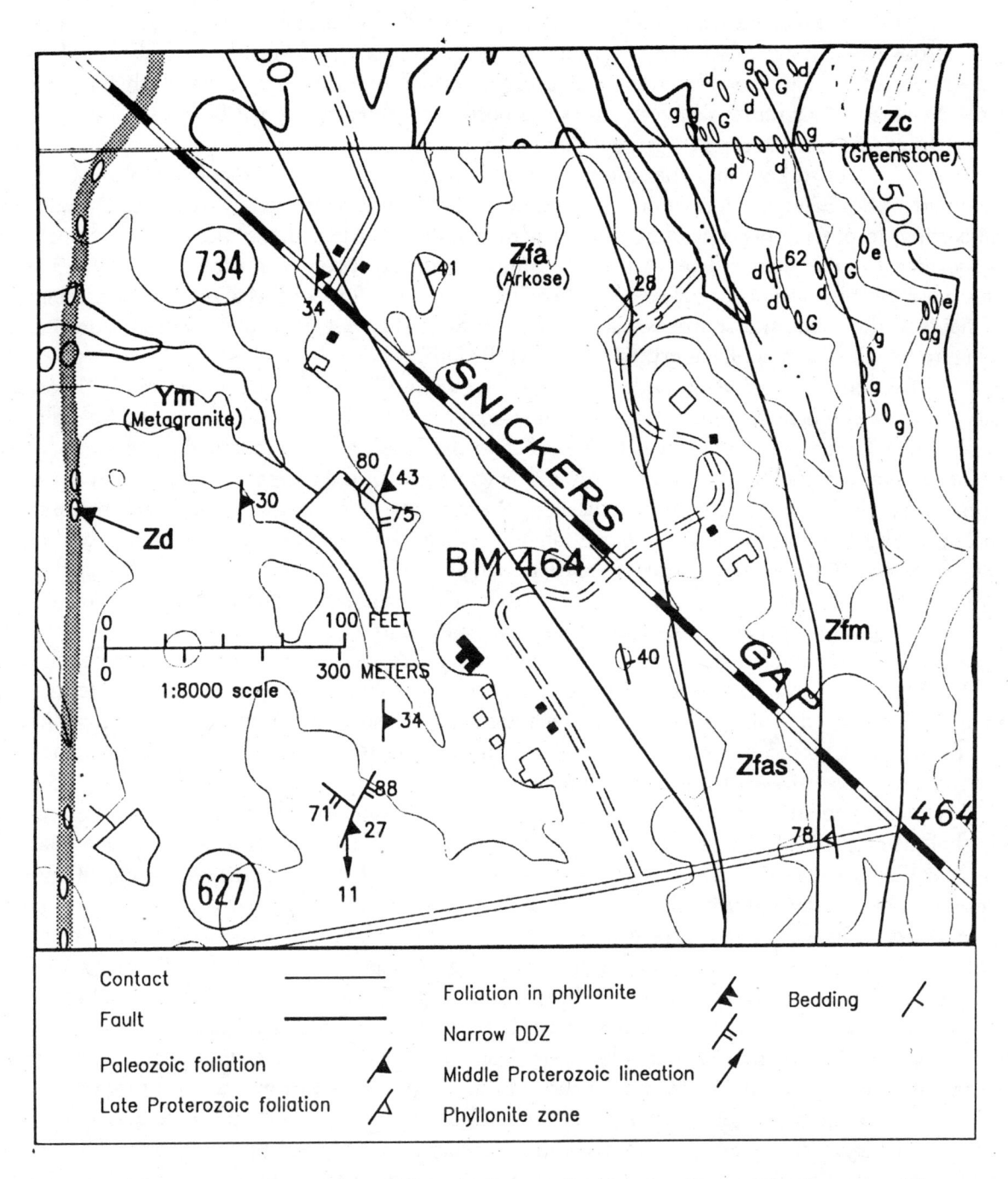

Figure 8. Geologic map in the vicinity of Stop 4 (optional). Dresden Farm. The letters in Zfm mean: d=dolomite; g=greenstone; G=metagabbro; e=brecciated epidosite; p=phyllite. Zd=metadiabase sills; Zc=Catoctin Formation; Zf=Fauquier Formation: Zfm=metamudstone unit, Zfas=meta-arkose and metasiltstone unit, Zfa=meta-arkose unit; Ym=Marshall Metagranite.

STOP 5. Meredyth Vineyards—Late Proterozoic Faulting. Lunch Stop and Wine Tour. Figures 9 and 10. Middleburg 7.5-minute quadrangle.

At this stop we will see 1) meta-arkoses typical of the lower part of the Fauquier Formation; 2) a northwest-trending offset of the contact between the Fauquier Formation and the Marshall Metagranite that we interpret as a Late Proterozoic normal fault; and 3) a zone of phyllonitization in the Marshall Metagranite which predates the northwest-trending normal fault.

In examining outcrop A (Figure 9), make note of the good preservation of the original clastic texture of the meta-arkose and the slabby bedding. These features are characteristic of rocks in the lower part of the Fauquier Formation. In some places bedding is expressed by variation in grain size and sorting; in others only a bedding-parallel parting indicates the bedding orientation. It might be a good idea to take along a few small hand samples for comparison with other rocks at this stop, since many of the rocks occur only as float. Small outcrops and float of this arkose occur throughout the fields to the south and southeast, and in the woods to the west beyond the winery (see outcrop B, Figure 9), all with a similar northeast strike and southeast dip.

After lunch (at picnic grounds, C on map) we will walk to the hilltop indicated by D on the map and walk through the vineyard toward the outcrop labeled E. The average strike of bedding in the Fauquier Formation in this area is subparallel to the vine rows. Abundant float of meta-arkose can be seen on the knoll as we enter the rows. The contact with the basement is inferred by presence of float. Abundance of coarse granite rocks along the vine rows increases as we approach the outcrop labeled E. The map for this area was based largely on float, with control of a few outcrops and auger holes. This mapping has documented an abrupt change from the prominent northeast regional trend of the Fauquier/Marshall contact that occurs north of this locality, to an anomalous northwest trend.

The pavement exposure of coarse megacrystic granite (E) is less deformed than most occurrences of this unit in the nearby vicinity. A few DDZs can be seen in this exposure. In thin section, the coarse granite has both a primary and metamorphic mineralogy essentially identical to the granite described at Stop 3. The megacrysts are alkali feldspar. A penetrative foliation similar to that described at Stop 3 is common in the local granite, though nearly absent at E. Development of this fabric was coeval with retrogressive alteration.

Proceed to the outcrops labeled F and G on the map. Float between E and F consists of Marshall Metagranite and weathered greenstone from dikes. Outcrop F consists of a coarse phase of the Marshall Metagranite. On the northwesternmost edge of that outcrop is a zone of intense foliation. The nearby outcrop G consists of an intensely foliated rock with quartz augen. These outcrops are on the southeast edge of a belt of similar rocks about 170 m wide that extends about 2 km to the northeast. These rocks are interpreted to be retrogressively altered and variably sheared Marshall Metagranite. We refer to this belt as a zone of phyllonitization, although not all rocks are phyllitic. Thin sections of various rocks in this belt have a wide range of textures. The quartz "augen" of outcrop G are large and commonly have very complex amoeboid shapes that appear to float in a sea of variably aligned fine white mica, quartz, epidote, sphene, green biotite, and magnetite. The outline of many quartz augen exactly mimics the anhedral interlocking shapes that quartz has with respect to feldspars in the granites. In a few places, relict feldspars remain, mostly altered to sericite plus quartz. Opaques rimmed by sphene similar to those in unaltered granites occur also, but euhedral (metamorphic) magnetite is also very common. Biotite is very minor in abundance in these rocks, and magnetite is the principal iron-bearing mineral, making the shear zone at F very magnetic. In the phyllitic rocks, the large amoeboid quartz grains are generally converted to a finer-grained aggregate, strung out parallel to foliation.

The development of these fabrics clearly was accompanied by an influx of water, which reacted with feldspar to produce white mica and (with the release of iron from the titanomagnetite-

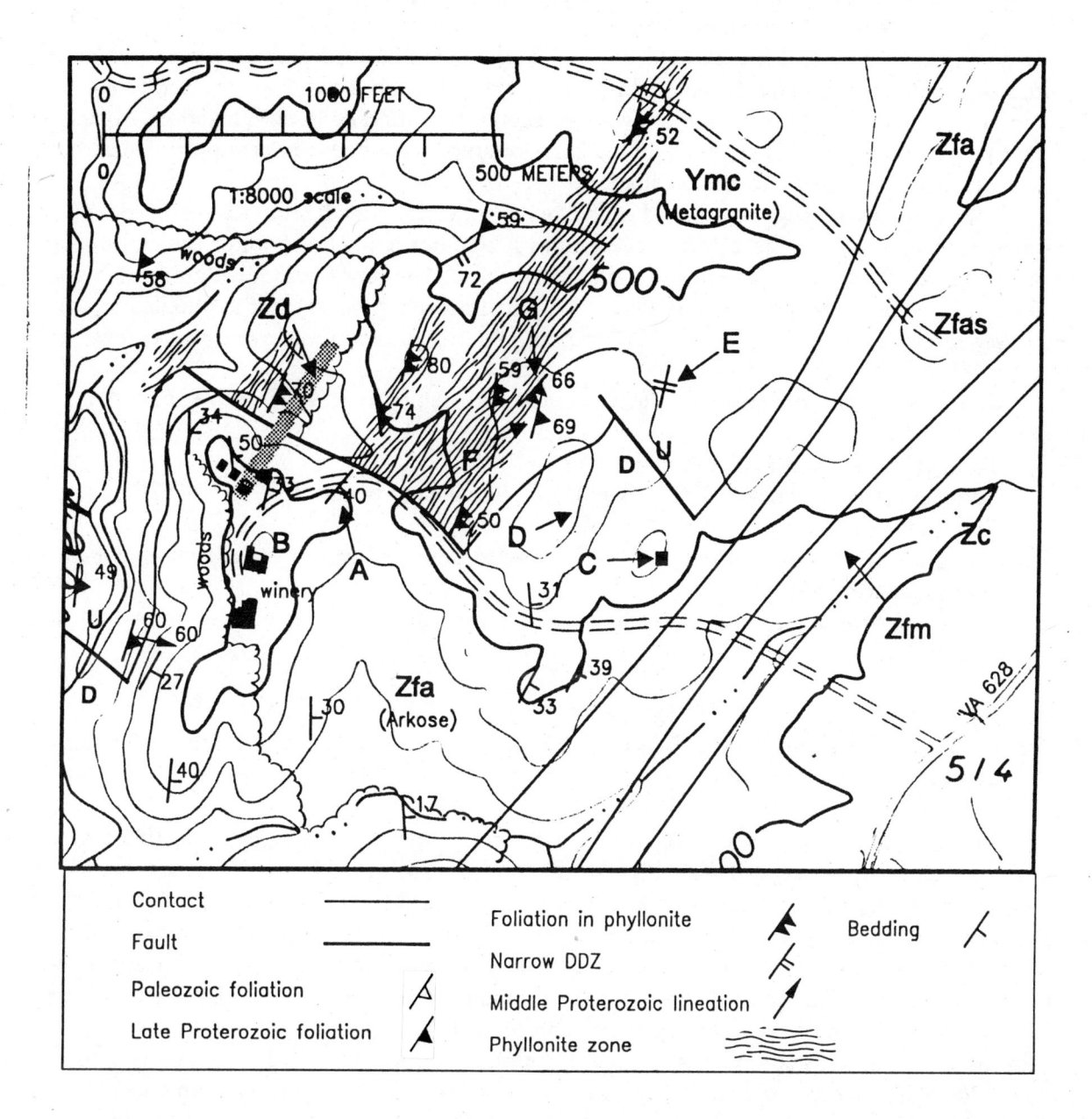

Figure 9. Geologic map in the vicinity of Stop 5. Meredyth Vineyards. Zc=Catoctin Formation; Zf=Fauquier Formation: Zfm=metamudstone unit, Zfas=meta-arkose and metasiltstone unit, Zfa=meta-arkose unit; Ym=Marshall Metagranite (fine- to medium-grained phase); Ymc=Marshall Metagranite (coarse-grained phase).

to-sphene reaction) biotite, epidote, and magnetite. The abundance of metamorphic magnetite in these rocks indicates that the fluid was more oxidizing than fluids involved in producing the penetrative foliation in unaltered granite. Feldspars altered mostly to mica, while quartz was preserved in augen. Although some rocks in the zone of phyllonitized granite resemble phyllites of the Fauquier Formation, thin sections have no evidence of sedimentary clastic textures.

The truncated phyllonite zone is the most important feature to be seen here. It begins 2 km to the northeast of here and is truncated in the vicinity of the farm road that leads to the winery (southwest of outcrops F and G). The fields south of the road on strike with the phyllonite zone contain arkose of the Fauquier Formation, with no sign of hydration or intense foliation. In fact, on the next property to the southwest, some of the best-preserved cross-bedded arkose of the quadrangle occurs directly along strike with the phyllonite zone.

This relationship suggests that the phyllonite zone developed prior to deposition of the Fauquier Formation, indicating a tectonic event with significant influx of water under low-grade metamorphic conditions sometime in the Late Proterozoic. If the phyllonitic foliation formed under the same conditions as the phyllonite at Glenwood Racetrack to the north in the Middleburg quadrangle (see discussion under Structural Geology section), then this tectonic event was extensional.

Retrace route out of Meredyth Vineyards and turn right (southwest) on VA 628.

21.0   Turn right on paved road VA 686.
22.1   Turn right (north) on VA 626. If skipping Optional Stop 6, turn left on VA 626 and go to 7.0 miles in roadlog.
23.5   Along this stretch of road the geology is complex due to a horst of granitic basement in the over sequence (Figure 9).
24.5   Pull off on right side of road onto grass shoulder just before small bridge over Burnt Mill Run. Stop 6.

STOP 6. Complex Basement/Cover Relations. (Optional Stop) Figure 10. Middleburg 7.5-minute quadrangle.

The map pattern of the contact between basement and cover is complex in this area and tightly constrained by outcrops (Figure 10). At this stop, the Fauquier Formation occurs in a narrow zone between metagranite to the east and west. The Fauquier Formation here consists mainly of matrix-supported cobble conglomerate and some phyllite. Two different granites are exposed; west of VA 626 the granite is coarse with augen-shaped K-feldspar megacrysts and a significant amount of biotite. East of VA 626 the granite is a fine- to medium-grained leucogranite. Both are rock types commonly found in the Marshall Metagranite. To the west we interpret the Fauquier Formation to be resting on the Marshall Metagranite nonconformably as evidenced by several lenses of metaconglomerate. To the east we interpret the meta-arkoses of the Fauquier Formation to be in fault contact with the Marshall Metagranite. To complicate relations even more, there is a narrow inlier of Marshall Metagranite within the Fauquier Formation to the south of this stop. All bedding in the Fauquier Formation dips east consistently on both flanks of this basement inlier, so we map a normal fault along its western edge.

A large exposure of Fauquier Formation metaconglomerate occurs on the east side of the road, south of the creek. Besides quartz clasts, most of the clasts are of medium-grained leucogranite that in hand specimen looks like the nearby leucogranite of the Marshall Metagranite. A few clasts appear to be the coarse granite.

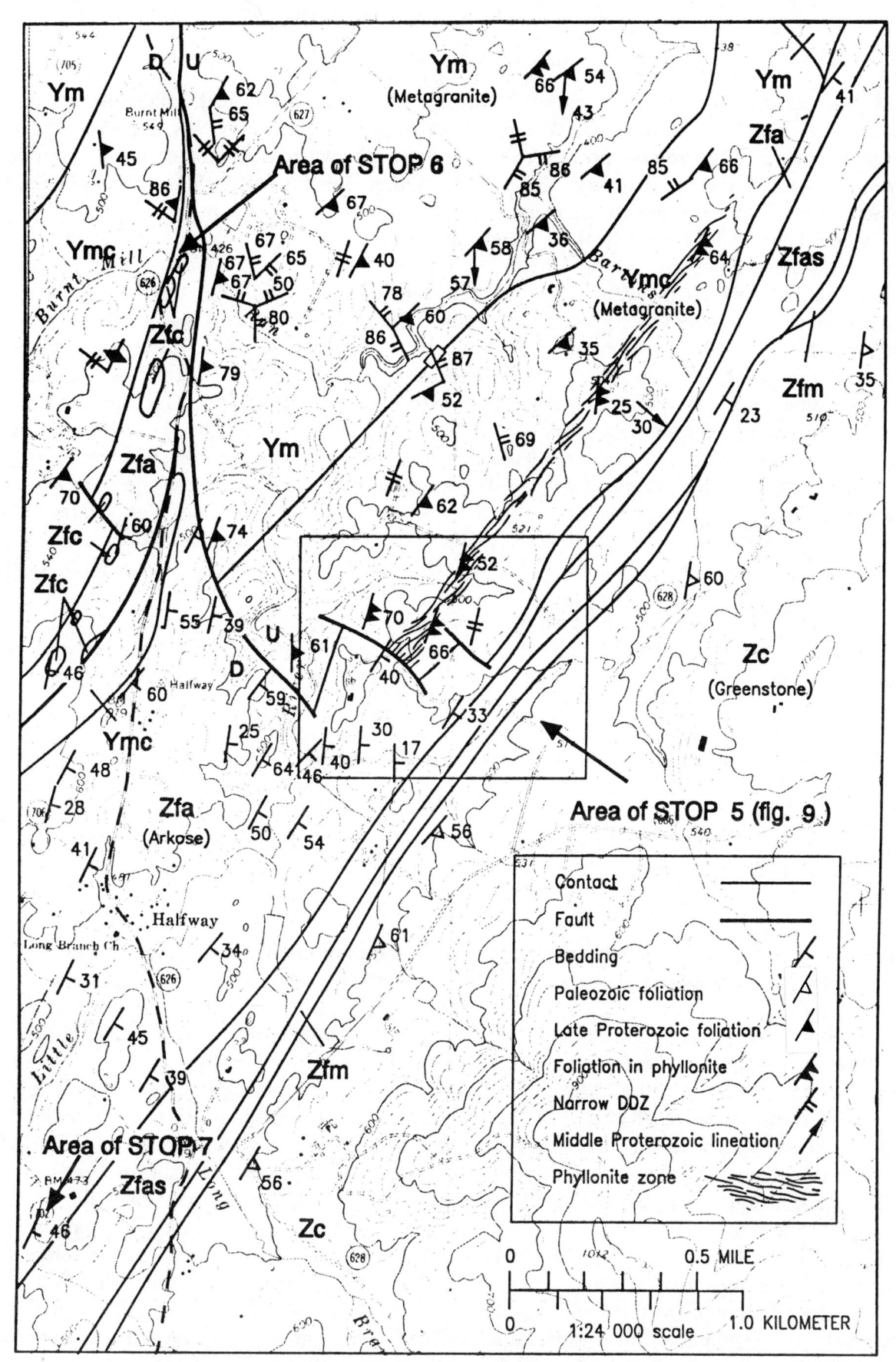

Figure 10. Geologic map in the vicinity of Halfway, Virginia, including Stops 5, 6, and 7. Zc=Catoctin Formation; Zf=Fauquier Formation: Zfm=metamudstone unit, Zfas=meta-arkose and metasiltstone unit, Zfa=meta-arkose unit, Zfc=cobble conglomerate unit; Ym=Marshall Metagranite (fine- to medium-grained phase); Ymc=Marshall Metagranite (coarse-grained phase).

Across the road, on the north side of the creek, are outcrops of metagranite (south of the driveway) and phyllite (road bank north of the driveway) in close proximity. Between them is greenstone, which could be a sill intruded along the contact between the basement and the overlying phyllite. It cannot be traced beyond this outcrop. Other outcrops of phyllite, interlayered with metasandstone and near metaconglomerate, occur along strike a few hundred meters to the southwest. The phyllite probably is an overbank deposit associated with stream channels that deposited the conglomerates.

The hill slope on the west side of VA 626, across from VA 627, has a cluster of outcrops. Small pavement exposures are of Fauquier Formation metaconglomerate. One is especially good, with clasts of medium-to-fine leucogranite and coarse biotite granite similar to the local coarse granite. A few meters to the west is a clump of bushes. At its northern end is another exposure of metaconglomerate, separated from basement granite by only a meter of cover. The coarse basement granite is moderately deformed, with biotite-rich shear planes that anastomose between K-feldspar megacrysts, imparting an augen texture. A leucocratic granite dike cuts the coarse granite. <u>Please do not sample the dike; such exposures are rare.</u> The leucocratic dike looks similar to the granite on the east side of the road. Other dikes of leucogranite cutting coarse granite have been located 2 km south of here. Dikes of similar coarse granite cutting medium-grained leucogranite occur north of Middleburg, supporting Espenshade's (1986) hypothesis that the coarse granite is a phase of the Marshall Metagranite.

A large outcrop of leucocratic granite, about 460 m (1500 ft) south of the intersection of VA 626 and 627, exemplifies the breccia that seems to be spatially associated with the inferred Late Proterozoic faults. In this outcrop, two sets of DDZs, striking N14°W and N42°E, are present, and appear to be related to the brecciation.

Cautiously turn around in nearby driveway and retrace route south along VA 626.

27.0   Intersection with VA 686; continue straight (south) on VA 626.
27.5   Turn right (west) on dirt road VA 702. Private training track on left.
28.0   Turn left into driveway leading to horse barn. Stop 7.

STOP 7. Crossbedded Metasandstone of the Fauquier Formation. Figure 10. Middleburg 7.5 Minute quadrangle.

A very large outcrop here is an excellent exposure of bedding features of metasandstone in the lower part of the Fauquier Formation. The metasandstone is mostly medium-grained, with some coarse-grained layers, generally feldspathic, and has abundant cross-stratification. These rocks are equivalent to the meta-arkose unit of the Fauquier Formation (Espenshade, 1986) and the Bunker Hill Formation of Wehr (1985). Finer grained metasandstone and metasiltstone crop out 180 m (590 ft) to the southeast; these rocks are equivalent to the meta-arkose and metasiltstone units of the Fauquier Formation of Espenshade (1986) and the Monumental Mills Formation of Wehr (1985).

On the nearly vertical face of the outcrop (roughly parallel to strike), many tangential cross beds are present. A few places appear to be trough cross beds. Thickness between bottomsets is generally 2–6 cm, but some sets are thicker. There is insufficient exposure to determine the exact paleo-current direction, but the orientation of foresets consistently indicates flow to the southwest. This current direction is common in this area and consistent with the dominant current direction found in the Mechum River Formation to the west (Schwab, 1974).

Return to VA 702 and turn right (east) retracing route to VA 626.

28.6   Turn right (south) on VA 626.
31.9   Turn right (west) on US 55.
34.0   Bunker Hill with outcrops of coarse arkose of lower Fauquier Formation. Type locality of Bunker Hill Formation of Furcron (1969).
36.15 Cross railroad tracks.
36.2   Turn left on VA 1001.
36.5   Turn left (south) on US 17.
37.2   Turn right on VA 691.
37.9   Turn right into private driveway and park in front of metal gate as close to the woods as possible. Stop 8.

STOP 8. Horner Run Fault. Figure 11. Marshall 7.5-minute quadrangle.

At this stop we will visit four or five outcrops in quick succession to examine basement-cover relations and a northwest-trending fault of Late Proterozoic age, named the Horner Run fault by Espenshade (1986). We will begin in a small pit about 120 m (400 ft) west of Rt. 691 and then work our way back to the highway and look at several outcrops on the east side of Rt. 691.

An excellent geologic map of the Marshall quadrangle has been published by Gilbert H. Espenshade (1986), and what we will show you at this stop is derived almost entirely from his work (Figure 11). This fault extends for approximately 7.2 km along a northwest trend across the northwest part of the Marshall quadrangle and northeast part of the Orlean quadrangle (Clarke, unpublished data). The fault juxtaposes the Marshall Metagranite on the northeast against the entire stratigraphic section of the Fauquier Formation (at least 790 m [2600 ft] thick) on the southwest. The fault is presumed to be a steeply southwest-dipping normal fault with movement down to the southwest. At this stop the uppermost part of the Fauquier Formation (metarhythmite unit of Espenshade, 1986) crops out in a shallow syncline in fault contact with the Marshall Metagranite. Bedding in the rhythmite and the nearby underlying meta-arkose and metasiltstone generally, but not always, trends north-northeast directly into the fault. Except for offsets like the Horner Run fault and faults seen at Stops 2, 5, and 6, the eastern contact of the Marshall Metagranite trends north-northeast for tens of miles through at least five 7.5-minute quadrangles. At the southeast end of the Horner Run fault, conglomerate and arkose found in the lowermost Fauquier Formation are juxtaposed with metarhythmite found in the uppermost Fauquier Formation. In that area a cross section suggests at least 550 m (1800 ft) of offset (Espenshade, 1986, Plate 1, cross section D-D').

The upper age limit of the Horner Run fault is constrained by geologic relations at its eastern end approximately 3 km to the southeast of this stop. A very thin belt of discontinuous lenses of metasediment in the lower part of the overlying Catoctin Formation clearly is not offset by the fault. It is not clear whether the lowermost part of the Catoctin (low-Ti metabasalt breccia of Espenshade, 1986) below these metasediments is offset by the fault; however, Espenshade points out that this breccia is much thicker to the southwest of the fault than it is to the northeast. New U-Pb zircon ages have been obtained for a rhyolite flow and a rhyolite dike approximately 29 and 56 km north of here, respectively (Aleinikoff et al., 1991). The metarhyolite flow is near the base of the Catoctin Formation on the west limb of the Blue Ridge anticlinorium near Bluemont, Virginia, and has an age of about 600 Ma. The metarhyolite dike intrudes a basement granite in the core of the anticlinorium and is dated at 561 ± 7 Ma. These dates agree well with Rb/Sr ages obtained by Badger and Sinha (1988) for metabasalts of the Catoctin Formation. This suggests that the lower part of the Catoctin

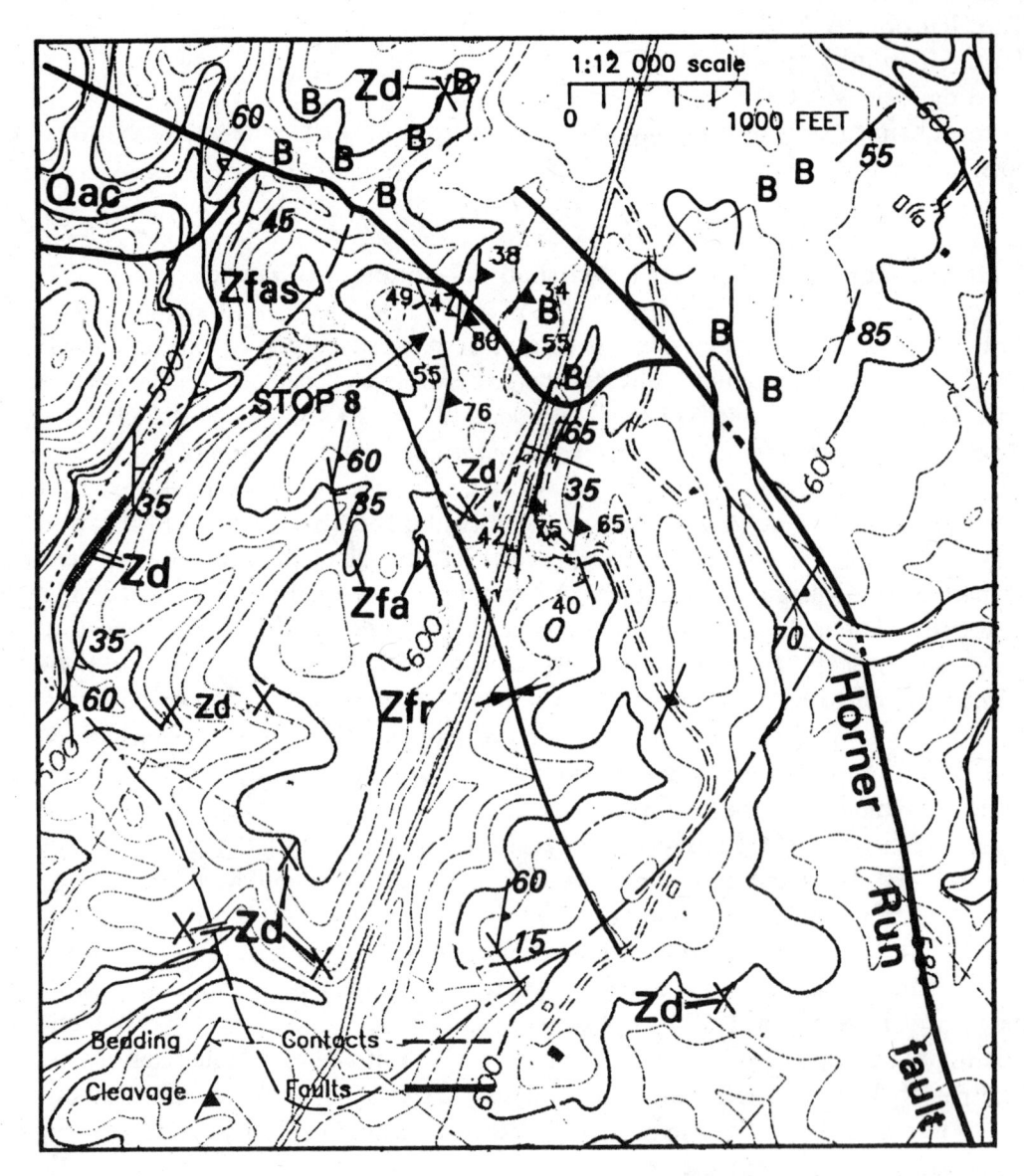

Figure 11. Geologic map in the vicinity of Stop 8, modified from Espenshade (1986). Horner Run fault. Qac=alluvium and bottom land colluvium; Zd=metadiabase and metagabbro sills; Zf=Fauquier Formation: Zfr=rhythmite unit, Zfa=meta-arkose unit, Zfas=meta-arkose and metasiltstone unit; Ym=Marshall Metagranite.

Formation is latest Late Proterozoic in age. Therefore, the upper age limit of the Horner Run fault is latest Late Proterozoic. The age of the uppermost Fauquier Formation that is cut by the Horner Run fault must also be Late Proterozoic, but cannot be determined as precisely.

In two outcrops of Marshall Metagranite at this stop, there is a tectonic breccia of rounded to subangular granite clasts in a very dark, fine-grained matrix. Espenshade (1986) noted the presence of this breccia in outcrops of Marshall Metagranite that are within roughly 760 m (2500 ft) of the fault. Farther from the fault the breccia is very rare.

The uppermost unit of the Fauquier Formation here is a very fine grained, thinly bedded (to laminated) siltstone that Espenshade (1986) called metarhythmite. The light-colored layers or laminae are mainly quartz silt, and the dark layers consist of fine sericite, with lesser chlorite and biotite. Magnetite is common and clearly grows across fine layers, confirming metamorphic derivation. Pyrite is also common and can form cubes up to 0.6 cm across. If you look carefully, many, if not most, samples of this laminated rock have syndepositional sedimentary structures such as slump folds, normal faults, very low angle bedding truncations, and scours. Generally the intrafolial folds are extremely small (less than 3 mm in wavelength); however, the last outcrop at this stop has mesoscopic recumbent slump folds whose wavelengths are 15 to 20 cm. The axes of these slump folds are approximately subhorizontal, trend roughly east-west, and suggest movement to the south. Could these slumps be related to movement on the Horner Run fault? In the same outcrop the regional cleavage is oriented N12°E; 42°SE and crosscuts the folds at a very high angle. Wehr (1985) put these rocks in the upper part of the Monumental Mills Formation of his Lynchburg Group, and interpreted their depositional environment to be a pro-delta slope. Thiesmeyer (1939) called these rocks varved slates that were deposited in freshwater lakes under glacial conditions. Thiesmeyer was careful to note that there is no direct evidence to support glaciation. Espenshade (1986) felt that the metarhythmite unit might be lacustrine or estuarine. We feel that a moderately deep-water environment is suggested and favor the model of Wehr (1985). However, a deep lake environment cannot be ruled out.

Continue south along VA 691.

45.5   Turn right on VA 689 (west).
46.3   Turn left (south) on VA 738 at Dudie.
48.1   Turn right (west) on VA 678.
48.65 Turn left (south) on VA 681 just before Carter Run.
48.9   Pull off on left side of road. Stop 9.

STOP 9. Cliff Mills Section of Upper Fauquier Formation. (Optional Stop). Jeffersonton 7.5-minute quadrangle.

These cliff exposures on the east bank of Carter Run are the occasional rare example of an excellent outcrop in the upper part of the Fauquier Formation (Monumental Mills Formation of Wehr, 1985). This outcrop consists of laminated to thin-bedded metasiltstone interbedded with thin to medium beds of fine-grained metasandstone. Very thin beds or laminae of metasiltstone, identical to those at Stop 8, are best seen a few tens of feet higher in the stratigraphic section and are exposed at mileage 48.6. Some of the fine-grained metasandstone weathers to a punky brown and gives the rock a pitted appearance. Graded bedding, occasional shallow channels, and bedding/cleavage relations all suggest that bedding is right-side-up. Large round biotites are common and may be related to contact metamorphism caused by intrusion of fairly thick sills of metadiabase found on the hill to the east. Bedding trends N20°W; 42°NE, and a well-developed Paleozoic slaty cleavage is oriented

N20°E; 71°SE.

Uphill to the east are small outcrops of medium- to coarse-grained meta-arkose which mark the top of the metasediments of the Fauquier Formation. Farther east at the base of Piney Mountain is float of parallel laminated siltstone, blue quartz pebble conglomerate, and dull, medium-gray quartzite. These metasediments may correlate with the Ball Mountain Formation of Wehr (1985). More mapping is needed to confirm this. Yet farther east on Piney Mountain is a thick succession of metabasalt breccia of the basal Catoctin Formation (low-Ti metabasalt breccia of Espenshade, 1986). This breccia is very distinctive, with a layering defining bedding that is very rare in the Catoctin. It may be hyaloclastite breccias that erupted subaqueously (Kline et al., 1990a).

Continue southeast on VA 681.

50.0	Excellent outcrop of metabasalt breccia showing crude layering. Layers of poorly sorted large subrounded clasts of scoriaceous and massive metabasalt in a matrix of angular small fragments alternate with layers dominated by matrix with very few larger clasts.
50.7	Turn right (south) on VA 691.
51.8	Turn right (west) on VA 688.
52.1	Turn left at fork on VA 613 toward Waterloo Bridge.
52.3	Cross Waterloo Bridge.
52.4	Immediately after bridge continue straight (south) on VA 622.
53.7	Turn right (west) on US 211 and immediately get in left lane.
53.8	Turn left (south) on VA 229.
54.6	Turn left into South Wales Corporation complex and golf course. Stop 10.

STOP 10. South Wales. Regional Stratigraphic Correlations. Figure 12. Jeffersonton 7.5-minute quadrangle.

All of the previous stops on this field trip were metasediments that have been mapped as part of the Fauquier Formation of Espenshade (1986), which correlate with the Bunker Hill and Monumental Mills Formations of Wehr (1985). At this stop coarsely laminated to thin-bedded metasiltstone interbedded with medium- to thick-bedded sandstone is present. These lithologies are part of a more heterogeneous stratigraphic sequence that includes sooty-black metasiltstone and slate; quartz pebble conglomerate with a punky, black matrix; dull-gray, medium-grained quartzite; and pyrite-biotite-chlorite schist interbedded with biotite metasandstone. Most of these rocks contain numerous concordant tabular bodies of fine-grained metadiabase and medium- to coarse-grained metapyroxenite(?), which we interpret as sills. All of the coarser-grained sills are extremely altered, and while the protoliths of some of them appear to be ultramafic rocks, very limited unpublished chemical data suggests a mafic composition strikingly similar to the low-titanium suite of the Catoctin Formation (Espenshade, 1986).

Although some of the sedimentary lithic types mentioned above can be found in the Fauquier Formation, the entire sequence is strikingly different from what we have already seen. This area in the southeastern corner of the Jeffersonton quadrangle has been mapped as "Swift Run Formation(?)" by Wehr (1985). Rocks in much of this area are particularly poorly exposed, making stratigraphic and structural analysis problematic. This area is the northeastern limit of Wehr's (1983; 1985) excellent study, but he had difficulty deciding what to do with these rocks. We feel the use of the name "Swift Run Formation" for rocks on the east limb of the Blue Ridge anticlinorium raises as many problems as it was intended to resolve. The use of the name should be restricted to the western limb of the

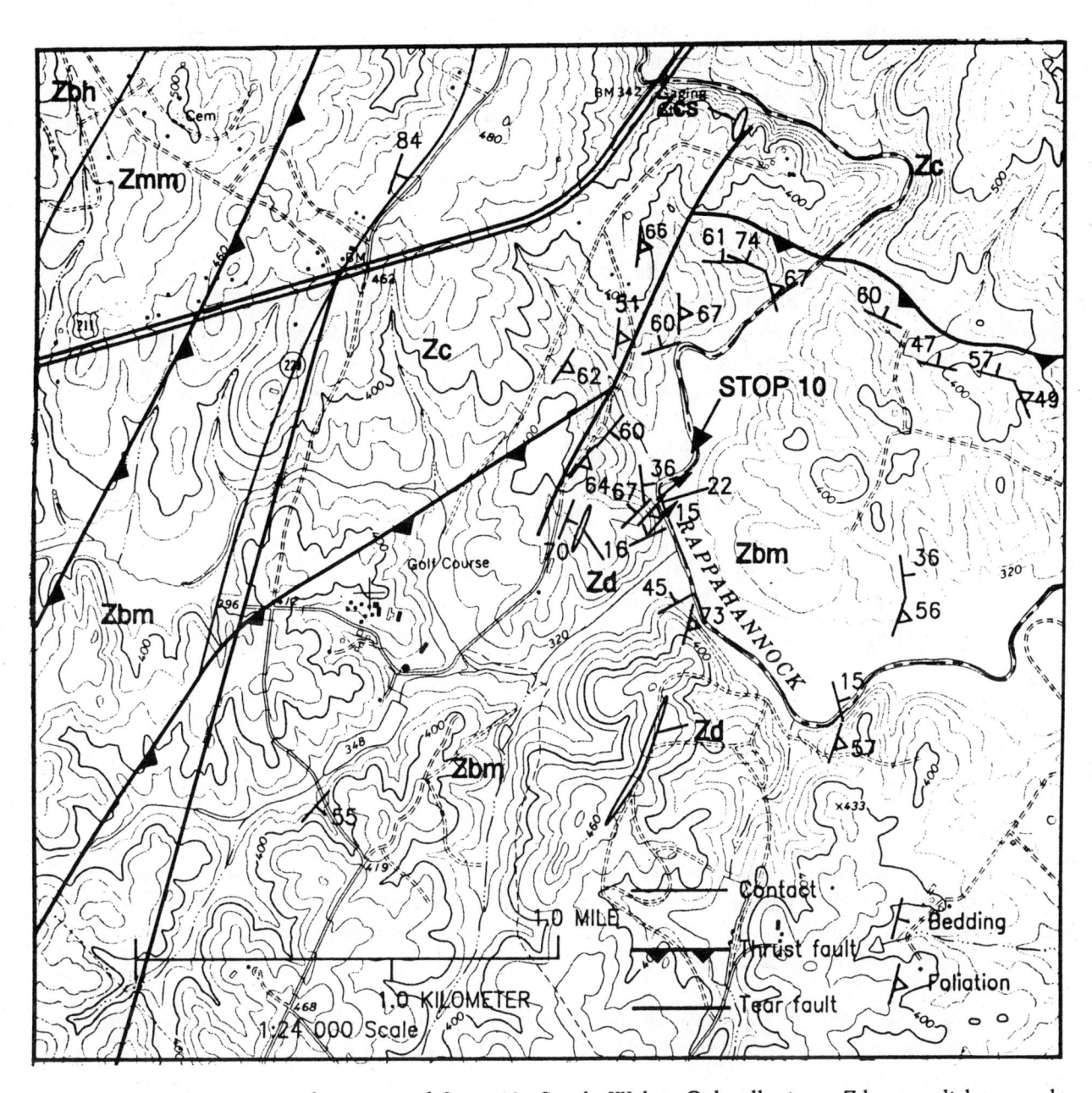

Figure 12. Geologic map of vicinity of Stop 10. South Wales. Qal=alluvium; Zd=metadiabase and metagabbro sills; Zc=metabasalts of Catoctin Formation; Zcs=lenses of metasediments in Catoctin Formation; Zbh=Bunker Hill Formation of Wehr (1985); Zmm=Monumental Mills Formation of Wehr (1985); Zbm=Ball Mountain Formation of Wehr (1985).

anticlinorium. Instead, these rocks appear to correlate better with the Ball Mountain Formation as defined by Wehr (1985). Wehr (1985) noted that many of the lithic types in the area mapped as Swift Run(?) were similar to lithic types in the Ball Mountain Formation. We make this correlation tentatively until more mapping has been completed in the Brandy Station and Remington quadrangles to the south.

The rocks at this stop are primarily coarsely laminated to thin-bedded metasiltstone that have graded bedding, some flame-shaped structures probably formed by load casting, minor slump folds, and possible ripples. Interbedded with these siltstones are some fine- to medium-grained feldspathic metasandstone beds ranging from 15 cm to 2 m thick. It is possible that the thicker sandstone beds are a combination of thinner beds. Wehr (1985) suggests that the Ball Mountain Formation was deposited in fairly deep water as sediment gravity flows. Primary structures preserved are consistent with this interpretation. These rocks have a well-developed foliation and several asymmetric tight northeast-plunging folds. The Ball Mountain Formation in the southeast corner of the Jeffersonton 7.5-minute quadrangle crops out over a rather broad area at least three miles across. In much of this area bedding dips gently and the rocks are folded into broad open structures. The Ball Mountain Formation in this area is probably part of a fairly flat thrust sheet that is transported northward and westward over the Catoctin Formation. The Ball Mountain Formation at Stop 10 (Figure 12) and to the east dips to the north or northwest near this thrust. Mapped relations to the west of Stop 10, and in the Culpeper area to the south, suggest that the Ball Mountain and the Monumental Mills Formations of Wehr (1985) (equivalent with the upper part of the Fauquier Formation) are also separated by a thrust fault (Wehr, 1985). It is this thrust fault that Rankin et al. (in press) suggest separates ancient North America (Laurentia) on the west from their Jefferson terrane on the east.

Retrace route out of South Wales to VA 229. Turn right on VA 229. Go 0.8 miles, turn right (east) on US 211 to Warrenton. The intersection of US 211 and (Business) US 29 in Warrenton is shown on any standard Virginia state roadmap for your return trip.

REFERENCES

Accademia Nazionale Dei Lincei, 1980, Geodynamic Evolution of the Afro-Arabian Rift System: Rome, 705 p.

Aleinikoff, J. N., Zartman, R. E., Rankin, D. W., Lyttle, P. T., Burton, W. C., and McDowell, R. C., 1991, New U-Pb zircon ages for rhyolite of the Catoctin and Mount Rogers Formations—More evidence for two pulses of Iapetan rifting in the central and southern Appalachians: Geological Society of America Abstracts with Programs, in press.

Badger, R. L., and Sinha, A. K., 1988, Age and Sr isotopic signature of the Catoctin volcanic province: Implications for subcrustal mantle evolution: Geology, v. 16, p. 692–695.

Barberi, F., and Varet, J., 1970, The Erta Ale volcanic range (Danakil Depression, northern Afar, Ethiopia): Bulletin Volcanologique, v. 3B-4, p. 848–917.

Bartholomew, M. J., and Lewis, S. E., 1984, Evolution of Grenville massifs in the Blue Ridge geologic province, southern and central Appalachians, in Bartholomew, M. J., Force, E. R., Sinha, A. K., and Herz, N., eds., The Grenville Event in the Appalachians and Related Topics: Geological Society of America Special Paper 194, p. 229–254.

Carlisle, D., 1963, Pillow breccias and their aquagene tuffs, Quadra Island, British Columbia: Journal of Geology, v. 71, p. 48–71.

Clarke, J. W., 1981, Billion-year-old rocks of the Blue Ridge anticlinorium of northern Virginia: A review: Virginia Journal of Science, v. 32, p. 127.

——, 1984, The core of the Blue Ridge anticlinorium in northern Virginia, in Bartholomew, M. J., Force, E. R., Sinha, A. K., and Herz, N., eds., The Grenville Event in the Appalachians and Related Topics: Geological Society of America Special Paper 194, p. 155–160.

Conley, J. F., 1978, Geology of the Piedmont of Virginia—Interpretations and problems, in Contributions to Virginia Geology — III: Virginia Division of Mineral Resources Publication 7, p. 115–149.

Dimroth, E., Cousineau, P., Leduc, M., and Sanschagrin, Y., 1978, Structure and organization of Archean subaqueous basalt flows, Rouyn-Noranda area, Quebec, Canada: Canadian Journal of Earth Sciences, v. 15, p. 902–918.

Espenshade, G. H., 1986, Geology of the Marshall quadrangle, Fauquier County, Virginia: U.S. Geological Survey Bulletin 1560, 60 p.

——, and Clarke, J. W., 1976, Geology of the Blue Ridge anticlinorium in northern Virginia: Geological Society of America Northeast-Southeast Sections Joint Meeting Field Trip Guidebook No. 5, 26 p.

Evans, N., 1984, Latest Precambrian to Ordovician metamorphism in the Virginia Blue Ridge: An alternative explanation for the origin of the contrasting Lovingston and Pedlar basement terranes: unpublished Ph.D. dissertation, Virginia Polytechnic Institute and State University, Blacksburg, 313 p.

Furcron, A. S., 1939, Geology and mineral resources of the Warrenton quadrangle, Virginia: Virginia Division of Mineral Resources Bulletin 54, 94 p.

——, 1969, Late Precambrian and early Paleozoic erosional and depositional sequences of northern and central Virginia, in Precambrian-Paleozoic Appalachian Problems: Georgia Geological Survey Bulletin, v. 80, p. 57–88.

Gooch, E. O., 1958, Infolded metasedimentary rocks near the axial zone of the Catoctin Mountain–Blue Ridge anticlinorium in Virginia: Geological Society of America Bulletin, v. 69, p. 569–574.

Kline, S. W., Conley, J. C., and Evans, N., 1990a, Hyaloclastite pillow breccia in the Catoctin metabasalt of the eastern limb of the Blue Ridge anticlinorium in Virginia: Southeastern Geology, v. 30, p. 241–258.

——, Lyttle, P. T., and Froelich, A. J., 1990b, Geologic map of the Loudoun County part of the Middleburg 7.5-minute quadrangle, Virginia: U.S. Geological Survey Open-file Report 90–641.

Lukert, M. T., and Banks, P. O., 1984, Geology and age of the Robertson River Pluton, in Bartholomew, M. J., Force, E. R., Sinha, A. K., and Herz, N., eds., The Grenville Event in the Appalachians and Related Topics: Geological Society of America Special Paper 194, p. 161–166.

———, and Halladay, C. R., 1980, Geology of the Massies Corner quadrangle, Virginia: Virginia Division of Mineral Resources Publication 17, map with accompanying text.

Mitra, G., 1979, Ductile deformation zones in the Blue Ridge basement and estimation of finite strains: Geological Society of America Bulletin, v. 90, p. 935–951.

Nelson, W. A., 1962, Geology and mineral resources of Albemarle County: Virginia Division of Mineral Resources Bulletin 77, 92p.

Parker, P. E., 1968, Geologic investigation of the Lincoln and Bluemont quadrangles, Virginia: Virginia Division of Mineral Resources Report of Investigations 14, 23 p.

Pavlides, L., 1990, Geology of part of the northern Virginia Piedmont: U.S. Geological Survey Open-file Report 90–548.

Rankin, D. W., 1975, The continental margin of eastern North America in the southern Appalachians: The opening and closing of the proto-Atlantic ocean: American Journal of Science, v. 275–A, p. 298–336.

———, Drake, A. A., Jr., and Ratcliffe, N. M., in press, Proterozoic North American (Laurentian) rocks of the Appalachian orogen, in Reed, J. C., Jr., Bickford, M. E., Houston, R. S., Link, P. K., Rankin, D. W., Sims, P. K., and Van Schmus, W. R., eds., Precambrian: Conterminous U. S.: Geological Society of America, The Geology of North America, v. C–2.

Reed, J. C., Jr., and Morgan, B. A., 1971, Chemical alteration and spilitization of the Catoctin greenstones, Shenandoah National Park, Virginia: Journal of Geology, v. 79, p. 526–548.

Schindler, J. S., 1990, Geologic map of the Lincoln 7.5-minute quadrangle: U.S. Geological Survey Open-file Report 90–640.

Schwab, F. L., 1974, Mechum River Formation: Late Precambrian (?) alluvium in the Blue Ridge province of Virginia: Journal of Sedimentary Petrology, v. 44, p. 862–871.

Simpson, E. L., and Eriksson, K. A., 1989, Sedimentology of the Unicoi Formation in southern and central Virginia: Evidence for Late Proterozoic to Early Cambrian rift-to-passive margin transition: Geological Society of America Bulletin, v. 101, p. 42–54.

Thiesmeyer, L. R., 1939, Varved slates in Fauquier County, Virginia: Virginia Geological Survey Bulletin 51-D, p. 105–118.

Tollo, R. P., Aleinikoff, J. N., Gray, K. J., 1991, New U-Pb zircon isotopic data from the Robertson River igneous suite, Virginia Blue Ridge: Implications for the duration of Late Proterozoic anorogenic magmatism: Geological Society of America Abstracts with Programs, in press.

Wehr, F., 1983, Geology of the Lynchburg Group in the Culpeper and Rockfish River areas, Virginia: unpublished Ph.D. dissertation, Virginia Polytechnic Institute and State University, Blacksburg, 254 p.

———, 1985, Stratigraphy of the Lynchburg Group and Swift Run Formation, Late Proterozoic (730–570 Ma), Central Virginia: Southeastern Geology, v. 25, no. 4, p. 225–239.

11
SIDELING HILL ROAD CUT AND VISITORS CENTER: AN EDUCATIONAL OPPORTUNITY COMBINING OUTCROP AND CLASSROOM

Kenneth A. Schwarz
Maryland Geological Survey
Baltimore, MD 21218

INTRODUCTION

Sideling Hill, six miles west of Hancock, Maryland, in Washington County, is one of a series of northeast-trending, subparallel ridges capped by erosion-resistant sandstone and conglomerate in the Valley and Ridge physiographic province (Figure 1).

During 1983, an excavation for the new I-68/U.S. 48 highway was cut through Sideling Hill and exposed a textbook example of a synclinal ridge in the Mississippian-age Purslane and Rockwell Formations. In late 1984, the Maryland Geological Survey (MGS) proposed that this feature be given geological recognition by the state for educating the public about the impressive geology of this site and of the Appalachian Mountains in general. To advertise this feature, the MGS was instrumental in publishing an article in *Maryland Magazine* (Winter 1987), as well as cover photos in *Geotimes* (July 1985) and the American Association of Petroleum Geologists Bulletin (Sept. 1987). In mid-1985, the Maryland Department of Transportation's State Highway Administration decided to build a rest stop at the site incorporating a Visitors Center, staffed by full-time park rangers and auxiliary summer staff from the Maryland Department of Natural Resources' Forest, Park, and Wildlife Service. These rangers will be trained to provide information to motorists by staff from the Maryland Department of Economic and Employment Development's Office of Tourism Development, as well as general geologic information by staff from the Department of Natural Resources' Maryland Geological Survey. Numerous educational opportunities will be planned, including many hands-on programs in the 65-seat assembly room at the Center.

Construction contracts were awarded to the Holloway Companies of Wixom, Michigan, and excavation work began in earnest in April 1983. Project completion is scheduled for August 1991. Although this trip may not allow a visit inside the center, the extent of its construction should be readily apparent. Restroom facilities for eastbound traffic will be built under a separate construction contract at a later date.

ROAD LOG

MILEAGE

Total	Incr	
0.0	0.0	Leave from Omni Hotel, Liberty St. entrance. Go south for one block to Lombard St.
0.1	0.1	Turn right (west) on Lombard St. for one block.
0.2	0.1	Turn left (south) on Howard St. Follow signs to I-395 southbound.
1.0	0.8	Follow signs for I-95 southbound for Washington, D.C.
1.8	0.8	Merge right with I-95 southbound.
4.9	3.1	Turn right onto exit lane for I-695 westbound for Towson.

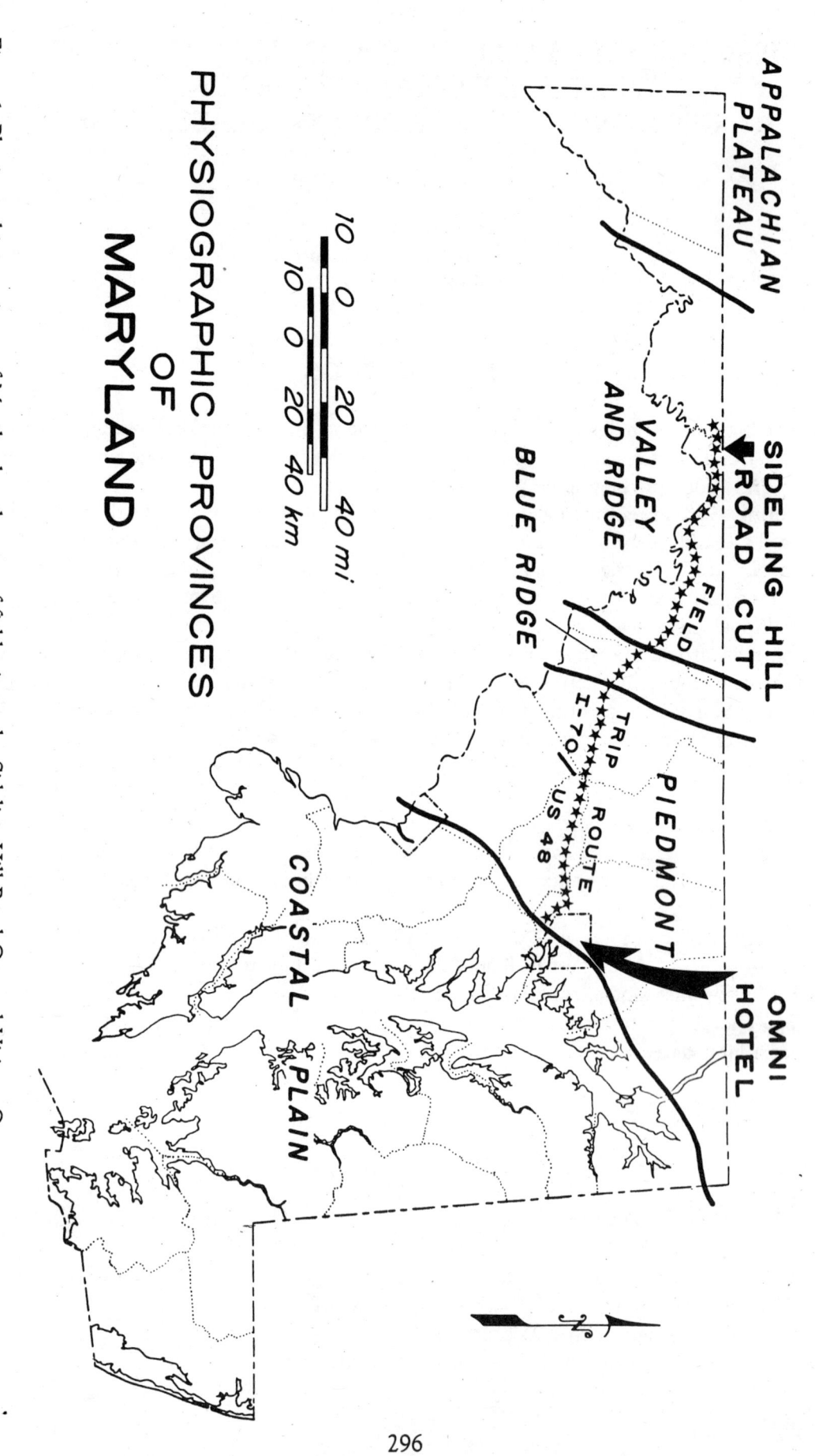

Figure 1. Physiographic provinces of Maryland and route of field trip to the Sideling Hill Road Cut and Visitors Center.

6.3	1.4	Fall Line. This incline marks the western edge of the Coastal Plain physiographic province (Figure 1). The Coastal Plain is composed of relatively flat-lying sedimentary rocks containing easily eroded gravels, sands, and shales of Lower Cretaceous through Pleistocene age and covering the eastern half of Maryland on both sides of the Chesapeake Bay. West of the Fall Line, the Piedmont physiographic province is composed of Precambrian and early Paleozoic age igneous and metamorphic rocks containing metasedimentary and metaigneous sequences of markedly different susceptibilities to erosion, resulting in a hilly terrain.
10.2	3.9	Turn right onto exit lane for I-70 westbound for Frederick, then stay in left lane for Frederick traffic.
13.5	3.3	Cross Patapsco River, leave Baltimore County, enter Howard County.
15.0	1.5	U.S. Rte. 29 exits left to Columbia and Washington. Keep straight ahead on I-70 westbound toward Frederick.
33.4	18.4	Leave Howard County, enter Carroll County.
34.9	1.5	Intersection with Maryland Rte. 27, Mt. Airy. Leave Carroll County, enter Frederick County. The Four County Farm at Parrs Spring contains the springheads of the Patapsco River, which flows east into the Chesapeake Bay. Maryland Rte. 27 follows Parr's Ridge, a major drainage divide separating basins whose waters flow east into the Chesapeake Bay from those of the Monacacy River whose waters flow south and west into the Potomac River.
36.1	1.2	On a clear day, Catoctin Mountain in the Blue Ridge physiographic province can be seen at the western horizon, about 20 miles away.
42.9	6.8	Good view of Catoctin Mountain across Frederick Valley.
44.2	1.3	Intersection with Maryland Rte. 144.
45.8	1.6	Long roadcut in siltstone of the Lower Cambrian Araby Formation. At the eastern end of the exposure is a diabase dike which has weathered to rusty-orange-colored, rounded boulders.
46.6	0.8	Cross Monocacy River. Stay on I-70 through numerous exits.
48.3	1.7	The Frederick Quarry of the Genstar Corporation on the left (south) extracts carbonate rocks from the Cambrian Frederick and Grove Limestones for masonry cement and crushed stone. Continuous pumping of water from the deep pit may lower the ground-water table locally and accelerate solution collapse in surrounding areas. The open, flat Frederick Valley is developed primarily on Paleozoic-age carbonates.
48.8	0.5	Sinkholes in the Frederick Limestone (not visible from I-70). In the field to the right (north) the thick clayey residual soil has collapsed into a solution cavity in the Upper Cambrian Frederick Limestone near the base of the second power pole. Diversion of the surface drainage into this sinkhole continues to flush out the clay filling and enlarges the sinkhole. Pinnacles of the underlying limestone bedrock can be seen in the deepest parts of the hole.
52.7	3.9	Cross east-dipping Mesozoic basin border fault, covered by colluvium from Catoctin Mountain. In certain areas of the Piedmont, as here near Frederick, rift basins containing Mesozoic sedimentary and igneous rocks cause a somewhat subdued, rolling topography. These basins are distinguished by the brick-red sedimentary rocks and soils.
55.8	3.1	Roadcut at crest of Catoctin Mountain. Basaltic lava flows of the Upper Precambrian Catoctin Formation. The rock has been strongly deformed and altered to greenstone or metabasalt (metamorphosed basalt) composed of the metamorphic minerals chlorite

and epidote. In places, flow tops containing a concentration of the yellowish-green mineral epidote and large pods of white quartz can be seen. Locally, magnetite in small octahedrons or irregular grains appears as a speckling on surfaces of the sheared greenstone. Large pillow-like structures may have formed when lavas invaded a lake or shallow sea to form subaqueous pillow lava, although other explanations are possible.

56.1	0.3	Continue west on I-70. The wide valley between Catoctin Mountain and South Mountain is known as the Middletown Valley. To the south, the valley is underlain by coarse-grained Precambrian-age gneiss and surrounded by Catoctin metabasalt; to the north it is composed of metabasalt and various pyroclastic units.
63.8	7.7	Exit for rest area at crest of South Mountain. Comfort stop and tourist information stop. Bedrock here is porphyritic rhyolite flows in the Catoctin Formation.
64.8	1.0	Appalachian Trail crosses I-70.
64.9	0.1	Crest of South Mountain, leave Frederick County, enter Washington County. The view to the west and northwest is the Great Valley. This valley, which is floored with bedrock of more than 8400 ft. of Cambrian and Ordovician-age limestone and shale, is known as the Hagerstown Valley in Maryland and the Shenandoah Valley in Virginia. The ridge-former on South Mountain is a quartzite member of the Cambrian Weverton Formation.
65.5	0.6	On a *very* clear day, the cut at Sideling Hill may be seen 38 miles to the northwest by parking at the "Speed Limit 55—Enforced from aircraft" sign and looking at far horizon to left of Fairview Mountain, which is only 21 miles away.
72.9	7.4	Ordovician Stonehenge Limestone outcrops in field to north (right).
73.4	0.5	Cross Antietam Creek.
79.4	6.0	Interchange with Maryland Rte. 63. Shale and siltstone of the Upper Ordovician Martinsburg Formation are exposed along the north (right) side of the highway just west of the interchange. This unit is about 3000 feet thick.
80.7	1.3	Cross Conococheague Creek.
82.2	1.5	Exposed along the roadcut at the overpass are strata of the Upper Ordovician Chambersburg Limestone. Continuing westward, the exposures are Middle Ordovician-age limestone and dolomite of the underlying St. Paul Group and Pinesburg Station Formation.
83.0	0.8	Exposures of interbedded dolomite and limestone of the Lower Ordovician Rockdale Run Formation in field to south (left). Ribbon karren, karst features (belts of limestone ledges alternating with bands of soil), are present.
86.7	3.7	Limestone and dolomite of the Upper Ordovician Conococheague Formation are exposed in pastures to the south (left) of the highway.
88.6	1.9	Leave Hagerstown Valley. Cross major thrust fault, the North Mountain fault, which places Cambrian rocks on Silurian and Devonian rocks. Leave Cambrian and Lower Ordovician–age carbonates, enter Upper Ordovician and Lower Silurian–age clastics. Note change in topography and vegetation.
94.0	5.4	Cross Licking Creek.
95.5	1.5	Exposures along the north (right) side of the highway are of interbedded sandstone, siltstone, and shale of the Upper Devonian Chemung Formation.
96.1	0.6	First unobstructed view of Sideling Hill cut at horizon to west (straight ahead).
97.4	1.3	Interchange with Maryland Rte. 615.
97.9	0.5	Series of views of Sideling Hill cut straight ahead.
101.4	3.5	Cross Tonoloway Creek.

101.9	0.5	Exposure of Devonian-age shale from the Woodmont, then Romney Formations.
104.1	2.2	Bear left onto U.S. Rts. 40-48 west toward Cumberland. Leave I-70, which goes north (right) into Pennsylvania.
104.4	0.3	Exposure of siltstone and shale of the Upper Silurian Wills Creek Formation.
104.9	0.5	Exposed along the access road on the north (right) side of the highway are red-brown siltstone and sandstone of the Middle Silurian Bloomsburg Formation.
106.3	1.4	The Upper Silurian Keyser Limestone is exposed in an old quarry in the hill to the south (left). At the south (left) end of the overpass of Sandy Mile Road is a large outcrop of the Lower Devonian Oriskany Sandstone. We will stop there on the way back.
106.8	0.5	Along the west (right) side of the highway is dark-gray shale of the Marcellus Member of the Devonian Romney Formation.
107.8	1.0	Interchange with Scenic U.S. Rte. 40 and Woodmont Road. Shale and siltstone of the Hamilton Member of the Romney Formation are exposed.
110.2	2.4	Bear right on to deceleration lane.
110.4	0.2	East end of roadcut at crest of Sideling Hill. Park in parking lot for westbound traffic.

STOP 1. SIDELING HILL ROADCUT

This spectacular deep roadcut (Figure 2) through the axis of a downfold or syncline at the crest of the mountain exposes over 800 feet of strata of the Lower Mississippian Rockwell and Purslane Formations. A number of different sedimentary rock types are present. From these rock types and fossils, the depositional environments have been determined. Minor tectonic structures produced during the folding of the layers are also present. The effects of lithology and structure upon movement of groundwater are well illustrated.

Construction

The cut is 370 ft deep, 740 ft wide at the top, and 200 ft wide at the base. There are four benches, each 80 ft higher than the previous one and about 20 ft wide, to control rock falls. About 5.7 million cubic yards of rock were removed from the crest at 1615 ft elevation, down to the present pass of 1280 ft. Practically all the rock material was used as asphalt aggregate, base-and-subbase-course material on the road, and riprap for road fill and steep slope fill, such as underlies the westbound parking area. This new 4.5 mile road, at only 6 percent grade, replaces the more than 7 miles of old road, which has steeper grades and three-lane switchbacks.

The $20.1 million project was $10 million less than the next competitive bidder. All new Caterpillar equipment was used to reduce down-time and simplify maintenance, parts storage, and field service. Completion was on-schedule, reflecting the military-like supervisory organization during the entire project.

Geology

The Rockwell Formation was probably deposited in an alluvial plain environment, east of a submerged area from eastern Ohio to western Maryland, during the early Mississippian, about 345 million years ago (Figure 3A) (Bjerstedt, 1986). A diamictite, which is an unusual, very poorly sorted rock composed of clay, silt, sand, and pebbles or cobbles of granite, graywacke, chert, and quartzite, outcrops about 70 feet above the base of the formation on the west side of the cut ("A" in Figure 2). About 600 feet of the Rockwell is exposed and consists of interbedded tan and gray-green, clay-rich sandstone, gray-green to dark-gray silty shale, and gray to dark-gray sandy siltstone with several

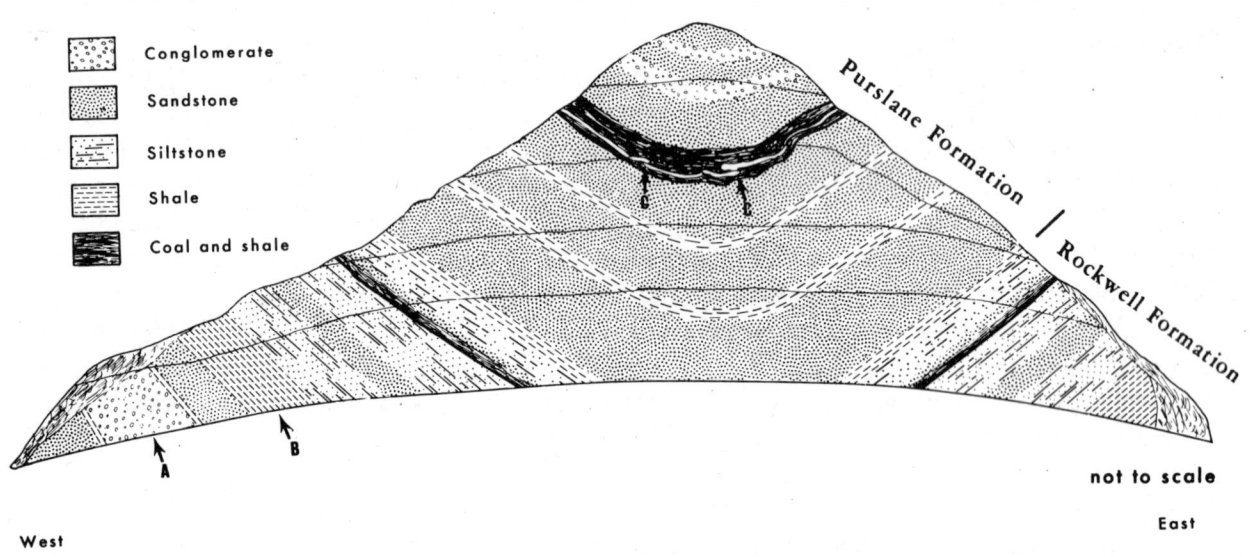

Figure 2. Photograph and interpretive geology of the north side of U.S. Rte. 48 road cut through Sideling Hill (Brezinski, 1989a). Semi-horizontal lines are benches; extreme east and west exposures of the Rockwell Formation are obscured by cover.

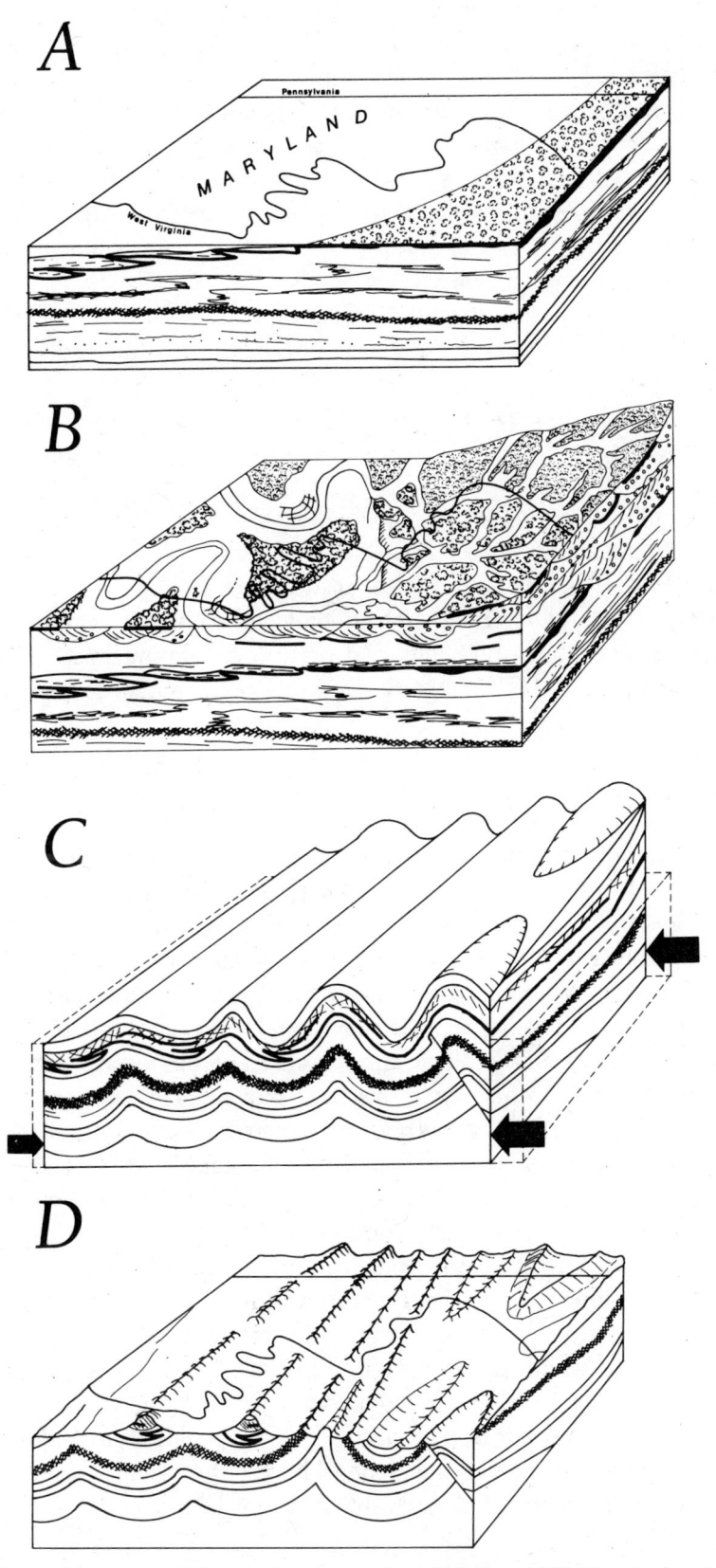

Figure 3. Sequence of development of the rocks exposed at Sideling Hill (Brezinski, 1989a). A, Shallow marine waters and adjacent shoreline swamps of the Riddlesburg sea. B, River systems of the Purslane. C, Folding during the Alleghanian mountain-building episode. D, Post mountain-building erosion to the ridges and valleys seen today.

intervals of red-brown claystone near the top. In places, thin shaly coal and coaly shale are interbedded with shale and siltstone.

Fossils are common, and are mostly plant fragments and imprints. Marine fossils are present within the black silty shale about 170 feet above the lowest exposure ("B" in Figure 2).

The overlying Purslane Formation consists of gray-green, tan, and white crossbedded sandstone and quartz-pebble conglomerate with interbedded gray siltstone, shale, and coaly shale. About 200 feet of sandstone is sufficiently resistant to erosion to create a topographic inversion: a synclinal mountain of resistant rock surrounded by anticlinal valleys which are underlain by older soluble limestone and easily eroded shale.

Near the top of the exposure is 45 feet of dark gray siltstone and semianthracite-grade coaly shale in which numerous plant remains are found. These are the only fossils in the Purslane Formation.

The thick sandstone and conglomerate are channel deposits laid down by rivers. The coal beds may have formed in swamps on flood plains adjacent to the fluvial channels (Figure 3B) (Brezinski, 1989b).

A mountain-building episode, the Alleghanian Orogeny, occurred in late Permian or early Triassic time about 240 million years ago. The compressional stresses during this orogeny were caused by the collision of North America and Africa, resulting in folding of the sediments (Figure 3C). These same stresses produced differential slippage between competent sandstone and relatively less competent carbonaceous siltstone and shale, and often locally mimicked the regional thin-skinned tectonics of the Appalachians ("C" in Figure 2). Erosion has cut into the folds, resulting in the landforms we see today (Figure 3D).

New Visitors Center

The Sideling Hill Visitors Center will have three levels and a mezzanine (Figure 4). The first level will have a multipurpose room, which can be used as a hands-on laboratory, a classroom, a meeting hall, or an auditorium with a seating capacity of 65 people.

The second level will be the main entry level. Located here will be the restrooms, the tourist information desk staffed by rangers (trained in geology and tourism) from the Department of Natural Resources' Forest, Park, and Wildlife Service, a tourism pamphlet rack similar to that at the South Mountain Visitor Center, a vista silhouette identifying features seen from the panoramic view to the east, and the base of a multistory geologic-time spiral. The spiral will extend to the roof of the building and identify geologic events, life forms, and relative ages pertinent to Sideling Hill. The plans also include wall murals showing methods of transportation and their effects on the land.

Exhibits on the mezzanine level will include a videotape of the construction of the road cut. The evolution of transportation and road-building technology in Maryland will be depicted in a series of button-activated panels geared for young inquisitive minds.

The third level will be the main geological exhibit floor, housing 12 separate exhibits. A repeating computer animation, projected on a screen, will show the sequences of sediment deposition, folding, and erosion that formed the Sideling Hill syncline. Examples of the three basic rock types (igneous, metamorphic, and sedimentary) form one exhibit. Other exhibits will focus on how geologists obtain or display information from: 1) sedimentary features and fossils (using actual rocks from the cut); 2) evidence of stream deposits; 3) tools used by the geologist, including air photos and space imagery; 4) a three-dimensional model of Sideling Hill for the visually impaired; and 5) numerous other topics to help explain the educational value of the roadcut.

Reenter westbound US. Rtes 40-48.

Figure 4. Artist's rendition of completed Visitors Center at Sideling Hill. View to northwest from westbound traffic lane.

110.7	0.3	Pass axis of syncline.
111.1	0.4	Mountain in foreground is Town Hill, also a syncline. Clearing in trees at right is for the Columbia Gas Transmission Corporation pipeline, which runs along the state line. Pennsylvania is to the north (right), Maryland to the south (left) of the clearing through the trees.
112.7	1.6	Maroon shale of the Devonian Hampshire Formation on north side of road (right).
113.0	0.3	Good exposure of steeply dipping maroon and green shale of the Devonian Hampshire Formation on right.
113.3	0.3	Cross Sideling Hill Creek, leave Washington County, enter Allegany County.
113.7	0.4	Bear right, exit U.S. Rts. 40-48, cross over for lunch stop at food store by Exxon station.

LUNCH

113.8	0.1	Reenter U.S. Rtes. 40-48 eastbound.
114.1	0.3	Cross Sideling Hill Creek, leave Allegany County, enter Washington County.
115.2	1.1	Bear right onto ramp for Scenic U.S. Rte. 40 and Mountain Road.
117.4	2.2	Highway makes a sharp U-turn to left at crest of Sideling Hill. Black carbonaceous shale of Lower Mississippian Purslane Formation is exposed in roadcut.
118.1	0.7	Outcrop on left is the Rockwell Formation (Lower Mississippian).
121.9	3.8	Woodmont Road and interchange with U.S. Rtes. 40-48. Scenic U.S. Rte. 40 ends. Keep straight on Maryland Rte. 144.
122.4	0.5	Turn left onto Sandy Mile Road.
123.4	1.0	Outcrop of Oriskany Sandstone on right. Pull in to dead-end road at right of bridge and park.

STOP 2. FOSSIL-BEARING DEVONIAN ORISKANY SANDSTONE AT SANDY MILE ROAD

About 320 ft of light tan to white, friable quartzose sandstone is exposed in near-vertical to steeply-dipping beds along the west side of Tonoloway Ridge. Molds of brachiopod shells are abundant throughout the exposure; less common are fossil horn corals and parts of crinoids (Perry and DeWitt, 1977).

Solution weathering commonly removes the carbonate cement and shell material to produce a porous bed with many cavities.

Uplimb thrusts, at low angles to the original bedding, have overthickened the Oriskany Sandstone about 20 percent by wedging. A combination of deep leaching of the calcite cement in the sandstone and local faulting and fracturing has obscured bedding in much of this exposure. This faulting and associated fracturing has generated important fracture porosity for natural gas accumulations in Allegany and Garrett Counties and in other producing states of the Appalachian Basin.

The Oriskany Sandstone is mined for glass sand at Berkeley Springs, West Virginia, approximately 6 miles to the south.

Turn around, retrace route south on Sandy Mile Road to Maryland Rte. 144.

124.4	1.0	Turn right (west) on Maryland Rte. 144.
124.9	0.5	Turn right (north) onto Woodmont Road and right (east) onto ramp for U.S. Rtes. 40-48 eastbound.
126.5	1.6	Pass under bridge for Sandy Mile Road.
128.5	2.0	Bear right onto ramp for I-70 east. Be careful —don't get into far right ramp for U.S. Rte. 522 south to Hancock.
164.8	36.3	Crest of South Mountain.
173.3	8.0	Crest of Catoctin Mountain. Basaltic lava flows of the Upper Precambrian Catoctin Formation are exposed on both sides of cut. Late-afternoon sun should cause the yellow-green mineral epidote to stand out in the north (left) side of the cut where pillow-like structures occur in the flow-tops of the greenstone. Also, large pods of white quartz are conspicuous.
177.9	4.6	Sugarloaf Mountain, a monadnock of resistant and late Precambrian quartzite, is a prominent topographic feature in the western Piedmont physiographic province to the southeast (right).
228.2	49.5	Retrace trip eastward on I-70/I-695/I-95/I-395 to Baltimore.

REFERENCES

Bjerstedt, T. W., 1986, Regional stratigraphy and sedimentology of the Lower Mississippian Rockwell Formation and Purslane Sandstone based on the new Sideling Hill road cut, Maryland: Southeastern Geology, v. 27, p. 69–94.

Brezinski, D. K., 1989a, Geology of the Sideling Hill road cut: Maryland Geological Survey Pamphlet.

———, 1989b, The Mississippian System in Maryland: Maryland Geological Survey Report of Investigations 52, 75p.

Perry, W. J., Jr., and de Witt, W., Jr., 1977, Field guide to thin-skinned tectonics in the central Appalachians: 1977 Annual AAPG/SEPM Convention, Washington, D.C., Post-meeting field trip 4, p. 30–33.